PERSPECTIVES IN ETHOLOGY

Volume 8

Whither Ethology?

CONTRIBUTORS

George W. Barlow
*Department of Zoology and Museum
of Vertebrate Zoology
University of California, Berkeley
Berkeley, California 94720*

Lynda I. A. Birke
*Animal Behaviour Research Group
Biology Department
The Open University
Milton Keynes MK7 6AA, England*

Marian Stamp Dawkins
*Department of Zoology
University of Oxford
Oxford, England*

Donald A. Dewsbury
*Department of Psychology
University of Florida
Gainesville, Florida 32611*

John L. Gittleman
*Department of Zoology
and Graduate Programs in Ethology
and Ecology
University of Tennessee
Knoxville, Tennessee 37916*

Paul Martin
*Sub-Department of Animal Behaviour
University of Cambridge
Madingley, Cambridge CB3 8AA, England*

J. E. R. Staddon
*Departments of Psychology and Zoology
Duke University
Durham, North Carolina 27706*

Milton D. Suboski
*Department of Psychology
Queen's University
Kingston, Ontario K7L 3N6, Canada*

Carel ten Cate
*Sub-Department of Animal Behaviour
University of Cambridge
Madingley, Cambridge, CB3 8AA, England
and
Zoological Laboratory
University of Groningen
Postbus 14, 9750AA Haren, The Netherlands*

A Continuation Order Plan is available for this series. A continuation order will bring delivery of each new volume immediately upon publication. Volumes are billed only upon actual shipment. For further information please contact the publisher.

PERSPECTIVES IN ETHOLOGY

Volume 8

Whither Ethology?

Edited by
P. P. G. Bateson
Sub-Department of Animal Behaviour
University of Cambridge
Cambridge, England

and

Peter H. Klopfer
Department of Zoology
Duke University
Durham, North Carolina

PLENUM PRESS • NEW YORK AND LONDON

The Library of Congress has cataloged this title as follows:

Perspectives in ethology. — Vol. 1- — New York: Plenum Press, 1973-
　　v.: ill.; 24 cm.

　Irregular.
　Includes bibliographies and indexes.
　Editors: v. 1-　　P.P.G. Bateson and P. H. Klopfer.
　ISSN 0738-4394 = Perspectives in ethology.

　1. Animal behavior — Collected works.　　I. Bateson, P. P. G. (Paul Patrick Gordon), 1938-　　. II. Klopfer, Peter H. [DNLM: W1 PE871AN]
QL750.P47　　　　　　　591.5′1 — dc19　　　　　　86-649219
　　　　　　　　　　　　　　　　　　　　　　　　AACR 2　MARC-S
Library of Congress　　　　　　[8610]

ISBN 0-306-42948-9

© 1989 Plenum Press, New York
A Division of Plenum Publishing Corporation
233 Spring Street, New York, N.Y. 10013

All rights reserved

No part of this book may be reproduced, stored in a retrieval system, or transmitted in any form or by any means, electronic, mechanical, photocopying, microfilming, recording, or otherwise, without written permission from the Publisher

Printed in the United States of America

PREFACE

It is possible to agonize too much about the true nature of ethology. Even so, people who call themselves ethologists have to live in a harsh competitive world in which funding agencies need recognizable fields to which they can allocate money and universities want subjects around which courses can be organized. In addition, people generally like to feel that they are part of an identifiable group. These facts of life must be recognized, but such pragmatism need not be confused with the way imaginative science is conducted. If questions are asked of the natural world, deep understanding is rarely going to come from one method or a single theoretical framework. Indeed, in our view, the attractiveness of ethology has been the willingness of its practitioners to draw inspiration from people trained in other disciplines. Any attempt to circumscribe ethology is likely to be intellectually stultifying. The dilemma is how to be competitive with other fields in order to be in a position to do research, while at the same time cooperating with them in order to maintain the vitality of the study of behavior. Ethology must have an identity and yet be open to new ideas and methods from "outside." This volume has been assembled as a contribution toward resolving the paradox of ethology as a subject that is both coherent and eclectic.

A critical question is whether ethology is held together by a single theory or simply characterized by a distinctive approach to the study of behavior. Ethologists clearly have different answers to these questions. We take the view that ethology as a coherent body of theory ceased to exist in the 1950s. The ideas were most clearly articulated by Konrad Lorenz and Niko Tinbergen in concepts and models such as the sign stimulus, the fixed action pattern, reaction specific energy, and hierarchically organized motivational systems with appetitive and consummatory phases. Although the "classical" ethologists, as they are now called, were keenly interested in the plasticity of behavior (imprinting being a famous case in point), their notion of instinct as an unchangeable component of behavior came under strong attack in the 1950s. One by one the concepts and theories succumbed to critical analysis and, by the beginning of the 1960s, any vestiges of common belief in an ethological theory of behavior had disappeared. For close to 30 years, people calling themselves ethologists have been pursuing radically different goals. While some were faithfully upholding the patterns of thought established by Konrad Lorenz, the pragmatically minded entered what was at first a largely atheoretical phase in their work. It was not the case that they worked without hypotheses, but at the time a common slogan was "good data!" rather than "grand ideas!" It was thought that the explanations for behavior would stare the ethologist in the face when enough information had been collected. We were uncomfortable with this positivist trend, and, indeed, one of our reasons for establishing *Perspectives in Ethology* was to encourage a fresh round of conceptual development. In our first editorial we wrote: "The requirement to be quantitative can mean that easy measures are chosen at the expense of representing the complexly patterned nature of a phenomenon. All too easily the process of data collection becomes a trivial excercise in describing the obvious or the irrelevant." We went on to urge our colleagues to search for new theoretical approaches in the study of behavior.

Our wishes for the subject were overtaken by events. The trends toward studies of function and behavioral ecology, which had been strongly fostered by ethologists, were suddenly incorporated into the new sociobiology movement of the 1970s. In other words, there was a nearly successful attempt to incorporate *all* research on behavior under this banner. Imaginations were captured by the new ideas from evolutionary biology and by the ways in which behavioral biology had been attractively married to population biology. E. O. Wilson portrayed the intended incorporation of all aspects of the study of behavior under one banner with his breathtaking prediction for the future of the subject. His bubble-gum (or dumbbell) drawing showed ethology largely replaced by sociobiology by the end of the century. Like many famous prophecies, this one was self-fulfilling for a while, and, in an astonishingly short period of time, the majority of aspiring graduate students wanted to work on a problem in sociobiology or behavioral ecology. The takeover will fail in the end (if it has not already done so) because large chunks of behavioral biology, which had been central concerns of ethology, were deemed to be irrelevant or uninteresting in the so-called "New Synthesis." Few students want to work on how behavior develops or on how it is controlled. It has gradually become apparent that this neglect of an important part of the biology of behavior has been crucial in the eventual failure of sociobiology to become a dominant mode of thought.

It is clear from this account that we believe that a broad ethology does have something distinctive to offer behavioral biology—not a theory but an approach. The four problems for ethologists posed by Niko Tinbergen, namely the evolution, function, development, and control of behavior, were seen as being distinct, as indeed they are. However, the view that they should not be too strongly divorced from each other emerges again and again in this book. By placing a particular question in a broader conceptual context, we achieve greater understanding, whatever the central question may be. We do not think that the caricature of ethology as the study of behavior in a natural setting was ever justified. Even the classical ethologists made some of their most striking discoveries in their homes or laboratories. Nonetheless, the readiness to consider what a particular behavior pattern might be for in the natural environment has been and should be distinctive of the subject. When this approach is coupled with comparisons between animals, the easy assumption that all animals solve the same problem in the same way is quickly shown to be false. The comparative approach continues to be an important characteristic of broad ethology.

Another feature of the subject, which is likely to be important in the coming decades, is the implicit systems approach of ethologists. People who study behavior are often portrayed as doing something rather easy and amateurish—"something that anyone could do in their holidays." This jibe is made so frequently that ethologists often believe, in the absence of a grand theory, they have little to offer science other than a few good stories. However, consider the neurophysiologists' touching confidence in the command neurons that provide *the* cause for the patterning of behavior, or the molecular biologists' (even great ones) belief that genes *for* behavior exist in the literal sense that one-to-one mapping exists in base pairs on DNA strands. These crudities are sometimes represented as attractive and necessary simplifications, but what they really demonstrate is a failure of intellect. The honest reductionists will admit that they cannot think in any other way about the neural basis of behavior or about gene influences on behavior. Ethologists should gain confidence from their ability to *easily* think in other ways, which reflects their experience with multiply influenced, free-running systems. They have genuine skills that enable them to understand the dynamics of behavioral processes and should make the most of them in promoting their own subject and in collaborating with people in other disciplines.

The contributions in this volume illustrate in greater detail the points we have made. Several distinguished practitioners in the study of behavior (not all of whom call themselves ethologists) offer their distinctive views of what has happened over the last few years and what could and should happen in the future. Specific proposals for the various ways in which the subject can move forward and collaborate with other disciplines are also examined. The renewed confidence in the value of broad ethology, as expressed in these chapters, was already apparent at the International Ethological Conference held in Madison, Wisconsin, in August 1987. The dilemma of providing an identifiable discipline that also interacts closely with others is being resolved by example. It is clearly possible to be both coherent and broad-minded. Ethology has re-emerged as a distinctive field that will continue to play an integrative role in the drive to understand the purpose of behavior patterns and how they evolved, developed, and are controlled.

P. P. G. Bateson
Sub-Department of Animal Behaviour
University of Cambridge
Madingley, Cambridge, CB3 8AA
England

P. H. Klopfer
Department of Zoology
Duke University
Durham, North Carolina 27706

CONTENTS

Chapter 1
HAS SOCIOBIOLOGY KILLED ETHOLOGY OR REVITALIZED IT?
George W. Barlow

I.	Abstract	1
II.	Introduction	2
III.	Ethology in Retrospect	5
	A. Experiential Line	5
	B. Homology/Analogy: The Comparative Approach	8
	C. Survival Value	10
	D. Causal Analyses	10
	E. Hierarchical Structure and Chains of Behavior	11
	F. The Ethogram and Chunks of Behavior	12
IV.	Contrasting Ethology and Sociobiology	13
	A. Formulation of Questions	14
	B. Methodology	14
	C. Reduction	15
	D. Genetics and Individuality	16
	E. Development	17
	F. Ecology	17
	G. Natural Selection	18
	H. Evolution	19
	I. Communication	19
V.	The Evolving Continuity between Ethology and Sociobiology	21
	A. The Relevance of Proximate Explanations	22
	B. Ultimate Causation: Giving Meaning	23
	C. The Role of Ethology in Current Behavioral Research	26

VI.	Whither Ethology?	32
	A. Issues Generated by Sociobiology	32
	B. Neglected Issues in Traditional Ethology	33
VII.	Closing Comments	35
VIII.	Acknowledgments	37
IX.	References	38

Chapter 2

THE FUTURE OF ETHOLOGY: HOW MANY LEGS ARE WE STANDING ON?

Marian Stamp Dawkins

I.	Abstract	47
II.	Introduction	47
III.	The Question of Mechanism	48
	A. Neuroethology	49
	B. Ethology, Sociobiology, and Behavioral Ecology	50
	C. Applied Ethology	52
IV.	Conclusions	53
V.	References	54

Chapter 3

THE COMPARATIVE APPROACH IN ETHOLOGY: AIMS AND LIMITATIONS

John L. Gittleman

I.	Abstract	55
II.	Introduction	56
III.	When Is the Comparative Approach Appropriate?	56
IV.	Aims of the Comparative Approach	58
	A. Defining and Describing New Variables	59
	B. Identifying Forms of Relationships	60
	C. Generating Hypotheses	61
	D. Testing General Hypotheses	62
	E. Testing Specific Hypotheses	63
V.	Limitations of the Comparative Approach	64
	A. The Data	64
	B. Biases in Distribution of Data	66
	C. Intraspecific Variation	67

	D.	Methodological Decisions in Data Usage	69
	E.	Establishing Causation	71
VI.	Future Directions		73
VII.	Acknowledgments		76
VIII.	References		77

Chapter 4
A BRIEF HISTORY OF THE STUDY OF ANIMAL BEHAVIOR IN NORTH AMERICA
Donald A. Dewsbury

I.	Abstract		85
II.	Introduction		85
III.	Pioneers		86
IV.	1890–1917		89
	A.	Zoology	89
	B.	Comparative Psychology	93
	C.	Interdisciplinary Interactions	98
V.	1917–1945		100
	A.	Zoology	101
	B.	Comparative Psychology	103
	C.	The Status of the Study of Animal Behavior in North America circa 1945	108
VI.	1945–Present		110
	A.	Recruits to the Population, 1945–1955	110
	B.	The Influence of European Ethology on North America	111
VII.	Conclusion		116
VIII.	Acknowledgments		117
IX.	References		117

Chapter 5
ANIMAL PSYCHOLOGY: THE TYRANNY OF ANTHROPOCENTRISM
J. E. R. Staddon

I.	Abstract	123
II.	Introduction	123
III.	Animals in Psychology: A Brief History	124
IV.	Animal Models	129

V.	Conclusion	132
VI.	Acknowledgments	134
VII.	References	134

Chapter 6
RECOGNITION LEARNING IN BIRDS
Milton D. Suboski

I.	Abstract	137
II.	Introduction: Purpose and Scope	138
III.	The Acquisition of Food Recognition	138
	A. Food Recognition from Parent–Offspring Interactions	139
	B. Social Transmission of Food Recognition	142
	C. Food Recognition without Reinforcement	144
IV.	Released-Image Recognition	146
	A. The Released-Image Recognition Model	147
	B. Comparison with Pavlovian Conditioning	150
V.	Recognition Learning Phenomena	152
	A. Imitation or Observational Learning of Food Recognition	152
	B. Search-Image Formation	153
	C. Avoidance-Image Formation	154
	D. Predator Recognition	155
	E. Imprinting	156
VI.	Assessment and Conclusions	163
VII.	Acknowledgments	164
VIII.	References	165

Chapter 7
PSYCHOIMMUNOLOGY: RELATIONS BETWEEN BRAIN, BEHAVIOR, AND IMMUNE FUNCTION
Paul Martin

I.	Abstract	173
II.	Introduction	174
III.	Types of Evidence for Links between Brain, Behavior, and Immune Function	176
	A. Clinical and Epidemiological Evidence	177
	B. Experimental Evidence	177
	C. Conditioning Effects	177

	D.	Anatomical Evidence	177
	E.	Brain Stimulation and Lesioning	178
	F.	Other Evidence	179
IV.	Clinical and Epidemiological Evidence		179
	A.	Associations between Psychological Factors and Illness	179
	B.	Associations between Life Events and Illness	180
	C.	Direct Measures of Immune Function in Humans	182
V.	Experimental Studies of Animals		185
	A.	Stress and Disease Susceptibility	185
	B.	Stress and Immune Function	186
	C.	"Control" as a Determinant of the Physiological Effects of Stress	187
	D.	Immunoenhancing Effects of Stress	189
VI.	Conditioning Effects		191
VII.	Mechanisms of CNS–Immune System Interactions		195
	A.	The Role of Corticosteroid Hormones	196
	B.	Other Possible Mediators	197
VIII.	Does the Immune System Influence Behavior?		200
IX.	A Mediating Role of Social Relationships?		201
X.	Future Research: The Role of Ethology		203
XI.	Appendix: Some Immunological Terms		204
XII.	Acknowledgments		205
XIII.	References		205

Chapter 8
HOW DO GENDER DIFFERENCES IN BEHAVIOR DEVELOP? A REANALYSIS OF THE ROLE OF EARLY EXPERIENCE
Lynda I. A. Birke

I.	Abstract	215
II.	Introduction	216
III.	Hormonal Effects on Behavioral Organization	218
IV.	Blurring the Dichotomy: Organization in Context	220
V.	Assumptions of the Hypothesis: Organizing Typical Animals	224
VI.	The Dichotomy Revisited: The Role of Hormones in Developmental Pathways	229
VII.	The Importance of Being Social: Some Implications For Research	235
VIII.	Acknowledgments	237
IX.	References	237

Chapter 9
BEHAVIORAL DEVELOPMENT: TOWARD UNDERSTANDING PROCESSES
Carel ten Cate

I.	Abstract	243
II.	Introduction	244
III.	Imprinting as a Phenomenon	245
IV.	Imprinting as a Developmental Process	246
	A. The Acquisition Process	248
	B. The Timing Process	253
	C. An Active Role for the Developing Individual?	261
V.	Discussion	263
VI.	Acknowledgments	266
VII.	References	266

INDEX 271

Chapter 1

HAS SOCIOBIOLOGY KILLED ETHOLOGY OR REVITALIZED IT?

George W. Barlow

Department of Zoology and Museum of Vertebrate Zoology
University of California
Berkeley, California 94720

I. ABSTRACT

I employ the definition of ethology that embraces biological studies of animal behavior, and that of sociobiology that includes both social interactions and behavioral ecology. The principal difference is the emphasis on behavior per se in ethology. I then sketch a personal view of the development of ethology, in the Lorenz–Tinbergen sense, and go on to point out its salient differences from sociobiology. The two fields, in their earlier states, can be separated in a number of ways. These include, most importantly, the source of hypotheses, which emanate from different evolutionary and ecological perspectives. The emphasis in sociobiology is on individuals and the costs and benefits to them of their actions. This is most sharply illustrated in explanations of how communication works. The methods employed also set the fields apart.

Now, however, the distinctions are blurring. Increasingly, those who study behavior at the level of mechanisms appreciate the importance of ultimate explanations. And the more evolutionarily minded are finding it necessary to move down to the level of causal mechanisms to fully explain the system. Thus mate choice in lekking birds can be illuminated by the details of the males' displays, and the decoupling of sex hormones and sexual behavior is understandable in relation to its adaptive significance.

Sociobiology has provided a rich supply of questions, many of which are best answered in an ethological setting. This is particularly the case with the increasing awareness of contingent strategies, alternative forms of reproductive behavior, kin recognition, mate choice, optimal foraging, and the like. In addition, many important issues in ethology were never well resolved; they need reexamination using the more sophisticated contemporary approach to experimentation together with improved technology.

II. INTRODUCTION

Ethology is dead, or at least senescent. That is, it is if you think of ethology in the narrow sense—the study of animal behavior as elaborated by Konrad Lorenz, Nikolaas Tinbergen, and Karl von Frisch. It has been quiescent for some time. No exciting ideas were emerging, and data gathering on key issues had lost its direction, despite the fact that many of the central precepts of ethology remain poorly tested. The pioneers of ethology would probably agree, though I imagine they would prefer to have it put less bluntly. After all, they made such great contributions to the study of animal behavior that ethology came to be identified with their particular approach, so they must feel a strong sense of union. But science is an unsentimental, impersonal creature that rumbles on, driven by the ambition of its practitioners. Lorenz and Tinbergen understand that better than do those who ponder ethology's mortality today.

This is best illustrated, in my mind, by an episode at the International Ethological Conference in the Hague, in 1965. The late Eckhard Hess proposed and campaigned for a radical change in the governing body of that biennial conference. He wanted Lorenz and Tinbergen made permanent members of the executive committee. Hess feared the field of ethology would be usurped by interlopers who would not maintain the faith. It was Tinbergen who spoke out in no uncertain terms, rejecting the unwise proposal. His argument was, in essence, that the seemingly protective cocoon would mummify ethology.

Any field of science must be open to change, and change is usually brought about by those who are regarded as sacrilegious upstarts. Particularly annoying, in all such revolts, is the perceived need of the bearers of the new truth to distance themselves from what has gone before, to create the impression that what is new is totally new and has little or no continuity with that which preceded it. Lorenz (1981) recently put it another, more colorful, way: ". . . the development of a science resembles that of a coral colony. The more it thrives and the faster it grows, the quicker its first beginnings . . . the vestiges of the founders and the contributions of the early discoverers . . . become overgrown and obscured by their own progeny."

In the 20 years since Hess' proposal, we have seen ethology grow from a coral colony into a virtual coral reef, accompanied by abundant speciation. Probably the most jarring event was the proclamation by E. O. Wilson (1975) that ethology, indeed the study of animal behavior in its broadest sense, would cease to exist—the dead coral at the base of the reef.

The demise of animal behavior, as a subject fit for study, would be wrought by sociobiology. Animal behavior was to fall through a yawning crack between sociobiology and population biology on the one side, and neurobiology on the other (Wilson, 1975, p. 698). Some of my colleagues refer to this as the dumbbell model: Neurobiology forms a large blob at one end and sociobiology and population biology the blob at the other; ethology is the narrow strip connecting them.

The objective of my essay is to argue to the contrary. The study of animal behavior had indeed begun to stagnate by 1975, and the advent of sociobiology was just the kick in the pants the field needed to get moving again. To make my point, I will review some of the more important contributions of ethology. Then I will contrast ethology and sociobiology to illustrate how they differ. But since this essay is in response to the specific charge of examining the state of ethology, sociobiology will be treated in only enough detail for a profitable contrast with ethology.

I must also make abundantly clear that this is a personal perspective that reflects my own educational ontogeny. I spent the years 1958–1960 with Lorenz, soaking up his views on ethology. Except for one year spent in Tinbergen's lab in Oxford, and one with Klaus Immelmann in Bielefeld, my views have been shaped by events in North America. I emphasize this because it has become so clear to me that history is always expressed through a personal filter the pores of which are shaped by one's environment. I was particularly impressed by this upon hearing John Durant lecture on the history of ethology, a remarkably British perspective (since published: Durant, 1986). I purposely avoided reading in this area, e.g., Thorpe (1979), because I want to present a personal view. The North American scene has been more controversial than that in Europe, where the distinction between ethology, sociobiology, and behavioral ecology has always been less acute.

After contrasting ethology and sociobiology, I will demonstrate how biology is best served by considering both the proximate and ultimate levels when explaining animal behavior. That means bringing ethology back into the picture to better understand how behavior is adaptive and how it has evolved, purposely blurring the distinction between ethology and sociobiology.

Most of the time I use the term *ethology* in its best and broadest sense. Ethology is the study of behavior itself and in an evolutionary framework, supported by a method that characterized the works of Lorenz and Tinbergen—empirical observations of naturally occurring behavior, guided by a few simple assumptions about the structure of behavior, producing testable hypotheses. Tinbergen (1963) delimited ethology in a broad way that is well suited to the contemporary situation: It concerns itself with the control and development of behavior, which might be equated to the proximate level, and with the function (adaptiveness) and evolution of behavior, roughly the ultimate level.

I also use the word *sociobiology* in its broadest sense. Krebs (1985) employed the term sociobiology interchangeably with behavioral ecology. He pointed out that sociobiology is frequently applied in a narrower sense just to social interactions, and that behavioral ecology pertains to behavior used to exploit and compete for resources. Behavioral ecology is indeed employed this way in recent literature. But behavioral ecology once had a different connotation, referring mostly to the behavioral mechanisms underlying the pattern of distribution of animals, as in seeking the optimal temperature or humidity, and related phenomena (e.g., Barlow, 1958; Porter *et al.*, 1973). In this essay, I follow Krebs' wider definition of sociobiology. Today the data of this sociobiology are by no means limited to behavioral events but ramify to embrace morphology and life history.

To return to the main theme, the course of science is often influenced by chance, or by mere accident. Charles Darwin, for instance, had a copy of Mendel's work in his office, which he obviously never read because it was uncut (Allen, 1975). If Darwin had been aware of the particulate nature of heredity his mind surely would have raced ahead, applying to evolutionary problems the algebra that derives from segregating units of inheritance. Freed of the paradox of dilution through blending in successive generations, he might have come yet closer to formulating many of the major issues of contemporary sociobiology.

The course of ethology also might have been different if Lorenz or Tinbergen had read R. A. Fisher (1930) and grasped the significance of population genetics to behavior. Instead, and perhaps due to the distraction of the nature/nurture controversy, they were concerned with the reality of behavioral genetics in the Mendelian tradition, and with the comparative evolutionary approach.

Another accident of history that contributed to the bias for Mendelian genetics in ethology was Lorenz's stint at Columbia University, where his father sent him as a young man to disrupt his romance with his future wife, Gretl. At Columbia, Lorenz came under the influence of T. H. Morgan, who made a profound impression on him (Konrad Lorenz, personal communication, 1959).

Had Lorenz and Tinbergen recognized the power of Fisher's concept of genetics, sociobiology might have developed before ethology. Imagine that took place, and that we only recently discovered some of the more exciting concepts from ethology such as releasers, stereotyped chunks of behavior, stimulus filtering, causal systems and displacement behavior, appeasement behavior, and phase specificity of development, as in imprinting. Such discoveries would excite the biological community even more than they did originally. Their relevance to evolutionary theory, now developed in the context of sociobiology, would be more obvious.

Imagine, further, the enthusiasm that would greet the development of the quantitative analytical methods of contemporary ethology. Those who had been grappling with sociobiological hypotheses would welcome such a precise analytical tool for testing the hypotheses that had been accumulating.

But that is not the way history has run. In fact, it is unlikely sociobiology could have developed as it has without ethology as a precursor. This historical fantasy, however, helps us appreciate the role ethology has played as a precursor to sociobiology, and how useful ethology can be to sociobiology today.

III. ETHOLOGY IN RETROSPECT

Most ethologists trace the beginning of their field to 1872 with the appearance of Darwin's *The Expression of the Emotions in Man and the Animals,* but it really was not until Lorenz (1935) and Tinbergen (1951) began actively publishing that the field took form. Lorenz and Tinbergen contributed to theory in a major way while advocating a strong basis for that theory in empirical observations. Using this balanced approach, they unearthed problems that radiated in many directions.

While reflecting on some of the issues in ethology I make no attempt to be comprehensive or even balanced. Rather, I simply select issues that prepare the reader for the comparison of ethology with sociobiology that follows, as well as for suggestions concerning future developments. As Harry S. Truman put it, "You can't know where you're going until you know where you've been."

A. Experiential Line

One of the most persistent misconceptions ethologists had to overcome was the belief that they concern themselves almost exclusively with "instinctive" behavior. While "instincts" did play a central role (Tinbergen, 1951), ethologists have always been interested in the role of experience and how it enters into the development of behavior. Lorenz was inspired here by his colleague Oskar Heinroth, who explored the early development of behavior by hand-raising a large number of species of birds. Through this exposure (e.g., Heinroth, 1911), Lorenz became aware of the phenomenon he later called *sexual imprinting,* the importance of which he recognized and communicated so well.

Lorenz was also fascinated with the "rigidification" that sometimes succeeds the process of learning. His most famous example (1952) was that of the water shrew that continued to jump over a nonexistent rock that had been removed from the shrew's learned path (see also Thorpe, 1956).

Lorenz (e.g., 1965) also theorized considerably on what he called *instinct–learning intercalation*. This was an attempt to integrate stereotyped species-typical behavior and plastic aspects of behavior into a coherent system (see also Eibl-Eibesfeldt, 1961).

Tinbergen did his earliest work on the behavior of wasps in the field (references in Baerends *et al.*, 1975). Much of it was concerned with the way wasps garner information when provisioning their nests. Those studies revealed the complexities of learning in such "lower" organisms, including one-trial learning. The findings presented challenges for theorists of learning behavior.

For some time, ethologists have been intrigued by another aspect of learning called "search image" (von Uexküll, in Hinde, 1966). Lukas Tinbergen (1960), Niko's brother, established the concept as an active focus of research through his field studies of prey capture by nest-provisioning titmice. He discovered that the kinds of insect prey the birds brought to the nest differed in frequency from the numbers predicted from the availability of prey, particularly when the prey were either sparse or superabundant. Through further analysis, he related differential reinforcement to selective perception, a precursor to optimal foraging.

One of the most flourishing areas in ethology has been the study of the acquisition of song in birds. Its historical roots are truly old (see Wickler, 1982), but much of the momentum of modern studies came from Thorpe (e.g., 1961; see also Marler and Mundinger, 1971) and his students. The tradition has been to analyze how song is acquired during early development and to relate this on the one hand to evolutionary problems such as the significance of song dialect (Kroodsma and Miller, 1982), and on the other hand to its neurological substrate (e.g., Konishi and Nottebohm, 1969).

Despite this array of research in learning and early development, ethologists were vigorously attacked and characterized as advocating genetic determinism. The assault was initiated by Kuo (summarized in 1967) and a school of American psychologists associated with T. C. Schneirla. Most of the criticism was directed at the writings of Lorenz, and particularly that which appeared before World War II (see Lehrman, 1953). The target, the parts of ethology that depicted instincts as little influenced by experience and under tight genetic control, was relatively narrow. The ethologists were vulnerable because of the sharp distinction they made at that time between instinct and learning.

Viewed in a modern context, the early ethologists' perception of how the genetic constitution of an animal is translated into the behaving phenotype does seem unsophisticated. Nonetheless, biologists, and that includes ethologists, have historically appreciated that the relationship between genotype and phenotype is not a neat one-to-one affair (Oppenheim, 1982). For their part, the ethologists pointed to this perspective and accused their critics of radical environmentalism, a not entirely unjustified criticism (Oppenheim, 1982).

One of the lasting contributions of the Schneirla group was their insistence that to comprehend behavior one must study the developmental processes that produce it. Likewise, the ethologists' argument that learning can be fully understood only in the context of its natural setting has gained ever wider acceptance among psychologists (e.g., Shettleworth, 1984).

This vigorous debate, the nature/nurture argument, led to an exchange of views that culminated in a profitable synthesis—the book by Robert Hinde entitled *Animal Behaviour: A Synthesis of Ethology and Comparative Psychology* (1966). That Hinde was able to produce this book derives in important ways from his close association in Cambridge with William Thorpe. The continuity of Hinde's book with Thorpe's (1956) *Learning and Instinct in Animals* is obvious.

Hinde's book produced increasing and broader interaction between ethologists and comparative psychologists that resulted in a more biological perception of learning processes. Perhaps most important was the growing realization that learning itself is subject to natural selection, that the ability to learn evolves. Such thinking led to a conceptual framework sometimes characterized as "constraints on learning" (Hinde and Stevenson-Hinde, 1973; Shettleworth, 1972), although that area of research arose relatively independently (Garcia et al., 1973).

We might well ask, then, where this line of ethological research has led. The study of imprinting blossomed into a large literature. Attention rapidly shifted away from sexual imprinting, which requires waiting until an animal is reproductively mature to test it, to filial imprinting, which requires only a brief interval after the experimental treatment (scientists, especially when not tenured, tend to be impatient).

At about the same time, the bulk of research on filial imprinting moved to psychology as psychologists came to appreciate the ramifications of imprinting for learning theory. The shift was also away from evolutionary considerations to ones concerning behavioral and physiological processes. However, some recent developments have put sexual imprinting back into an evolutionary setting (e.g., Bateson, 1982).

In contrast to imprinting, instinct–learning intercalation has not fared well. This concept was interwoven with the assumption of a distinction between innate and acquired components. Bateson (e.g., 1976) rejected the concept on that ground and he has been for the most part followed in the primary literature, though some introductory textbooks persist in the dichotimization into innate and acquired behavior.

Unfortunately, an important point has been lost in the rejection of instinct–learning intercalation. The concept may still have merit in a more sophisticated formulation that does not make such a clear separation between the origins of the components. One conceptualization that obviates the dichotomy is the cake-bake model of Bateson (1976). He correctly pointed out that though the starting ingredients of the cake can be identified, once stirred and baked it is impossible to classify parts of the cake as, say, learned or innate. This model has left the unintended impression, in some quarters, of a uniformity of behavioral events with regard to their development and control. However, Bateson pointed out that behavioral events are arrayed on a spectrum in this respect (see also Barlow, 1977).

It is this spectrum that interests me here, and that I want to relate to the intercalation as proposed by Lorenz. At one extreme, behavioral events are buffered from environmental inputs. Thus, the modal action patterns of many species show scarcely any noteworthy variation within and among individuals (Barlow, 1977). The environment, however, influences these events in profound ways, regulating when they are expressed, their sequence of occurrence, and their orientation. That is the intercalation that Lorenz originally expressed. Such behavior has been called *consummatory* by ethologists and *unconditioned responses* by psychologists.

At the other extreme is behavior so variable and diverse as to defy characterization except in terms of goals and consequences. Ethologists originally referred to behavior of this sort as *appetitive*. Lorenz's objective, I believe, was to clarify how various components of behavior fit together. The more stereotyped actions, those most resistant to experiential modification of their form, serve almost as tools for the more dynamic aspects of the system that regulate the timing and "aiming" of the recognizable chunks of behavior.

Oddly, psychologists seem to have had no problem with this, probably because they never imposed a nature/nurture dichotomy onto the formulation, and because they seemed comfortable with a spectrum of behavior as they worked out the basis of learning in a formal system. That formal system received a number of unexpected inputs from ethology. One-trial learning, for instance, has moved into the realm of learning psychology. Search image continues to get attention, largely because of its consequences for optimal foraging (e.g., Kamil and Sargent, 1981; Krebs, 1978). And ethologists still examine the acquisition of bird song, searching for its meaning in relation to dialect (e.g., Kroodsma and Miller, 1982), and coming ever closer to understanding its neurological basis (Notebohm, 1984).

The major fallout of the nature/nurture debate has been the increasing importance of behavioral development. The study of development, however, has not been blessed with coherent, provocative theories the like of which have driven sociobiology (Immelmann et al., 1981).

B. Homology/Analogy: The Comparative Approach

As traditional evolutionary biologists, schooled in systematics and comparative anatomy, Tinbergen and Lorenz, and their students, naturally concentrated on comparative approaches to behavior. A landmark paper by Lorenz (1941) was a comparative study of displays by ducks that had major implications for their classification as well as bearing on theoretical issues about homology and analogy, and how behavior evolves. Tinbergen (1959) and his students, for their part, engaged in a prolonged comparative study of gulls with parallel implications.

Because the focus of those investigations was displays (which were treated almost as morphological traits) it follows that the concepts to emerge dealt in large measure with communication. The central concept was that of some signal, called a releaser, whose counterpart receptor in the (central) nervous system was called a releasing mechanism. [In his enthusiasm for the progress being made by ethology, Julian Huxley (1963) proclaimed the concept of a unitary releaser–releasing mechanism *the* major advance in ethology.]

At the time of these discoveries, European ethologists were recovering from an overdose of subjectivity in the study of behavior (Tinbergen, 1951). Behavior was now to be viewed "objectively"; its external manifestation was the central piece in any analysis, even though intervening variables such as drive became important constructs. As a consequence, the concept of releaser–releasing mechanism–response came to convey a relatively fixed and invariant view of behavior. These integrated components of behavior were regarded as fundamental parts of the individual, equivalent to its limbs or tongue.

I do not argue that such behavior does not exist. Many examples were provided by Eibl-Eibesfeldt (1970). But the result was that, in our mental images, animals almost became automatons with little scope for alternative behavior. This vision of behavior may also have been influenced by what was perceived as a logical necessity for homologizing behavior, as in comparative studies of displays.

Such studies led inevitably to a consideration of the origin and evolution of displays. By tracing gradual changes in displays and associated structures across related species, ethologists were able to infer homology of behavior patterns (e.g., Lorenz, 1941, 1965; Tinbergen, 1959). Their studies of derived behavior patterns raised the problems of convergence and parallel evolution, which are endemic to all analyses that seek homologies.

Homology and analogy are central concepts in evolutionary theory. While most biologists believe they have a good grasp of those concepts, application can be difficult in specific cases. This is nowhere more evident than in comparisons between the behavior of humans and animals (Barlow, 1980; Blurton Jones, 1975; Chagnon and Irons, 1975; von Cranach, 1976; Lorenz, 1966; Mason and Lott, 1976). Lorenz's attempts (e.g., 1966) to interpret human behavior through such comparisons drew the fire of many critics then and for years afterward (e.g., Klopfer, 1977).

Atz (1970), in a memorial to Schneirla, concluded that it is impossible to apply homology to animal behavior, but the essay by Baerends (1958) is a persuasive argument to the contrary. In his later active years Lorenz (e.g., 1974) emphasized the value of comparison through analogy. The focus was on repeated evolution of solutions to common problems. But there theoretical development in ethology stopped. It is moot whether one considers the further elaboration of what came to be called "design features" a continuation of this line of thought within ethology, or a part of the development of sociobiology.

C. Survival Value

A fundamental assumption underlying ethological studies was that any reasonably regular piece of behavior has positive survival value. Otherwise, it would not be there. Apparently Tinbergen (e.g., 1965) became uneasy with that assumption. He and his students started doing experiments in the field to test whether certain behavior patterns are in fact adaptive. Among the pioneering studies involving gulls were ones on removal of eggshells from the nest (Tinbergen et al., 1967), and on breeding in synchronous colonies (Kruuk, 1964; Patterson, 1965).

In performing those studies, Tinbergen opened up an area of research for ethologists. Adaptiveness of behavior was now considered in both the positive and negative sense, what is now called *costs and benefits*. For instance, although carrying away an egg shell reduced the chance that the nest would be detected by a predator, it simultaneously increased the chance that the unguarded chick would be eaten. Nesting close together within the colony resulted in excessive disturbance from neighbors, but nesting farther away toward the perimeter of the colony increased the risk of predation. In each of these examples, the critical factor became the differences between individuals: Some gulls nested in the center of the colony, some on its perimeter.

This type of ethological research continued to prosper. While other inputs have been important, this genre of studies has contributed to sociobiological work emphasizing costs and benefits, with attention to individual differences.

D. Causal Analyses

Just as the comparative study of displays led naturally to analyses of communication, so did they inspire questions about their causation. Why is it that the same stimulus to a given individual causes different behavior at different times? To what extent can one predict the response as more and more inputs are regulated? How much of the differential responsiveness is determined by changing internal states, such as arousal in the short term or hormonal fluctuations in the long term?

Because the explanations involved hypothetical constructs that depend on conflicting tendencies to respond in one way or another, they were termed *motivational studies*. Behavior was resolved into systems and subsystems, such as aggression, courtship, incubation, and the like. These studies tried to account for the evolutionary origins of displays in terms of shifting causation and motivational conflicts (e.g., Baerends, 1975; Baerends and Cingel, 1962; Tinbergen, 1959). To those familiar with the literature on derived activities, terms such as *intention movement, displacement behavior, appeasement behavior, ritualization*, and *typical intensity* will bring to mind an area of research that was once exceedingly active but is quiescent today.

An issue that excited considerable controversy was the proposal by Lorenz (1950, 1965, 1966) that behavior may be performed spontaneously. The longer an "instinctive" behavior is not expressed, the more likely it is to be performed, until finally it "goes off" in the absence of stimulation. His irrepressible sense of humor led him to formalize this relationship by portraying a center in the central nervous system as a toilet. The dynamics of the stool and its tank became the psychohydraulic model. Water, representing reaction-specific energy, constantly trickles into the tank. As the tank fills, less stimulation is required to "release" the response. When completely full, the tank discharges without stimulation. Peter Marler once commented to me that the model would have gained wider acceptance if Lorenz had used an electronic capacitor as the metaphor explaining the changing threshold of response.

Lorenz's application of the psychohydraulic model to aggression excited the strongest opposition, and still does (Klopfer, 1985). Unfortunately, the baby has been thrown out with the bathwater: Studies of the causal mechanisms of aggression, including motivation, are now neglected by biologists (but see Maynard Smith and Riechert, 1984).

E. Hierarchical Structure and Chains of Behavior

In the context of ethology, the hierarchical structure of behavior usually means the organization of behavior into branching trees of groups and subgroups, running from complex variable events to simple muscle contractions (Tinbergen, 1951; Fentress and Stilwell, 1973). However, the principle has broader application and plays a central role in most analyses of behavioral structure (Dawkins, 1976a).

An early approach to behavioral sequences was the study of "chains" of stimuli and responses involving two organisms. This was beautifully demonstrated by Baerends (1941) in his classic analysis of hunting behavior in a wasp. A stimulus A1 triggers a response A2, which leads the wasp to attend to stimulus B1, resulting in response B2, and so on. In practice, the concept of a simple interaction chain, as when it involves two animals mating, was an oversimplification. Animals sometimes skip steps, and redundancy is common (Hinde, 1966). However, the important message here is that two organisms start interacting and the situation determines which path of the hierarchy will be followed. The organisms then work their way toward some end point. To those concerned only with outcomes of behavior, it would be easy to concentrate on just the last step, regarding it as the salient behavior for an ultimate explanation.

F. The Ethogram and Chunks of Behavior

Jennings (1906) described the behavior of animals in terms of recognizable movements he called their *action system*. Makkink (1936) later used the term *ethogram* in describing the behavior of a shorebird (see the brief historical summary in Schleidt et al., 1984). The ethogram developed into a style of description in which the recognizable displays and other stereotyped movements of an animal were recounted in detail. The components were then put together in a functional context.

It is easy to forget that this approach is in reality a research strategy, a prescription for how to start an investigation. It was not always so obvious. But the spirit of the ethogram is embedded in most contemporary field projects, and many laboratory programs have their roots in ethograms. The first step is still obvious—selecting the chunks of behavior to quantify.

Despite his tongue-in-cheek disdain for quantification, Lorenz (1950) recognized the necessity of resolving behavioral components into units to deal with them quantitatively. To many ethologists, the unit became the *fixed action pattern*, an unfortunate term arrived at by committee (Thorpe, 1951); I prefer to call such chunks of behavior *modal action patterns*, or MAPs (Barlow, 1977).

The essential point, however, is not the terminology; it is that qualitative ethograms progressed to quantitative descriptions of behavior. One popular quantitative representation of the ethogram was a flow diagram in which the motor events were disks interconnected by lines of varying thickness; the thickness was proportional to the number of times that particular sequence was recorded (e.g., Baerends et al., 1955). At a more sophisticated level, Wiepkema (1961) borrowed factor-analytic techniques from psychology, including a graphic representation, in his study of the spawning of a fish, the mussel-guarding bitterling (*Rhodeus amarus*). Delius (1969) approached skylark maintenance behavior using a number of quantitative techniques.

These quantitative studies were aimed at illuminating the continuous stream of overt behavior (e.g., Nelson, 1964). Often MAPs were used to test ideas about motivation (e.g., Baerends et al., 1955; Wiepkema, 1961). Another application was in communication, quantifying the sequence of MAPs expressed by two interacting animals (e.g., Hazlett and Bossert, 1965; Dingle, 1972). The culmination of these quantitative studies was the research of Golani (e.g., 1976), in which he used Labanotation (ballet notation) to record in exquisite detail the pattern and sequence of movements of interacting individuals. Schleidt et al. (1984) have recently proposed a procedure for making a standard ethogram that resembles Golani's method in spirit and is much simpler (but here the ethogram is again reduced to just a catalog of the motor events of an individual).

Ethologists have commonly employed another technique for quantifying behavior. It is an indirect measure, the consequences of the behavior (see Hinde, 1966). Thus, one studies not how a nest is built but how many twigs are moved, not how an animal drinks but how much, and so on. Consequences are removed various degrees from the actual behavior.

Today one hardly ever encounters ethograms in mainline behavior periodicals, though some are published in animal-group journals. The contribution that endures is the working method of regarding the stream of ongoing behavior as consisting of relatively continuously variable behavior (the old appetitive behavior) and chunks of relatively stereotyped behavior (the descendent of consummatory behavior). It is the naming of the latter, the chunks, MAPs, or whatever, and treating them as units, that still characterizes much of the current work in behavior. And this remains so despite the fact that MAPs grade smoothly into continuously variable behavior (Barlow, 1977).

The ethogram is not an end in itself. It is a tool, a method, a first step in describing behavior patterns and their organization. Ethograms gained meaning only as they became something more than a static catalog of behavioral events within a sparse motivational skeleton. As causal relations became better understood, ethograms assumed alternative forms, precursors of contingent strategies, and took on a dynamic status. As they did so, ethograms became a source of hypotheses about the organization and control of behavior. And MAPs remained the primary units of quantification in the testing of those hypotheses.

IV. CONTRASTING ETHOLOGY AND SOCIOBIOLOGY

When I read Wilson's eloquent book, *Sociobiology*, I often found it difficult to keep in mind that it was meant to be something different from ethology. Much of the material fitted comfortably into traditional ethology. There were indeed differences, and those differences have become clearer as sociobiology has matured.

I will point out a number of salient differences between ethology and sociobiology, most of which are inherent in the literature but are seldom articulated. This should sharpen our awareness of differences and thereby of similarities, and in turn help create a fruitful working relationship between the two disciplines.

A. Formulation of Questions

Consider the ways questions have been formulated within the two disciplines. The differences are not pronounced—we are dealing with shades of gray—but the emphasis is real. In my experience, Lorenz stressed the importance of deriving hypotheses from sound, detailed empirical observation, that is, induction. Tinbergen also seemed to favor that approach in the beginning, as did von Frisch when he wondered why the bees were waggling on the surface of the honeycomb. Later, Tinbergen became more deductive as he questioned the adaptive significance of behavior instead of assuming it. But the main thread was clear in the beginning.

Commonly an ethologist picked an animal for reasons that were almost as variable as the animal itself. The behavior chosen to note and to quantify was actually guided by hypotheses, but these were seldom stated; some observers may even have been unaware of them. The observations themselves, however, led to more precise, focused hypotheses (see Caws, 1969). During the period from about 1950 to 1965 ethology experienced a flurry of testing hypotheses. And most of this revolved around the central concepts in ethology, which had been arrived at inductively. Contemporary ethologists have become more hypotheticodeductive, but some workers are still disposed toward empirical observation as an end in itself.

Sociobiologists have been more deductive. Predictions have been deduced from a larger theoretical framework, such as the anisogamous relationship between the sexes (e.g., Williams, 1966). Typically, a hypothesis is stated right at the outset, such as the expectation of a biased sex ratio of 3:1 when mating competition is likely among sibling males (Trivers and Hare, 1976; further examples in Wilson, 1975). The most advanced expression of this way of thinking lies in formal modeling, as in the application of game theory (see Section V. C. 2).

B. Methodology

As pointed out in Section IIIF, ethologists characteristically start a project by making a functional ethogram of unmanipulated behavior. Surprisingly, a good deal of this research has been done in the laboratory. The investigator usually has a good grasp of the situation in nature, however, and often tries to simulate the important features of the natural habitat in captivity, especially in the beginning.

The sociobiologist also observes normal behavior, but more typically that of free-living animals in the field. Moreover, the investigator more often takes along clearly articulated preformulated hypotheses that guide data gathering. The data frequently are some steps removed from the actual behavior, being instead the consequences or outcomes of behavior. "Pure" sociobiologists, for instance, are seldom interested in the form of a display. Rather, examples of measures are typically outcomes such as owning a display site, number of mates inseminated, benefits conferred upon kin, and number of eggs thrown out of a nest.

For some time little experimentation was done within sociobiology. Rather, correlations were sought to test hypotheses. Increasingly, however, the experimental approach is employed; this is especially welcome and fitting when done in the field (e.g., Andersson, 1982; Ebersole, 1980; Krebs, 1982; Power, 1975).

C. Reduction

How reduction is construed is a telling difference between ethology and sociobiology. Ethologists have traditionally sought reductionist explanations of behavior at two proximate levels. One level involves the behavior itself—as in causal or system analyses. The other seeks the physiological substrates of behavior. But, since physiology and behavior are ultimately the same (they are simply measured differently), these all may be considered physiological explanations.

Sociobiological explanations are motivated by their relevance to ultimate factors—the adaptiveness, or evolution, of behavior. Thus, when a sociobiologist does turn to reduction it is typically formulated in the framework of optimality (Krebs and McCleery, 1984), whether stated or not. Costs are weighed against benefits. Ideally, the common currency is reproductive success, but this is often difficult to quantify. Cost–benefit analyses are applied to genetic explanations, as in kin selection, as well as to energetic ones.

One type of genetic reduction is the formulation of a model accounting for the way a particular strategy affects its bearer's genetic representation in succeeding generations (e.g., Parker, 1983; Grafen, 1984). Another type is the testing for kinship through biochemical methods, as when studying nepotism (Hanken and Sherman, 1981). [The bioenergetic approach, taken far enough, results in physiological explanations (e.g., Ryan, 1985).]

Ethology and sociobiology, then, have been separable: Ethology seeks reductionist explanations in physiology, and sociobiology would like to find them in genetics. In sociobiology, even energetic considerations reduce to differential reproductive success of different genotypes.

D. Genetics and Individuality

It is impressive how little concern for experimental genetics of behavior has been shown either by ethologists or sociobiologists. Moreover, ethologists and sociobiologists differ in the role they ascribe to genetics in behavior. The ethologists have confined their efforts to experimental genetics that utilize classical techniques such as crossing strains and even species (e.g., Franck, 1974; Manning, 1976). Although sociobiologists have carried out conventional genetic analyses (e.g., Maynard Smith and Riechert, 1984), their efforts have been directed mainly to theoretical models involving population genetics (e.g., Kirkpatrick, 1982; Lande, 1981). Oddly enough, most of behavioral genetics today is carried out by comparative psychologists or by geneticists. The findings have penetrated little into ethology or sociobiology.

Ethologists have shown scant interest in differences within species, regardless of their source. Such differences have been recognized, but they were treated as distracting variations around the mean. The basic idea was one of species universals, although with enough variation to allow evolution to occur. If competition was considered at all, it was with regard to interspecific interactions. Intraspecific competition was not seen as a driving force in evolution (though intraspecific interactions—for example, the evolution of signals—were). This type of thinking by ethologists led to a pervasive group-selectionist set of mind that was subconscious, though most ethologists would deny it (but see Eibl-Eibesfeldt, 1982).

Considerations of aggression best reveal this attitude. Julian Huxley (1963) wrote that escalating contests that result in injury "would mitigate against the survival of the species." Lorenz (1966) conveyed a similar view about damaging fights being contrary to the benefit of the species. Most surprising was Tinbergen's (1964) comment, in another context, that "one party—the actor—emits a signal, to which the other party—the reactor—responds in such a way that the welfare of the species is promoted." I say surprising because at the same time Tinbergen and his students were engaged in field studies on gulls that demonstrated how selection operates on the individual. All three of these ethologists professed adherence to the concept of individual selection. All three, nonetheless, lapsed into a convenient shorthand for explaining the evolution of behavior.

From the beginning, sociobiologists made an overt distinction between selection at the level of the group and selection at the level of the individual (Hamilton, 1964; Williams and Williams, 1957; Williams, 1966). Later, selection at the level of the gene was drawn into the picture (Dawkins, 1976b; Williams, 1966).

Most sociobiologists find it parsimonious to explain behavior first in terms of differences between individuals, and genetic differences are the very stuff of their arguments. Such differences are inherent in reasoning about fitness, inclusive fitness, and evolutionarily stable strategies. But most of all, the application of game theory (e.g., Maynard Smith, 1979) has propelled individuals into the forefront of sociobiological thinking (Slater, 1981; Krebs, 1985). Differences between individuals get the attention in evolutionary theories.

Given this, it is astonishing that the sources of behavioral differences between individuals are so seldom investigated by sociobiologists; commonly they are conveniently assumed to be genetic. Even when it can be shown that different experiences may lead to dissimilarities between individuals, the effect is still subsumed under the wisdom of the genes (Dawkins, 1976b; Wilson, 1975); the developmental plasticity itself becomes part of the individual difference. I accept this, but I also think it pushes important issues into the background.

E. Development

While ethologists have come to view the development of behavior as a crucial area, sociobiologists have been slow to concede the significance of development to their discipline. Sociobiologists recognize that selection operates at all stages of life history. However, development is not viewed as a process so much as a progression, a longitudinal expression of life history with selection operating at all stages: "As a rule, theoretical sociobiologists and behavioral ecologists treat the complex links between gene and behavioral phenotype . . . as modules that can be temporarily decoupled from the explanatory scheme . . . in order to provide the degree of simplification necessary for models in population biology" (Wilson, 1975, p. 717).

This attitude contrasts with theorizing in other areas of biology. Some morphologists, for example, are returning to the spirit of D'Arcy Thompson, Garstang, and De Beer in seeking to understand the evolution of differences through developmental processes. Alberch *et al.* (1979) formalized the dynamics of such processes, particularly timing and rate; they treated development as a central issue in evolutionary thought. In this view the morphologists and ethologists are in agreement. Here, the ethologists have been more progressive than have been the sociobiologists.

F. Ecology

Ethologists have long prided themselves on emphasizing that behavior can be understood only in its appropriate environment. And that has meant more than the study of communication or causal systems. They argued early and persuasively that any search for general laws of learning is hampered if one ignores the adaptive significance of differences between species (Lorenz, 1965). The ethologists' view of the environment, however, was a descriptive and static one, reflecting the status of ecology at the time. It amounted to the physical surroundings plus predators and members of one's own species—that is, autecology.

Sociobiologists have benefited from a more sophisticated concept of modern ecology. They have brought in the bioenergetic perspective, as in the importance of energy balance in a feeding territory, and demography. Such life-history variables as length of life, rate of predation, and reproductive effort have assumed central importance in determining such features as the mating system and timing of reproduction (Gadgil and Bossert, 1970; Sibly and Smith, 1985; Williams, 1966; Wilson, 1975). But, lest we forget, the demographic data are explained by the behavior of individuals.

G. Natural Selection

From the beginning, ethologists brought to their studies of behavior a presumption of natural selection and adaptation. A central theme has been that the behavior must be adaptive or else it would not persist. Just how the behavior is adaptive has usually been argued after the fact and has only seldom been put to the test. As noted, the British school of ethologists, led by Tinbergen, moved away from this complacent attitude. An appreciation of costs and benefits was emerging.

As mentioned in Section IV.C, sociobiologists zeroed in on costs and benefits in the context of natural selection. Alexander (1975), in a landmark paper, developed this way of thinking as applied to social behavior. His students, Hoogland and Sherman (1976), extended that line of thought in their analysis of the tradeoffs of group life in bank swallows. This further contributed to the shift away from interspecific to interindividual comparisons, or simply to "decisions" by individuals.

The roots of this conception of costs and benefits to the individual are several. One of the earliest grew from the calculations arising from considerations of kin selection. What cost should an altruistic individual bear in relation to the inclusive genetic benefit accruing through the recipient of the altruism (Hamilton, 1964)? Another source was the economics of optimality, stimulated by models of foraging (Krebs, 1978) but later extended to almost any type of behavior. Finally, game theory played an enormous role, concentrating on the costs and benefits of particular strategies within interactive contexts.

Sociobiology differs from ethology most strongly in its perception of adaptation. Its practitioners have perceived adaptation as a dynamic process in which individuals constantly evaluate the flux of environment. Selection produces a population of organisms, each individually and slightly differently attuned to the world around it. Such a viewpoint makes it easier to account for similar traits in different organisms and different traits within the same species, and for alternative tactics within a given individual. A population of organisms represents a suite of adaptations which may vary from generation to generation.

Ethology also perceived individuals as dynamic systems constantly evaluating the environment, but in a proximate sense. In ethology, however, the traditional view of adaptation was static. Although population genetics was understood, behavior was regarded as equally adaptive among all individuals, save the interesting exception.

H. Evolution

Ethologists have traditionally directed their efforts toward comparative studies of related species. That goal has its roots in systematics, reconstructing the evolutionary pathways of behavior of interest, or even working out taxonomic relationships (e.g., Selander, 1964). Because convergence or parallel evolution could vitiate such studies, analogy became an important consideration. But even though analogy was understood, there was no general appreciation of the need to integrate the findings of comparative studies with ecological principles. Consequently, mating systems continued to be accepted as given, and the forms of displays were regarded as arbitrary.

The chief concern of sociobiologists has been analogy, though this is seldom stated. The emphasis has been on evolutionary solutions to problems. For instance, given that resources in the environment are distributed patchily and are economically defensible, should the mating system be polygynous (Emlen and Oring, 1977; Horn, 1968)? Thus, recurring patterns are important not because they indicate relationships but because they reveal design features resulting from similar selective forces.

It might be argued that this difference between ethologists and sociobiologists is merely a consequence of the end points they study. Granted, MAPs are better material for studying phyletic relationships, hence homologies, whereas mating systems are more suitable for the study of analogous features. But it is also clear that the environment can shape MAPs through evolutionary time, sometimes in concert with the mating system (Crook, 1964). Therefore, the mating system can be viewed as a component of the environment.

I. Communication

Perhaps the most overt difference between ethology and sociobiology has arisen in the analysis of communication. At first, this difference in perspective seemed large, but it has shrunk considerably [compare Dawkins and Krebs (1978) with Krebs and Dawkins (1984)]. The difference, nonetheless, is important. In the traditional ethological model, communication is a cooperative endeavor—sender and receiver have coevolved to maximize the efficiency of exchanging information (Marler, 1959). The signals derive from various indicators of the sender's internal state, such as intention movements, displacement behavior, and epiphenomena associated with activation of the autonomic nervous system (Cullen, 1966; Morris, 1956). Thus, the behavior of the sender is a reliable sign of that which the sender is likely to do in the near future—communication is honest and informative (see also Zahavi, 1979).

It follows, since communication is cooperative, that displays have differing effects on the receiver depending on their form and the context in which they are given. This assumption was severely tested when Maynard Smith and Price (1973) proposed that, on the contrary, the particular displays are not so important. Rather, what matters is simply how long an animal persists. In this case, they were writing specifically about combat and the now famous game called the *War of Attrition*.

Dawkins and Krebs (1978) extended this way of thinking, in a delightfully provocative fashion, to further challenge ethological precepts about communication:

> The . . . attitude . . . we espouse emphasizes the struggle between individuals. If information is shared at all it is likely to be false information, but it is probably better to abandon the concept of information altogether. Natural selection favours individuals who successfully manipulate the behaviour of other individuals. . . . Of course, selection will also work on individuals to make them resist manipulation . . . . (Dawkins and Krebs, 1978)

The foregoing paragraph, taken out of context, appears to confound two concepts—manipulation and deceit. They are different (Richard Dawkins, personal communication, 1986). If a person tells someone to get off his property or he will shoot, it is one thing for him to have a pistol in plain view and yet another to claim that he has a pistol that he could use, but chooses to keep hidden in his pocket for the moment. Both forms of behavior are attempts to manipulate. But in the first case the receiver is sure the sender has the pistol. In the second case there is ample room for doubt because the sender could be lying. Even in the first case, there may be uncertainty about the sender's intent to use the pistol, or there could be reason to suspect the pistol is a fake; but testing that suspicion presents a greater risk to the receiver. Thus, there may be a continuum of situations ranging from obvious deceit through ambiguous risk to honest threat.

In the second edition of their chapter, Krebs and Dawkins (1984) modify their position. Communication is no longer couched in terms such as deceit and lying, and the interests of the receiver are brought in. Conflict of interest between sender and receiver is still important, but the authors point out the kinds of situations in which that should be expected; they include combat, some instances of male–female conflict, and parent–offspring conflict. They move closer to the mainstream ethological position by recognizing that "the evolution of ritualized signal movements or structures from their precursors is the product of coevolution between the *roles*." "Roles" here means that an individual is at one time the sender, at another time the receiver.

Predator and prey are always in conflict and try to manipulate and deceive one another: Consider the angler fish luring its victims, or nestling birds mimicking the hissing of a snake. But it confuses the issue to extend this line of thought to intraspecific interactions as when, referring to a female sea otter rearing the male's young, it is stated that "the effect is that the male manipulates the female in much the same way as he manipulates a stone [to break up mollusc]" (Krebs and Dawkins, 1984, p. 384). The sea otter example is misleading because it is also in the female's interest to rear the young. One could just as well argue that the female has manipulated the male to fertilize her ova. But, if so, the male happily manipulated.

So far, I have not seen an example of intraspecific deceit that could operate if all members of the population adopted that strategy. In the cases of female mimicry among fishes (Barlow, 1967; Dominey, 1980; Gross and Charnov, 1980), the deceived male is the parental caretaker. If all males simultaneously adopted the female-mimic role, none would reproduce successfully because their eggs and larvae depend on care by an honest, parental male. Another instance of negatively frequency-dependent deceit is bluff in a freshly molted mantis shrimp (*Gonodactylus bredini*) that is too soft to fight (Steger and Caldwell, 1983).

Thus, the sociobiological approach is set off by the recognition that communication often involves a conflict of interest between the sender and the receiver. For further progress along these lines, however, it is essential to maintain the distinction between interspecific and intraspecific interactions.

V. THE EVOLVING CONTINUITY BETWEEN ETHOLOGY AND SOCIOBIOLOGY

Ethology and sociobiology are changing rapidly, such that each increasingly has a foot in the other's camp. The articles one sees today reflect this blurring of boundaries. Consequently, much of the foregoing is history in the making. Nonetheless, I continue to encounter papers and seminars in which the work is framed in a blinkered format, ignoring either proximate mechanisms on the one hand, or ultimate explanations on the other.

In the following I first make the case for the importance of both levels of explanation. If a separation must be made, and I don't favor that, one could characterize, for ease of exposition, the ethological input as concentrating on proximate explanations, and the sociobiological contribution as seeking ultimate explanations.

A. The Relevance of Proximate Explanations

It is not preaching to the choir to urge more attention to proximate mechanisms when searching for ultimate explanations. Too often proximate mechanisms are not considered, despite the frequent proclamations that alternative hypotheses are a central consideration. The following example makes the point.

Power (1975) claimed to have demonstrated that altruism does not exist among mountain bluebirds. He did this by removing one mate of a pair caring for nestlings. If a newly arrived mate then helped care for the nestlings, altruism would have been demonstrated on the assumption that the new mate was not related to the nestlings. The new mates, however, ignored the nestlings, so Power rejected the altruism hypothesis.

Emlen (1976) rejoined that there is an abundant literature demonstrating that birds usually do not accept young unless they are hormonally prepared for them. This usually requires having passed through the sequence of events leading up to hatching and the presence of nestlings (see also Beach, 1978). The alternative and more plausible hypothesis, then, was that the failure of the new mate to provide care was due to the absence of the appropriate hormonal substrate. Thus, proximate mechanisms can invalidate an experiment, post hoc ultimate explanations notwithstanding (e.g., Power, 1975).

Removing and then returning a subject is a favorite experiment of field workers, but it can create pitfalls at the proximate level. To illustrate, when testing the effects of prior ownership the territory holder may be removed for periods of several days before being released to contest for its territory with the new occupant (e.g., Krebs, 1982). But the effect of removal is more than one of just being away from the territory. For instance, confinement can alter the levels of sex steroids and thereby interfere with aggressive responses independent of the effect of owning a territory. Failing to consider proximate mechanisms as alternatives or supplements to ultimate ones may therefore lead to unwarranted conclusions.

On a more positive note, addressing behavior *qua* behavior can shed light on questions about ultimate significance. Lekking, an aggregation of males in an arena from which females select mates, is an excellent example. Some males are selected more often than others, but the reasons for this are not clear. Is female choice based on quality of the male however assessed, his position in the lek, or the size of his display ground? This apparently straightforward issue has not been easy to resolve. Gibson and Bradbury (1985) sought the answer in the sage grouse (*Centrocercus urophasianus*). They quantified several aspects of the males' lekking behavior. Only vocalizations were analyzed in the kind of detail that has characterized ethology (probably because the available instruments permitted relatively easy and accurate quantification of sounds). They also measured the key features of territory dimensions, position, and male size.

One acoustic component of the strut display, as well as rate of display and lek attendance, correlated positively with mating success. And the measures of display differed among individual males. No aspect of territory size or location explained male success. Gibson and Bradbury expressed regret that visual aspects of the display had not been measured. Given that the acoustic signal was so informative, a comparably careful analysis of the visual aspect would be in order.

It is a truism that each biological phenomenon can be explained at both the proximate and ultimate levels. And proximate and ultimate explanations must be compatible. Proximate hypotheses are often more immediately falsifiable. Because of their primacy over ultimate mechanisms, they set the boundaries for ultimate explanations. Further, there is also the risk that the behavior is not an evolved adaptation in the sense of Williams (1966). Considerable effort can be wasted pursuing the ultimate explanation when there is none.

The question arises, then, of what constitutes an adequate explanation. Most behavioral phenomena are not satisfactorily explained at the proximate level until they are carried down some levels of organization. That having been said, there is no denying the utility of connecting proximate and ultimate explanations, to provide pattern and thus meaning to the findings.

B. Ultimate Causation: Giving Meaning

Sociobiologists have often been indifferent to explanations at the proximate level. In contrast, there is widespread and growing acceptance, among those working on behavior at the level of mechanisms, of the view that placing behavior in an evolutionary context is beneficial (e.g., Ingle and Crews, 1985). An instructive recent discovery in comparative endocrinology, for instance, is that reproductive activities rise and fall in inverse relation to sex hormones. Thus, a reptile might engage in courtship and copulation when sex steroids are at their lowest level, but show no sexual behavior when the levels are at their highest. Only an adaptive approach makes sense of the situation.

Crews and Moore (1986) characterized the connection between reproductive activities and titers of sex hormones as being of three types. The first is the familiar case with temperate-zone species, a close temporal association between gonadal recrudescence, sex steroids, and mating behavior.

The second type is one in which the gonads are continuously responsive and sex hormones remain at a high level. If a favorable climatic event occurs, as when the zebra finch (*Taeniopygia guttata castonotis*) experiences rain in its arid environment, gametes develop rapidly followed by reproduction.

The third case is the most remarkable. It is illustrated by the male red-sided garter snake (*Thamnophis sirtalis parietalis*). Upon emerging from winter hibernacula the males quickly mate with the females, but they have small gonads and low levels of circulating androgens. There is apparently not enough time for them to produce sperm in the spring because mating occurs so quickly. So the males undergo spermatogenesis in late summer when they have high titers of sex steroids, and they store the sperm through the winter. The result is temporal decoupling of sexual behavior and male sex hormones. Mating behavior evolved in response to a specific environmental situation—cold torpor followed by rising temperatures—and not to the internal clue of steroid levels (Crews and Moore, 1986).

At the level of manifest behavior, the literature has been growing on copulatory behavior of rodents, canids, and other mammals. Much of this has been written by comparative psychologists with little regard to evolutionary implications. Consequently, it has been difficult to make sense of the diversity that was discovered.

Dewsbury, however, has been seeking correlations between behavior, morphology, and environment. He found that rodents that lock during copulation have safer nesting sites than do those that do not lock (Dewsbury, 1975). Later, he reported that polygyny in rodents is positively correlated with numerous thrusts and ejaculations per mating bout, with ornate penises, and a number of other features (Dewsbury, 1981). The opposite conditions often predict monogamy. The adaptive significance is obvious and gives meaning to the variation.

It was crucial to relate the findings about reptile endocrinology and rodent mating behavior to adaptation to different environments. That greatly enriched our understanding. But note that the evolutionary models here are of the type employed by early ethologists. They apply at the level of species. The next step ought to be an examination of interindividual differences within, say, the red-sided garter snake or the prairie vole, and the costs and benefits associated with those differences.

The following example will surprise some readers. It concerns sexual imprinting. Because it has been such a major issue in ethology, one would assume that sexual imprinting is well rooted in evolutionary theory. Quite the contrary, ultimate explanations were post hoc. A commonly discussed hypothesis was that by learning its species, with the parent as the model, an organism saves neural connections that otherwise would have been devoted to coding species identification.

The suspicion has been growing, however, that when sexual imprinting is implicated in species identification, the individuals may actually have an inherent disposition to respond preferentially to their own species (Bateson, 1979; Gallagher, 1978; Gottlieb, 1976; Hess, 1973). In contrast, ten Cate (1984) feels the evidence for such a predisposition is not conclusive, at least from his studies.

Bateson (e.g., 1983) put forth an alternative and perhaps more plausible ultimate explanation for sexual imprinting: The individual learns its particular parent or sibling in order to optimize its degree of outcrossing when picking a mate. Whether this is the correct ultimate explanation is almost immaterial. Reevaluation of sexual imprinting creates a number of hypotheses to test and to relate to individual decisions within different mating systems to see if they make evolutionary sense. Some may jump at this concept as a way of connecting sexual imprinting with human behavior. It is tempting to equate optimal outcrossing with the incest taboo, but that is a risky extension at the very least (Patrick Bateson, personal communication, 1987).

The final examples were chosen because they may give meaning to a kind of display that drew the attention of early ethologists but remained a puzzle, and because they constitute such a commendable melding of ethology and sociobiology. They concern the occurrence of postcopulatory displays (Barfield and Geyer, 1972; Lorenz, 1966). Why should a male pigeon, immediately after copulation, perform a display flight with loudly clapping wings, often followed by the hen? No satisfactory answer was provided, probably because the answers were framed in the context of motivation models and had no reference to competition within the species.

A similar situation exists with the anthophorid bee *Centris pallida*. After copulation, the male of a mating pair emits an acoustic signal while holding and stroking the female (Alcock and Buchmann, 1985). The male also emits a different, brief, sound just prior to the breakup of the pair. Sperm is transferred almost at once when the male first grasps the female, so the sounds and palpating by the male are postcopulatory displays. The male does make occasional further genital contact, particularly just before releasing the female.

Alcock and Buchmann considered alternative hypotheses and found one most plausible: Vocalization and stroking by the male are meant to turn off female receptivity; the late genital contacting is a means of determining her receptivity. To test this, they presented two kinds of females to other males— females that had been inseminated but not stimulated tactilely and acoustically, and females that had been both inseminated and stimulated. New males accepted the unstimulated females as mates, but soon left the stimulated females. From the male's evolutionary perspective, it is to his advantage to turn off female receptivity and thereby avoid sperm competition or supplantation.

But why should the female submit to this? Alcock and Buchmann offered a highly speculative reason, drawing on the variation in size among males, and on the presumed preference of the female for the sperm of larger males. I leave it to the reader to go to the source and judge whether this is a just-so story or a stimulating call for further research.

These examples make the point. Putting physiological and behavioral findings into an evolutionary context that centers on differences among individuals can open doors to entirely unexpected areas of research. It can also give meaning to well-known but heretofore unexplained behavior.

C. The Role of Ethology in Current Behavioral Research

The most important contribution the ethological approach can make is to put behavior back into behavioral research, or to keep it there. At the 1986 meetings of the Animal Behavior Society in Tucson, Arizona, I heard a number of papers in which the data presented were morphological measurements. True, the hypotheses under test had been generated by behavioral phenomena, and the data were relevant to them. Further, the findings were of inherent interest. But they might just as well have been presented to audiences not concerned with behavior.

What has set the ethological approach apart has been attention to the performance of behavior, and its immediate causation—that is, to the skinny bar portion of Wilson's (1975) dumbbell. There would be no way of understanding the adaptiveness of decoupling sex steroids and mating behavior in garter snakes if one studied that species at only the population or neuroendocrinological levels. Likewise, one could speculate forever on why some male sage grouse are preferred by females; the answer would not be found in the neurophysiology of the males, nor in population studies. Only a direct analysis of the behavior itself contains the answer. Having gotten that answer, neurological explanations may then become relevant. Thus, there is a necessary and equal connection between population phenomena, individual behavior, and physiological substrates.

I turn now to a number of research areas that, for the most part, are considered sociobiological to illustrate how the ethological approach can be profitably applied. In many instances, this is already being done but is seldom recognized for what it is.

1. Energetics

Behavioral ecology relies heavily on the currency of energetic costs and benefits (and also on the risk of predation). Whether explicit or implicit, this often translates into a detailed study of the time–energy budget (e.g., McCann, 1983). The activities of the animal are divided up into behavioral components which range from continuously variable actions such as locomotion to discrete events such as displays. Time spent in each activity is converted to energy expended, usually employing some relatively crude estimate, such as the known cost of locomotion under specified conditions such as body weight, speed, and temperature (e.g., Feldmeth, 1983).

Seldom is there any attempt to obtain detailed information about the different displays and MAPs used in social interactions. Too often they are all assumed to have the same costs, and these are not measured but are variously inferred from resting or active metabolism. When such behavior is measured directly, it can show remarkably high metabolic costs (Ryan, 1985; Wells and Taigen, 1986). Doubtless, some displays in other organisms are done at virtually no cost. However, that displays can be energetically expensive has important ramifications for arguments about signaling as manipulation (i.e., Dawkins and Krebs, 1978). In general, activities such as attacks in territorial defense are sufficiently infrequent that they account for a relatively small amount of the energy budget.

2. Game Theory as Applied to Combat

Nowhere in sociobiology has theory moved further ahead of data than in the application of game theory. This is clearest for analyses of combat. The advent of the war-of-attrition model (Maynard Smith, 1974) ignited the field, stimulating a progressive elaboration and refinement of theory, seeking an ever-closer approximation to reality in the sense of accurately predicting unexpected findings (e.g., Enquist and Leimar, 1983; Parker and Rubenstein, 1981). What is lacking is sufficient studies testing these fascinating ideas. Initially, old ethological data were resurrected (e.g., Caryl, 1979). While some recent research has been carried out (e.g., Enquist *et al.*, 1985; Jacobsson *et al.*, 1979; Mosler, 1985; Paton and Caryl, 1986; Riechert, 1978, 1986; Turner and Huntingford, 1986), much remains to be done.

We need careful analyses illuminating the process of assessment. Are decisions to escalate the only critical variable? To what extent are fights a series of rounds, and is information assimilated and integrated from round to round if not immediately from displays? Taking as our guide the paper by Gibson and Bradbury (1985) on lekking sage grouse, would a more careful analysis of communication lead to a different interpretation, such as predicting the winner earlier in a contest (see especially Turner and Huntingford, 1986)? Can the contestants be manipulated experimentally to test directly the predictions made by theory?

One of the more productive aspects of game theory has come from regarding interaction as a tradeoff between conflicting interests. Should the individual cooperate or behave selfishly (see Vehrencamp, 1981)? This line of inquiry started with Trivers' (1971) article on reciprocal altruism, and led to a more formal statement of the prisoner's dilemma in the tit-for-tat model (Axelrod and Hamilton, 1981). That has caused investigators to pose questions that would not otherwise have been asked, as in tolerance and conflict among tree swallows (Lombardo, 1985). The appropriate data, in most instances, consist of quantifying behavioral interactions. This type of model provides a bridge to noncombative interactions in which a conflict of interest is nonetheless usually present.

3. Mating Behavior and Other Noncombative Interactions

How animals select their mates, and why, has been the center of attention in much sociobiological literature (e.g., Bateson, 1983). On the assumption that the only consideration is anisogamy, theory predicts that females ought to be the choosy sex, and that males ought to compete among themselves to fertilize the eggs of as many females as possible (Williams, 1966, 1975).

But what is it the female is choosing, if indeed she chooses, and why? Does she aim for the male with superior genes as Trivers (1972) proposed? Does she choose arbitrarily and start the process of runaway selection (Fisher, 1930; Kirkpatrick, 1982)? Does the female discriminate in favor of the most vigorous and hence heterozygous male (Jeffry Mitton and Randy Thornhill, personal communication, 1986), or the male least infected by parasites (Hamilton and Zuk, 1982), and thereby avoid the dead end of exhausting genetic variation? As another possibility, does she try to optimize the tradeoffs between outbreeding and inbreeding (Bateson, 1983)? The female might be responding to all of the above, or to some combination of them.

This essay is not the place to tackle ultimate explanations of mate choice. The reason for bringing them up is that theory is now well ahead of empirical findings. The array of hypotheses could be narrowed by establishing empirically the basis for mate choice in a number of species. And that can be a tricky task. The main difficulty, especially in the field, is that the female may be responding to a number of traits, not just the one the experimenter fixes on (see Halliday, 1983).

What is needed is manipulation of key stimuli. By coloring some males, Semler (1971) demonstrated that females of a polychromatic population of threespine sticklebacks (*Gasterosteus aculeatus*) prefer males with red bellies to those without. Similarly, Andersson (1982) manipulated the lengths of the conspicuous tail feathers of male longtailed widowbirds (*Euplectes progne*). Compared to baseline control groups studied earlier in the season, male widowbirds with normal or shortened tails experienced a drop in reproductive success whereas the males with artificially elongated tails maintained their success at the previous level.

Of considerable moment to theory would be experiments designed to test the extent to which the sensory systems by which mates are assessed are tuned to certain physical features of the environment, such as the prevailing background (Endler, 1978) or wavelength of light in a forest (Burtt, 1986; Hailman, 1977) or lake (Lythgoe, 1979, 1984), or sounds in a complex environment (Ryan, 1986). Those design features of perception, shaped by the organism's surroundings, are the primary adaptations of the sensory system. They should then constrain the nature of social stimuli that are effective in social communication. That ought to predict which display movements, colors borne, or vocalizations emitted by males are the most conspicuous and thereby elicit the strongest response in females. This could have considerable bearing on arguments about how runaway selection gets started, and also on such issues as rapid speciation (e.g., Dominey, 1984). There must be some identifiable basis to differential responsiveness among females other than some untestable construct such as aesthetics.

As the foregoing examples suggest, most of the literature has concentrated on female choice for the obvious reason that theory predicts females are the choosy sex. However, given the opportunity, even highly polygynous males may discriminate among potential mates. For example, males of the lekking pupfish *Cyprinodon macularius californiensis* choose the largest, hence most fecund, female of a group (Loiselle, 1982).

Further, females and males ought to be comparably selective when the male shares parental care with the female. She should pick a mate that is best equipped to provide care and is not apt to desert her. But how does she know that? As for the faithful male in a biparental pair, his reproductive success is identical to hers. Therefore, he should be just as choosy as the female—but is he?

Remarkably little research has been done on mate choice in biparental species, probably because most theory has developed around the more challenging issue of sexual selection. In a study of the biparental Midas cichlid fish (*Cichlasoma citrinellum*), Rogers (1985) found that the female selects the largest, most aggressive male with previous breeding experience. Those traits proved important in holding a territory and in defending the young from predators. Contrary to expectations, the male did not discriminate among females according to size, aggressiveness, or prior breeding experience. Thus, the male behaves as one might predict for a polygynous species.

Rogers suspects that female Midas cichlids set up dominance relationships and that the largest females get the largest males, obviating the need for males to choose. Additionally, all females are maximally aggressive when with young, so there is no need for males to discriminate among females on that basis.

The relevance of fine-grain behavioral studies extends well beyond these bread-and-butter issues of mate assessment and sexual selection. Documenting the nature of, and variation in, mate bonding in monogamous species in relation to ecology and phyletic history could throw light on the evolution of mating systems. Parent–offspring and sib–sib relations are increasingly analyzed in an ethological context with attention to the details of the behavior itself (Horsfall, 1984; Mock, 1984; Stamps *et al.*, 1985), inspired by theoretical papers on parent–offspring conflict (Trivers, 1972; Stamps and Metcalf, 1980).

4. Discriminating among Kin

The issues of kin selection and nepotism have also generated research on interactions between parents and offspring, and among siblings and more distantly related individuals (reviewed by Holmes and Sherman, 1982). Increasingly, attention has turned to ways in which individuals discriminate among others depending on degree of relationship. The closer the kinship, the more cooperative and less aggressive ought to be social interaction. But how do animals make this discrimination?

Holmes and Sherman (1982) offered four hypotheses for how animals might acquire information about relatedness: spatial distribution, association, phenotype matching, and "recognition alleles." Their review, and their own findings on two species of ground squirrels, suggest that phenotype matching is the most probable candidate.

Holmes and Sherman showed how a careful program of experiments involving cross-fostering of young squirrels can narrow down the possible behavioral mechanisms involved. The test procedures concentrated on direct measures of hostile behavior. The experimental design and measurements were typical of ethological research.

But studies of kin recognition extend well beyond friendly versus hostile responses, to behavior such as choice of mates. And here again sociobiology converges with ethology. Hoogland (1982) reported that prairie dogs avoid mating with close kin. And Holmes and Sherman (1982), while cross-fostering ground squirrels, obtained evidence for a brief period of learning kin, occurring some time around weaning and emergence from the natal burrow. The parallel with sexual imprinting and optimal outcrossing in birds is remarkable. Thus, the concept of kin selection, seen at first as so different from anything proposed in ethology, rapidly converged with ethology when proximate mechanisms were sought.

5. Plasticity of Behavior

A major trend in sociobiology seems paradoxical when compared with the publicity that surrounded sociobiology a decade ago. At that time, the press pounced on the issue of genetic determinism, especially as it related to human behavior (Gwynne *et al.*, 1978). Now contingent strategies and *alternative reproductive behaviors* (ARBs to some) seem increasingly important (e.g., Austad, 1984).

How the animal responds is regarded more and more in the context of its evaluating the situation and making a measured response. To be sure, authors hasten to add that they imply no cognition, at least not in the sense of the animal understanding the ultimate consequences. They mean merely that the system has built into it alternative strategies, options if you will. I have seen no note of how this contrasts with the traditional ethological position, which viewed the organism almost as an automaton. Can these different perceptions be resolved? Are we dealing with the blind men palpating an elephant?

There is some merit in Lorenz's (1965) instinct–learning intercalation, even if it is an oversimplification and perhaps even misleading with regard to how behavior develops (Bateson, 1976). Behavior is not a homogeneous cake, nor is it of two types, instinctive and learned. But the pieces and the matrix of behavior do vary enormously in a given individual.

Some behavior appears as chunks that are stereotyped, easily recognizable, recur, and are exceedingly resistant to environmental modification of their basic coordination (Barlow, 1977, 1981). Lorenz (1965) argued that conventional learning consists of making the association between the relevant stimulus and the production of these chunks of behavior. The psychologist may recognize these as unconditioned responses, and other behavior as consummatory behavior or MAPs. This is not to say that either is genetically determined in the strict sense. All behavior does indeed develop. Nonetheless, what we conventionally call learning has little influence on the form of MAPs, but a great influence on how such forms of behavior are used.

Learning has become central in some areas of sociobiology, most prominently in optimal foraging (Krebs, 1978). The experimental paradigms have shown remarkable convergence with those in psychology, from which there is much to learn (Kamil, 1983; Pulliam, 1981; Sherry *et al.*, 1981). By attending to the details of learning, sociobiologists are contributing importantly to understanding fundamental problems in ecology and hence to ultimate explanations. Moreover, learning is being studied in the field, as in Alejandro Kacelnick's (personal communication, 1986) analysis of nest provisioning in starlings. That will interest comparative psychologists because their research is becoming more naturalistic (Staddon, 1980). (Interestingly, many comparative and physiological psychologists now refer to themselves as biological psychologists.)

There is probably no area of sociobiology research in which plasticity of behavior is not important. I think most investigators are coming around to this position. The source of adaptive variation, however, might be in dispute depending on one's predilections. Whatever the bias, detailed study of behavior under different environmental circumstances is essential for understanding what is going on.

A quote from Waddington (1975, p. 170, originally 1959) is appropriate here:

> The relation of the behaviour of an animal to the evolutionary process is not solely that of a product. Behaviour is also one of the factors which determine the magnitude and type of evolutionary pressure to which the animal will be subjected. It is at the same time a producer of evolutionary change as well as a resultant of it, since it is the animal's behaviour which to a considerable extent determines the nature of the environment to which it will submit itself and the character of the selective forces with which it will consent to wrestle. The "feed-back" or circularity in relation between an animal and its environment is rather generally neglected in present-day theorizing.

Waddington's comments are just as timely today as when they were first written. An animal may by accident or by choice find itself in one social group or another, or none, and that can have profound consequences. It may behave differently in each situation. And its behavior can have a bearing on the survival of the group in which it resides. Thus, behavior can drive evolution. Bateson (1988) has prepared an essay elaborating on this point.

VI. WHITHER ETHOLOGY?

Most predictions are either extrapolations from what is already happening, or a plea for others to do what interests us. I will try to avoid emphasizing my own interests though I imagine they are apparent. There is no escaping foretelling in terms of the present. The strategy is to pick issues that need attention and that would profit from the ethologists' touch.

A. Issues Generated by Sociobiology

I have already touched on a number of these, but some need repeating here. Perhaps the most remarkable effect sociobiology has had is one that would not have been predicted from the controversy that marked its advent. Instead of an era of genetic determinism, closer study has resulted in a conception of animals as remarkably labile. They seem infinitely adaptive, capable of inventing games that theorists elaborate, though no one really implies that higher cognition is involved. What animals do is contingent upon what they are and what confronts them. It makes no difference that they are only good at what they have evolved to do. Nor is it important, for the moment, to ascribe their contingent strategies to genetic differences. What matters is that we increasingly view animals, in a limited way, as calculating machines. This has significant consequences for future research strategies, and opens the door to thinking about the cognitive abilities that animals may possess.

Why do individuals respond now one way and now another? Early ethological research answered this question in terms of immediate sensory input and recent history, as in motivational models. That may again prove useful. But now it needs integrating into a larger framework of natural history. For instance, a young male in a social group behaves submissively until it becomes large enough to challenge. Thus, an important determinant of decision making derives from where one lies on a developmental trajectory.

Given that, how is conflict resolved, or cooperation forged, when two individuals find themselves at the same point on the developmental trajectory and in the same place? What is the role of past history, of genetic differences, of context such as prior residency? Is prowess the overriding factor, or can differences in motivation outweigh it (Barlow et al., 1986)? Conflict resolution, because of the extensive theory underlying it, is ripe for this type of research. It is therefore amazing how little is being done. Empirical research would have a salutary effect on theory and stimulate revision.

The converse of conflict, cooperation, is poorly studied despite well-articulated theory. The general framework of reciprocal altruism (Trivers, 1971; Axelrod and Hamilton, 1981), or mutualism (Wrangham, 1982), has evoked considerable interest, though little in the way of experimentation meant to falsify hypotheses (but see Lombardo, 1985).

Underlying these issues and those not discussed is individuality (Slater, 1981). We are just beginning to see the fruits of our attention to individual differences. The main thing is to understand the significance of the differences. Later, we will need to know more about the sources of such differences. That will lead to more detailed studies of development and genetics.

B. Neglected Issues in Traditional Ethology

Are we overly critical of the early ethological research? I think not, and that is not to be taken as an ungrateful comment. Many of the most widely accepted and cherished precepts of ethology were never adequately tested. As time passes, the older papers are read less often and, instead, are cited from secondary sources, which insulates them from close scrutiny. In this way time gives validity. The methodology and lack of statistical sophistication in much of the earlier studies would not stand the scrutiny of modern journal referees. Confidence, instead, has been gained by the coherence of a body of findings, by the reasonableness of their conclusions, and by personal experiences tending to bear them out.

A good example of the need for much more work is the releaser concept. Virtually all introductory textbooks on animal behavior use the case of the stickleback. Pelkwijk and Tinbergen (1937), in a classic study, showed that the red belly of a crude model of another fish elicits attack from the male stickleback holding a territory. However, a number of workers have published largely ignored articles reporting their failure to reproduce those results (see Rowland and Sevenster, 1985).

Baerends (1985) has taken up this case in an attempt to understand why others have not been able to demonstrate this simple "red-belly" releaser. Using previously unpublished data collected by Nick Colias in 1949, Baerends concluded that the motivational situation in most of the earlier studies, particularly the presence of fear, interfered with the response to the red belly, and that experience played a large role in the fearful state. Rowland and Sevenster (1985) also tackled this issue, through experiments using both European and North American sticklebacks, and did not get a clear answer. The issue is still open. The more important question is: Does this represent the general state of affairs for what we have called releasers?

A sister concept in ethology is that of *heterogeneous summation*. It was introduced by Seitz (1940) in his analysis of the behavior of a cichlid fish, *Astatotilapia strigigena*. Though his analysis was not adequately quantified, Seitz proposed that various stimuli trigger responses by summating independently of each other.

The most widely cited demonstration of the simple additive nature of heterogeneous summation, at least in textbooks on animal behavior, comes from the exhaustive experimental analysis done by Leong (1969) on aggressive responses to dummies by another cichlid fish, *Haplochromis burtoni*. However, the statistical analysis may be flawed (Lee Machlis, personal communication, 1980) for the kinds of reasons outlined by Machlis *et al.* (1985). The outstanding demonstration of heterogeneous summation remains the elegant work of Baerends and his colleagues (e.g., Baerends and Drent, 1982). Curio (1963) has cautioned that even when stimuli appear to summate it is unlikely that the relationship is a simple additive one. Clearly, there is much to do before the concept is accepted as generally applicable.

A number of other issues in ethology stopped short of culmination, such as the spontaneity of behavior, particularly aggression, and how it relates to the ecology of the species in question. I suppose this is just human nature. Scientists, like fans of popular music, are to a degree neophiliacs. We run with the popular theme until the excitement wanes, then switch to the new "hit song" as it climbs to the top of the chart. In other instances the problems may have been perceived as simply not worthy of further attention. Whatever the reason, motivational models disappeared, as did descriptive comparative studies.

Other issues appear to have been dropped because the technology was not adequate to make their further pursuit feasible. Thus, workers gave up in frustration when attempting extensive quantitative analyses in which mountains of data were gathered in various formats such as voiced entries on a tape recorder, or square waves generated by a multiple-pen recorder. The neophilia phenomenon may have been operating here as well. Note that the application of measures of uncertainty to determine information transmitted (e.g., Dingle, 1972), which calls for large data fields, disappeared from ethology at about the same time it lost fashion in related fields such as ecology. Current analyses of combat among animals focus on information, but in an altogether different context (e.g., Paton and Caryl, 1986).

The conceptual issues here are several that revolve around complex motor output. They relate to the structure of behavior as it is thought to reflect the underlying organization of the central nervous system. Are there lawful properties that govern the transitions from one type of behavior to another, or the recurrence of a given behavior (e.g., Nelson, 1964)?

If two organisms are involved, then the stream of data bears on the problem of communication. If animal A performs a particular signal, how does that affect the probability that the other animal will perform a given act (Dingle 1972)? If such studies were now done in the context of game theory, would the models hold up?

The fine structure of MAPs has seldom been analyzed, and only by Golani (e.g., 1976) and his associates. The findings have been applied to interactions and to recovery from major insult to the nervous system (Golani, 1976, 1981). The results are promising. There has been, however, no attempt to examine the genetics of the structure of the MAPs themselves (Barlow, 1981).

Now the technology exists to do these things, though it is sometimes frightfully expensive. Many labs have access to microcomputers and to digitizing pads that make data acquisition relatively painless. The software has been developed for other types of analyses but could be modified for animal behavior. There are now high-speed television systems that can record at 1000 frames per second. Powerful computer graphics can even rotate the derived figure in space to compensate for movements not in the plane of the lens. If such equipment had been available 25 years ago we would now have a subculture of ethologists well versed in the manipulation of massive amounts of data.

The pure sociobiologist may find this not only uninteresting but not worth doing. But such studies are important. They provide behavior models for neuroethologists to aim for: "Besides the needs for further improvement in anatomical, chemical, and physiological techniques, these questions underline the urgent need for better methods for assessing behavior in ways that will permit its variety to be better correlated with brain variables" (Bullock, 1984, p. 477). Analyzing the dynamics of the expression of MAPs provides such a bridge.

Neuroethologists are moving increasingly in the direction of analyzing complex, naturally occurring behavior (Ingle and Crews, 1985). One of the most challenging problems remaining in biology is explaining how the brain works. If in fact E. O. Wilson's dumbbell model were to prevail, it would make the connection between behavior and brain exceedingly difficult. We need to study behavior *qua* behavior to bridge the gap, and we need to provide the quantitative dimension so that behavior can be better translated into neurobiology and endocrinology.

VII. CLOSING COMMENTS

I hope I have been persuasive in my argument that there is no sharp distinction between ethology and sociobiology, although they are clearly different at the extremes. A major difference between the fields lies in the backgrounds of the practitioners. And even that difference breaks down because many prominent sociobiologists were trained originally as ethologists. The essential difference is in the crisp focus of ethologists on *behavior:*

> Ethologists and behavioral ecologists need to remember that the subject under investigation is behavior, not adaptation, optimization or even evolution. Concepts such as optimization are theoretical constructs that we use to help us understand and predict behavior, but are not themselves the objects of study. (Kamil, 1983, p. 292)

I would count as a devout sociobiologist someone who would disagree with Kamil's assertion, whereas an ethologist would agree.

Another difference that is sometimes proposed is that sociobiologists avoid explanations that rest on lower levels of organization. The view is that by moving down levels the essential aspects of design features will be lost (Krebs, 1985). True enough. But why stop there? Understanding design features produces clearer models of how behavior is organized. And that helps in the study of behavior in and of itself, and of its neuroendocrinological substrates.

Study of behavior at the proximate level feeds back on analysis at the ultimate level. It sets boundaries, tests predictions, and challenges assumptions underlying theory (e.g., Turner and Huntingford, 1986). In this way behavioral studies stimulate revision of existing theory or even creation of new theory. Because this reality of scientific inquiry is so often ignored I want to pause on one example recently pointed out to me by Judy Stamps (personal communication, 1987). It concerns the concept of ideal free distribution (Fretwell and Lucas, 1970) as it is so widely applied to establishing territories.

A pivotal precept of ideal free distribution in territoriality is that animals seeking territories balance off quality of territory with density of occupants. When quality is equal, densely settled areas are avoided. This basic assumption is central to derivative concepts such as the Beau Geste hypothesis (Krebs, 1977). Yasukawa and Searcy (1985), however, found that male red-winged blackbirds appear to be attracted to areas of numerous male territories. It might be argued that this is a special case of colonial nesters. However, Stamps finds that juvenile anoline lizards are drawn to territory holders; and she has found many examples of similar findings in the literature. The bottom line is that the basic assumption underlying the application of ideal free distribution to territoriality may not be general. If so, then its theoretical underpinnings require reevaluation, with possibly far-reaching consequences for the generality of the theory.

Another benefit of studying behavioral mechanisms is that they lend credence to findings obtained in higher-order studies by establishing the reality of a mechanism that can produce the hypothesized outcome. I might add, there is nothing about moving down levels of organization that requires giving up the broader view at higher levels. Reductionism, therefore, is complementary to holism, not antithetical. John Krebs obviously appreciates this or he would not be analyzing learning in fine detail to understand foraging strategies. Ethology and sociobiology are not separate enterprises, and the well-being of ethology depends intimately on progress in sociobiology. So it is important to ask about the health of sociobiology.

In a 10-year retrospective, Krebs (1985) looked for the kinds of new ideas that feed sociobiology and that have emerged since Wilson's (1975) treatise on sociobiology. They were surprisingly few and either not so new or not so general, despite the encouraging advances that have been made empirically. The most general advance mentioned was the growing appreciation of differences among individuals. Krebs also cited our changing perspective on communication, from a coevolved and cooperative system to one in which manipulation for selfish ends is often the case. The most specific advance mentioned, and one that remains untested, was the proposal by Hamilton and Zuk (1982) that females select mates on the basis of the males' resistance to disease, and the implications of that for maintaining genetic variation in the face of sexual selection.

Compared with advances sociobiology made in the decade prior to 1975, albeit quietly, the Krebs list is short and the ideas less global by far. Doubtless, the reader could provide his or her own list, but I suspect the advances thereon would be either relatively specific, or in actuality empirical. For further progress in the conceptual realm we need data that threaten the existing framework.

No, ethology in the broad sense is not dead. Sociobiology has breathed life into it by providing an array of stimulating questions that will keep it going for a long while. The future of research in animal behavior lies in a comprehensive, integrated approach that views behavior first and foremost in an evolutionary context. Investigators must then be willing to follow the reductionist path far enough to provide adequate answers to the questions they pose. That will lead to surprises, compelling reevaluation of the more encompassing theories. That should produce another burst of exciting ideas comparable to those in the births of ethology and sociobiology.

But what shall we call this field? Sociobiology is too narrow, even when it includes behavioral ecology, and so is socioecology. "Animal behavior" is fine but it is too long; besides, it runs the risk of being confused with the behaviorism of the logical positivists. The label animal behaviorist is, moreover, an inelegant term. As the reader might have predicted, I would prefer retaining ethology. It is brief. Its root refers simply to habits and the term is thus inclusive. The use of the word antedates Lorenz and Tinbergen, as you can quickly verify by examining the Zoological Record. It has grown in breadth and changed so much that it can no longer be equated with the original European approach to behavior. And ethology has gained wide usage in the social sciences as the evolutionary approach of the biologists to the study of behavior. For all these reasons, I vote for ethology.

VIII. ACKNOWLEDGMENTS

This article, in its nascent form, was presented as an invited lecture at the Third Regional Conference on the History and Philosophy of Science at the University of Colorado, Boulder, in 1979. A later draft of that article was read and criticized by David Barash, Colin Barnett, Robert Hinde, and Glendon Schubert, to whom I extend my warm thanks. The manuscript has been substantially altered to take into account subsequent developments. I am grateful to the following persons, whose comments on the ultimate version led to considerable improvement: Pat Bateson, Roy Caldwell, Richard Francis, Stephen E. Glickman, David L. G. Noakes, Susan E. Riechert, William Rogers, and Thelma E. Rowell. The writing was supported by Grant HD18496 from the National Institute of Child Health and Human Development, which I am delighted to acknowledge.

IX. REFERENCES

Alberch, P., Gould, S. J., Oster, G. F., and Wake, D. B. (1979). Size and shape in ontogeny and phylogeny. *Paleobiology* **5**:296–317.
Alcock, J., and Buchman, S. L. (1985). The significance of post-insemination display by male *Centris pallida* (Hymenoptera: Anthophoridae). *Z. Tierpsychol.* **68**:231–243.
Alexander, R. D. (1975). The search for a general theory of behavior. *Behav. Sci.* **20**:77–100.
Allen, G. (1975). *Life Science in the Twentieth Century*, Wiley, New York.
Andersson, M. (1982). Female choice selects for extreme tail length in a widowbird. *Nature* **299**:818–820.
Atz, J. W. (1970). The application of the idea of homology to behavior. In Aronson L. R., Tobach, E., Lehrman, D. S., and Rosenblatt, J. S. (eds.), *Development and Evolution of Behavior*, Freeman, San Francisco, pp. 53–74.
Austad, S. N. (1984). A classification of alternative reproductive behaviors, and methods for field testing ESS models. *Amer. Zool.* **24**:309–320.
Axelrod, R., and Hamilton, W. D. (1981). The evolution of cooperation. *Science* **211**:1390–1396.
Baerends, G. P. (1941). Fortplfanzungsverhalten und Orientierung der Grabwespe (*Ammophila campestris*). *J. Tijdschr. Ent.* **84**:68–275.
Baerends, G. P. (1958). Comparative methods and the concept of homology in the study of behaviour. *Arch. Neerl. Zool.* **13**(Suppl. 1):401–417.
Baerends, G. P. (1975). An evaluation of the conflict hypothesis as an explanatory principle for the evolution of displays. In Baerends, G., Beer, C., and Manning, A. (eds.), *Function and Evolution in Behaviour*, Clarendon Press, Oxford, pp. 187–227.
Baerends, G. P. (1985). Do the dummy experiments with sticklebacks support the IRM-concept? *Behaviour* **93**:258–277.
Baerends, G. P., and Cingel, N. A. van der (1962). On the phylogenetic origin of the snap display in the common heron (*Ardea cinerea* L.). *Symp. Zool. Soc. London* **8**:7–24.
Baerends, G. P., and Drent, R. H. (eds.) (1982). The herring gull and its egg. Part II. The responsiveness to egg-features. *Behaviour* **82**:(entire issue).
Baerends, G. P., Brouwer, R., and Waterbolk, H. Tj. (1955). Ethological studies on *Lebistes reticulatus* (Peters). I. An analysis of the male courtship pattern. *Behaviour* **8**:249–332.
Baerends, G. P., Beer, C., and Manning, A. (eds.) (1975). *Essays on Function and Evolution of Behaviour*, Clarendon Press, Oxford.
Barfield, R. J., and Geyer, L. A. (1972). Sexual behavior: Ultrasonic post-ejaculatory song of the male rat. *Science* **176**:1249–1350.
Barlow, G. W. (1958). Daily movements of desert pupfish, *Cyprinodon macularius*, in shore pools of the Salton Sea, California. *Ecology* **39**:580–587.
Barlow, G. W. (1967). Social behavior of a South American leaf fish, *Polycentrus schomburgkii*, with an account of recurring pseudofemale behavior. *Amer. Midl. Nat.* **78**:215–234.
Barlow, G. W. (1977). Modal action patterns. In Sebeok, T. A. (ed.), *How Animals Communicate*, Indiana University Press, Bloomington, pp. 98–134.

Barlow, G. W. (1980). The development of sociobiology: A biologist's perspective. In Barlow, G. W., and Silverberg, J. (eds.), *Sociobiology: Beyond Nature/Nurture? Reports, Definitions and Debate,* Westview Press, Boulder, Colorado, pp. 3-24.

Barlow G. W. (1981). Genetics and the development of behavior, with special reference to patterned motor output. In Immelmann, K., Barlow, G. W., Petrinovich, L., and Main, M. (eds.), *Behavioral Development. The Bielefeld Interdisciplinary Project,* Cambridge University Press, Cambridge, pp. 191-251.

Barlow, G. W., Rogers, W., and Fraley, N. (1986). Do Midas cichlids win through prowess or daring? It depends. *Behav. Ecol. Sociobiol.* **19:**1-8.

Bateson, P. P. G. (1976). Specificity and the origins of behavior. *Advan. Stud. Behav.* **6:**1-20.

Bateson, P. P. G. (1979). How do sensitive periods arise and what are they for? *Anim. Behav.* **27:**470-486.

Bateson, P. P. G. (1982). Preferences for cousins in Japanese quail. *Nature* **295:**236-237.

Bateson, P. (1983). Optimal outbreeding. In Bateson, P. (ed.), *Mate Choice,* Cambridge University Press, Cambridge, pp. 257-277.

Bateson, P. (1988). The active role of behaviour in evolution. In Ho, M.-W., and Fox, S. (eds.), *Process and Metaphors in the New Evolutionary Paradigm,* Wiley, Chichester, pp. 191-207.

Beach, F. A. (1978). Animal models for human sexuality. Sex, hormones and behavior. *Ciba Found. Symp.* **62:**113-143.

Blurton Jones, N. G. (1975). Ethology, anthropology, and childhood. In Fox, R. (ed.), *Biosocial Anthropology,* Wiley, New York, pp. 69-92.

Bullock, T. H. (1984). Comparative neuroscience holds promise for quiet revolutions. *Science* **225:**473-478.

Burtt, E. H. (1986). An analysis of physical, physiological and optical aspects of avian coloration with emphasis on wood-warblers. *Amer. Ornithol. Union, Ornithol. Monogr.* **38:**1-126.

Caryl, P. G. (1979). Communication by agonistic displays: What can games theory contribute to ethology? *Behaviour* **68:**136-169.

Caws, P. (1969). The structure of discovery. *Science* **166:**1375-1380.

Chagnon, N. A., and Irons, W. (eds.) (1975). *Evolutionary Biology and Human Social Behavior. An Anthropological Perspective,* Duxbury, North Scituate, Massachusetts.

Crews, D., and Moore, M. C. (1986). Evolution of mechanisms controlling mating behavior. *Science* **231:**121-125.

Crook, J. H. (1964). The evolution of social organisation and visual communication in the weaver birds (Ploceinae). *Behaviour Suppl.* **10:**1-178.

Cullen, J. M. (1966). Reduction of ambiguity through ritualization. *Phil. Trans. Roy. Soc., Ser. B.* **251:**363-374.

Curio, E. (1963). Probleme des Feinderkennes bei Vögeln. *Proc. 13th Intern. Ornithol. Congr.,* pp. 206-239.

Dawkins, R. (1976a). Hierarchical organisation: A candidate principle for ethology. In Bateson, P. P. G., and Hinde, R. A. (eds.), *Growing Points in Ethology,* Cambridge University Press, Cambridge, pp. 7-54.

Dawkins, R. (1976b). *The Selfish Gene,* Oxford University Press, London.

Dawkins, R., and Krebs, J. R. (1978). Animal signals: Information or manipulation? In Krebs, J. R., and Davies, N. B. (eds.), *Behavioural Ecology: An Evolutionary Approach,* Blackwell Scientific, Oxford, pp. 282-309.

Delius, J. D. (1969). A stochastic analysis of the maintenance behaviour of skylarks. *Behaviour* **33:**137-178.

Dewsbury, D. A. (1975). Diversity and adaptation in rodent copulatory behavior. *Science* **190:**947-954.

Dewsbury, D. A. (1981). An exercise in the prediction of monogamy in the field from laboratory data on 42 species of muroid rodents. *The Biologist* **63:**138-162.

Dingle, H. (1972). Aggressive behavior in stomatopods and the use of information theory in the analysis of animal communication. In Winn, H. E., and Olla, B. L. (eds.), *Behavior of Marine Animals,* Vol. 1, Plenum Press, New York, pp. 126–156.

Dominey, W. J. (1980). Female mimicry in male bluegill sunfish—a genetic polymorphism? *Nature* **284:**546–548.

Dominey, W. J. (1984). Effects of sexual selection and life history on speciation: Species flocks in African cichlids and Hawaiian *Drosophila.* In Echelle, A. A., and Kornfield, I. (eds.), *Evolution of Fish Species Flocks,* University of Maine at Orono Press, Orono, pp. 231–249.

Durant, J. R. (1986). The making of ethology: The Association for the Study of Animal Behaviour, 1936–1986. *Anim. Behav.* **34:**1601–1616.

Ebersole, J. P. (1980). Food density and territory size: An alternative model and a test on the reef fish *Eupomacentrus leucostictus. Am. Nat.* **115:**492–509.

Eibl-Eibesfeldt, I. (1961). The interactions of unlearned behaviour patterns and learning in mammals. In Delafresnaye, J. F. (ed.), *Brain Mechanism and Learning,* Blackwell, Oxford, pp. 58–73.

Eibl-Eibesfeldt, I. (1970). *Ethology. The Biology of Behavior,* Holt, Rinehart and Winston, New York.

Eibl-Eibesfeldt, I. (1982). Warfare, man's indoctrinability and group selection. *Z. Tierpsychol.* **60:**177–198.

Emlen, S. T. (1976). Altruism in mountain bluebirds? *Science* **191:**808–809.

Emlen, S. T., and Oring, L. W. (1977). Ecology: Sexual selection and the evolution of mating systems. *Science* **197:**215–223.

Endler, J. A. (1978). A predator's view of animal color patterns. *Evol. Biol.* **11:**319–364.

Enquist, M., and Leimar, O. (1983). Evolution of fighting behaviour: Decision rules and assessment of relative strength. *J. Theor. Biol.* **102:**387–410.

Enquist, M., Plane, E., and Roed, J. (1985). Aggressive communication in fulmars (*Fulmarus glacialis*). *Anim. Behav.* **33:**1007–1020.

Feldmeth, C. R. (1983). Costs of aggression in trout and pupfish. In Aspey, W. P., and Lustick, S. I. (eds.), *Behaviorial Energetics: The Cost of Survival in Vertebrates,* Ohio State University Press, Columbus, pp. 117–138.

Fentress, J. C., and Stilwell, F. (1973). Grammar of a movement sequence in inbred mice. *Nature* **244:**52–53.

Fisher, R. A. (1930). *The Genetical Theory of Natural Selection,* Dover, New York.

Franck, D. (1974). The genetic basis of evolutionary changes in behaviour patterns. In van Abeelen, J. H. F. (ed.), *The Genetics of Behaviour,* North-Holland, Amsterdam, pp. 119–140.

Fretwell, S. D., and Lucas, H. L. (1970). On territorial behavior and other factors influencing habitat distribution in birds. *Acta Biotheoret.* **19:**16–36.

Gadgil, M., and Bossert, W. H. (1970). Life history consequences of natural selection. *Am. Nat.* **104:**1–24.

Gallagher, J. (1978). Sexual imprinting: Variations in the persistence of mate preference due to differences in stimulus quality in Japanese quail (*Coturnix coturnix japonica*). *Behav. Biol.* **22:**559–564.

Garcia, J., Clarke, J. C., and Hankins, W. G. (1973). Natural responses to scheduled rewards. *Perspect. Ethol.* **1:**1–41.

Gibson, R. M., and Bradbury, J. W. (1985). Sexual selection in lekking sage grouse: Phenotypic correlates of male mating success. *Behav. Ecol. Sociobiol.* **18:**117–123.

Golani, I. (1976). Homeostatic motor processes in mammalian interactions: A choreography of display. *Perspect. Ethol.* **2:**69–134.

Golani, I. (1981). The search for invariants in motor behavior. In Immelman, K., Barlow, G. W., Petrinovich, L., and Main, M. (eds.), *Behavioral Development. The Bielefeld Interdisciplinary Project,* Cambridge University Press, Cambridge, pp. 372–390.

Gottlieb, G. (1976). Early development of species-specific auditory perception in birds. In Gottlieb, G. (ed.), *Neural and Behavioral Specificity*, Academic Press, New York, pp. 237–280.

Grafen, A. (1984). Natural selection, kin selection and group selection. In Krebs, J. R., and Davies, N. B. (eds.), *Behavioural Ecology. An Evolutionary Approach*, Sinauer, Sunderland, Massachusetts, pp. 62–84.

Gross, M. R., and Charnov, E. L. (1980). Alternative male life histories in bluegill sunfish. *Proc. Natl. Acad. Sci. USA* **77**:6937–6940.

Gwynne, P., Begley, S., and Mayer, A. J. (1978). Our selfish genes. *Newsweek* **92**(16):118–123.

Hailman, J. P. (1977). *Optical Signals: Animal Communication and Light*, Indiana University Press, Bloomington.

Halliday, T. R. (1983). The study of mate choice. In Bateson, P. (ed.), *Mate Choice*, Cambridge University Press, Cambridge, pp. 3–32.

Hamilton, W. D. (1964). The genetical evolution of social behaviour. I and II. *J. Theor. Biol.* **7**:1–52.

Hamilton, W. D., and Zuk, M. (1982). Heritable true fitness and bright birds: A role for parasites? *Science* **218**:384–387.

Hanken, J., and Sherman, P. W. (1981). Multiple paternity in Belding's ground squirrel litters. *Science* **212**:351–353.

Hazlett, B. A., and Bossert, W. H. (1965). A statistical analysis of the aggressive communication systems of some hermit crabs. *Anim. Behav.* **13**:357–373.

Heinroth, O. (1911). Beiträge zur Biologie, nämentlich Ethologie und Psychologie der Anatiden. *Verh. Int. Orn. Kongr.* **5**:589–702.

Hess, E. H. (1973). *Imprinting*, Van Nostrand Reinhold, New York.

Hinde, R. A. (1966). *Animal Behaviour. A Synthesis of Ethology and Comparative Psychology*, McGraw-Hill, New York.

Hinde, R. A., and Stevenson-Hinde, J., (eds.) (1973). *Constraints on Learning. Limitations and Predispositions*, Academic Press, New York.

Holmes, W. G., and Sherman, P. W. (1982). The ontogeny of kin recognition in two species of ground squirrels. *Amer. Zool.* **22**:491–517.

Hoogland, J. L. (1982). Prairie dogs avoid extreme inbreeding. *Science* **215**:1639–1641.

Hoogland, J. L., and Sherman, P. W. (1976). Advantages and disadvantages of bank swallow (*Riparia riparia*) coloniality. *Ecol. Monogr.* **46**:33–58.

Horn, H. S. (1968). The adaptive significance of colonial nesting in the Brewer's blackbird (*Euphagus cyanocephalus*). *Ecology* **49**:682–694.

Horsfall, J. A. (1984). Brood reduction and brood division in coots. *Anim. Behav.* **32**:216–225.

Huxley, J. (1963). Lorenzian ethology. *Z. Tierpsychol.* **20**:402–409.

Immelmann, K., Barlow, G. W., Petrinovich, L., and Main, M. (1981). General introduction. In Immelmann, K., Barlow, G. W., Petrinovich, L., and Main, M. (eds.), *Behavioral Development. The Bielefeld Interdisciplinary Project*, Cambridge University Press, Cambridge, pp. 1–18.

Ingle, D., and Crews, D. (1985). Vertebrate neuroethology: Definitions and paradigms. *Ann. Rev. Neurosci.* **8**:457–494.

Jakobsson, S., Radesäter, T., and Jarvi, T. (1979). On the fighting behavior of *Nannacara anomala* (Pisces, Cichlidae). *Z. Tierpsychol.* **49**:210–220.

Jennings, H. S. (1906). *Behavior of the Lower Organisms*, Columbia University Press, New York.

Kamil, A. C. (1983). Optimal foraging theory and the psychology of learning. *Amer. Zool.* **23**:291–302.

Kamil, A. C., and Sargent, T. D. (eds.) (1981). *Foraging Behavior. Ecological, Ethological and Psychological Approaches*, Garland, STPM Press, New York.

Kirkpatrick, M. (1982). Sexual selection and the evolution of female choice. *Evolution* **36**:1–12.

Klopfer, P. H. (1977). Social Darwinism lives! (Should it?). *Yale J. Biol. Med.* **50**:77–84.

Klopfer, P. H. (1985). On central controls for aggression. In Bateson, P. P. G., and Klopfer, P. H. (eds.), *Perspectives in Ethology*, Plenum Press, New York, **6**:33–44.

Konishi, M., and Nottebohm, F. (1969). Experimental studies in the ontogeny of avian vocalizations. In Hinde, R. A., (ed.), *Bird Vocalizations in Relation to Current Problems in Biology and Psychology*, Cambridge University Press, Cambridge, pp. 29–48.

Krebs, J. (1977). The significance of song repertoires: The Beau Geste hypothesis. *Anim. Behav.* **25**:475–478.

Krebs, J. (1978). Optimal foraging: Decision rules for predators. In Krebs, J. R., and Davies, N. B. (eds.), *Behavioural Ecology. An Evolutionary Approach*, Sinauer, Sunderland, Massachusetts, pp. 23–63.

Krebs, J. (1982). Territorial defense in the great tit (*Parus major*): Do residents always win? *Behav. Ecol. Sociobiol.* **11**:185–194.

Krebs, J. (1985). Sociobiology ten years on. *New Sci.* (**1476**):40–43.

Krebs, J. R., and Dawkins, R. (1984). Animal signals: Mind-reading and manipulation. In Krebs, J. R., and Davies, N. B. (eds.), *Behavioural Ecology. An Evolutionary Approach*, 2nd ed., Sinauer Associates, Sunderland, Massachusetts, pp. 380–402.

Krebs, J. R., and McCleary, R. H. (1984). Optimization in behavioural ecology. In Krebs, J. R., and Davies, N. B. (eds.), *Behavioural Ecology. An Evolutionary Approach*, 2nd ed. Sinauer Associates, Sunderland, Massachusetts, pp. 91–121.

Kroodsma, D. E., and Miller, E. H. (eds.) (1982). *Acoustic Communication in Birds*, Academic Press, New York.

Kruuk, H. (1964). Predators and anti-predator behaviour of the black-headed gull (*Larus ridibundus* L.). *Behaviour Suppl.* **11**:1–29.

Kuo, Z. Y. (1967). *The Dynamics of Behavior Development. An Epigenetic View*, Random House, New York.

Lande, R. (1981). Models of speciation by sexual selection of polygenic traits. *Proc. Natl. Acad. Sci. USA* **78**:3721–3725.

Lehrman, D. S. (1953). A critique of Konrad Lorenz's theory of instinctive behavior. *Quart. Rev. Biol.* **28**:337–363.

Leong, C. Y. (1969). The quantitative effect of releasers on the attack readiness of the fish *Haplochromis burtoni* (Cichlidae: Pisces). *Z. Vergl. Physiol.* **65**:29–50.

Loiselle, P. V. (1982). Male spawning-partner preference in an arena-breeding teleost *Cyprinodon macularius californiensis* Girard (Atherinomorpha: Cyprinodontidae). *Am. Nat.* **120**:721–732.

Lombardo, M. P. (1985). Mutual restraint in tree swallows: A test of the TIT FOR TAT model of reciprocity. *Science* **227**:1363–1365.

Lorenz, K. Z. (1935). Der Kumpan in der Welt des Vogels. *J. Ornithol.* **83**:137–213, 289–413.

Lorenz, K. Z. (1941). Vergleichende Bewegungsstudien an Anatiden. *J. Ornithol.* **89**(Suppl. 3):194–293.

Lorenz, K. Z. (1950). The comparative method of studying innate behavior patterns. *Soc. Exper. Biol. Symp.* **4**:221–268.

Lorenz, K. Z. (1952). *King Solomon's Ring*, Methuen, London.

Lorenz, K. Z. (1965). *Evolution and Modification of Behavior*, University of Chicago Press, Chicago.

Lorenz, K. Z. (1966). *On Aggression*, Harcourt Brace Jovanovich, New York.

Lorenz, K. Z. (1974). Analogy as a source of knowledge. *Science* **185**:229–234.

Lorenz, K. Z. (1981). *The Foundations of Ethology. The Principal Ideas and Discoveries in Animal Behavior*, Simon and Schuster, New York.

Lythgoe, J. N. (1979). *The Ecology of Vision*, Clarendon Press, Oxford.

Lythgoe, J. N. (1984). Visual pigments and environmental light. *Vis. Res.* **24**:1–26.

Machlis, L., Dodd, P. W. D., and Fentress, J. C. (1985). The pooling fallacy: Problems arising

when individuals contribute more than one observation to the data set. *Z. Tierpsychol.* **68:**201–214.
Makkink, G. F. (1936). An attempt at an ethogram of the European avocet (*Recurvirostra avosetta* L.) with ethological and psychological remarks. *Ardea* **25:**1–60.
Manning, A. (1976). The place of genetics in the study of behaviour. In Bateson, P. P. G., and Hinde, R. A. (eds.), *Growing Points in Ethology*, Cambridge University Press, Cambridge, pp. 327–343.
Marler, P. (1959). Developments in the study of animal communication. In Bell, P. R. (ed.), *Darwin's Biological Work. Some Aspects Reconsidered*, Wiley, New York, pp. 150–206.
Marler, P., and Mundinger, P. (1971). Vocal learning in birds. In Moltz, H. (ed.), *Ontogeny of Vertebrate Behavior*, Academic Press, New York, pp. 389–450.
Mason, W. A., and Lott, D. F. (1976). Ethology and comparative psychology. *Ann. Rev. Psychol.* **27:**129–154.
Maynard Smith, J. (1974). The theory of games and the evolution of animal conflicts. *J. Theor. Biol.* **47:**209–221.
Maynard Smith, J. (1979). Game theory and the evolution of behaviour. *Proc. R. Soc. London B* **205:**475–488.
Maynard Smith, J., and Price, G. R. (1973). The logic of animal conflict. *Nature* **246:**15–18.
Maynard Smith, J., and Riechert, S. E. (1984). A conflicting-tendency model of spider agonistic behaviour: Hybrid-pure population line comparisons. *Anim. Behav.* **32:**564–578.
McCann, T. S. (1983). Activity budgets of southern elephant seals, *Mirounga leonina*, during the breeding season. *Z. Tierpsychol.* **61:**111–126.
Mock, D. W. (1984). Sibicidal aggression and resource monopolization in birds. *Science* **225:**731–733.
Morris, D. (1956). The feather postures of birds and the problem of the origin of social signals. *Behaviour* **9:**75–114.
Mosler, H.-J. (1985). Making the decision to continue the fight or to flee. An analysis of contests between male *Haplochromis burtoni* (Pisces). *Behaviour* **92:**129–145.
Nelson, K. (1964). The temporal patterning of courtship in the glandulocaudine fishes (Ostariophysi, Characidae). *Behaviour* **24:**90–146.
Nottebohm, F. (1984). Bird song as a model in which to study brain processes in relation to learning. *Condor* **86:**227–236.
Oppenheim, R. W. (1982). Präformation und Epigenese in der Entwicklung des Nervensystems und des Verhaltens: Probleme, Begriffe und ihre Geschichte. In Immelmann, K., Barlow, G. W., Petrinovich, L., and Main, M. (eds.), *Verhaltensentwicklung bei Mensch und Tier. Das Bielefeld-Projekt*, Verlag Paul Parey, Berlin, pp. 157–221.
Parker, G. A. (1983). Mate quality and mating decisions. In Bateson, P. (ed.), *Mate Choice*, Cambridge University Press, Cambridge, pp. 141–164.
Parker, G. A., and Rubenstein, D. I. (1981). Role assessment, reserve strategy, and acquisition of information in asymmetric animal conflicts. *Anim. Behav.* **29:**221–240.
Paton, D., and Caryl, P. G. (1986). Communication by agonistic displays: I. Variation in information content between samples. *Behaviour* **98:**213–239.
Patterson, I. J. (1965). Timing and spacing of broods in the black-headed gull (*Larus ridibundus* L.). *Ibis* **107:**433–460.
Pelkwijk, J. J. ter, and Tinbergen, N. (1937). Eine reizbiologische Analyse einiger Verhaltensweisen von *Gasterosteus aculeatus* L. *Z. Tierpsychol.* **1:**193–200.
Porter, W. P., Mitchell, J. W., Beckman, W. A., and DeWitt, C. B. (1973). Behavioral implications of mechanistic ecology: Thermal and behavioral modeling of desert ectotherms and their microenvironment. *Oecologia* **13:**1–54.

Power, H. W. (1975). Mountain bluebirds: Experimental evidence against altruism. *Science* **189**:142–243.
Pulliam, H. R. (1981). Learning to forage optimally. In Kamil, A. C., and Sargent, T. D. (eds.), *Foraging Behavior: Ecological, Ethological, and Psychological Approaches*, Garland STPM Press, New York, pp. 379–388.
Riechert, S. E. (1978). Games spiders play: Behavioral variability in territorial disputes. *Behav. Ecol. Sociobiol.* **4**:1–28.
Riechert, S. E. (1986). Spider fights: A test of evolutionary game theory. *Amer. Sci.* **47**:604–610.
Rogers, W. (1985). Monogamy, mate choice and breeding success in the Midas cichlid (*Cichlasoma citrinellum*). Ph.D. dissertation, University of California, Berkeley.
Rowland, W. J., and Sevenster, P. (1985). Sign stimuli in the threespine stickleback (*Gasterosteus aculeatus*): A re-examination and extension of some classic experiments. *Behaviour* **93**:241–257.
Ryan, M. J. (1985). Energetic efficiency of vocalization by the frog *Physalaemus pustulosus*. *J. Exp. Biol.* **116**:47–52.
Ryan, M. J. (1986). Neuroanatomy influences speciation rates among anurans. *Proc. Natl. Acad. Sci. USA* **83**:1379–1382.
Schleidt, W. M., Yakalis, G., Donnelly, M., and McGarry, J. (1984). A proposal for a standard ethogram, exemplified by an ethogram of the bluebreasted quail (*Coturnix chinensis*). *Z. Tierpsychol.* **64**:193–220.
Seitz, A. (1940). Die Paarbildung bei einigen Cichliden. *Z. Tierpsychol.* **4**:40–84.
Selander, R. B. (1964). Sexual behavior in blister beetles (Coleoptera: Meloidae). I. The genus *Pyrota*. *Can. Entomol.* **96**:1037–1082.
Semler, D. E. (1971). Some aspects of adaptation in a polymorphism for breeding colours in the threespine stickleback (*Gasterosteus aculeatus*). *J. Zool. London* **165**:291–302.
Shettleworth, S. J. (1972). Constraints on learning. *Advan. Stud. Behav.* **4**:1–68.
Shettleworth, S. J. (1984). Learning and behavioural ecology. In Krebs, J. R., and Davies, N. B. (eds.), *Behavioural Ecology. An Evolutionary Approach*, 2nd ed., Blackwell, Oxford, pp. 170–194.
Sherry, D. F., Krebs, J. R., and Cowie, R. J. (1981). Memory for the location of stored food in marsh tits. *Anim. Behav.* **24**:1260–1266.
Sibly, R. M., and Smith, R. H. (eds.) (1985). *Ecological Consequences of Adaptive Behaviour*, Blackwell Scientific, Palo Alto, California.
Slater, P. J. B. (1981). Individual differences in animal behavior. *Perspect. Ethol.* **4**:35–49.
Staddon, J. E. R. (ed.) (1980). *Limits to Action: The Allocation of Individual Behavior*, Academic Press, New York.
Stamps, J. A., and Metcalf, R. A. (1980). Parent–offspring conflict. In Barlow, G. W., and Silverman, J. (eds.), *Sociobiology: Beyond Nature/Nurture? Reports, Definitions and Debate*, Westview Press, Boulder, Colorado, pp. 589–618.
Stamps, J., Clark, A., and Arrowood, P. (1985). Parent–offspring conflict in budgerigars. *Behaviour* 94:3–40.
Steger, R., and Caldwell, R. L. (1983). Intraspecific deception by bluffing: A defense strategy of newly molted stomatopods (Arthropoda: Crustacea). *Science* **221**:558–560.
ten Cate, C. (1984). The influence of social relations on the development of species recognition in zebra finch males. *Behaviour* **91**:263–285.
Thorpe, W. H. (1951). The definition of terms used in animal behaviour. *Bull. Anim. Behav.* **9**:34–40.
Thorpe, W. H. (1956). *Learning and Instinct in Animals*, Methuen, London.
Thorpe, W. H. (1961). *Bird-song*, Cambridge University Press, Cambridge.

Thorpe, W. H. (1979). *The Origins and Rise of Ethology: The Science of the Natural Behaviour of Animals*, Praeger, New York.
Tinbergen, L. (1960). The natural control of insects in pinewoods: I. Factors influencing the intensity of predation by song birds. *Arch. Neerl. Zool.* **13**:265–343.
Tinbergen, N. (1951). *The Study of Instinct*, Clarendon Press, Oxford.
Tinbergen, N. (1959). Comparative studies of the behaviour of gulls (Laridae): A progress report. *Behaviour* **15**:1–70.
Tinbergen, N. (1963). On aims and methods of ethology. *Z. Tierpsychol.* **20**:410–433.
Tinbergen, N. (1964). The evolution of signaling devices. In Etkin, W. (ed.), *Social Behavior and Organization among Vertebrates*, University of Chicago Press, Chicago, pp. 206–230.
Tinbergen, N. (1965). Behavior and natural selection. In Moore, J. A. (ed.), *Ideas in Modern Biology*, Doubleday, New York, pp. 519–542.
Tinbergen, N., Impekoven, M., and Franck, D. (1967). An experiment on spacing out as a defence against predation. *Behaviour* **28**:307–321.
Trivers, R. L. (1971). The evolution of reciprocal altruism. *Quart. Rev. Biol.* **46**:35–57.
Trivers, R. L. (1972). Parental investment and sexual selection. In Campbell, B. (ed.), *Sexual Selection and the Descent of Man*, Aldine, Chicago, pp. 136–179.
Trivers, R. L., and Hare, H. (1976). Haplodiploidy and the evolution of the social insects. *Science* **191**:249–263.
Turner, G. F., and Huntingford, F. A. (1986). A problem for game theory analysis: Assessment and intention in male mouthbrooder contests. *Anim. Behav.* **34**:961–970.
Vehrencamp, S. L. (1981). Optimal degree of skew in cooperative societies. *Amer. Zool.* **23**:327–335.
von Cranach, M. (ed.) (1976). *Methods of Inference from Animals to Human Behavior*, Aldine Press, Chicago.
Waddington, C. H. (1975). *The Evolution of an Evolutionist*, Edinburgh University Press, Edinburgh.
Wells, K. D., and Taigen, T. L. (1986). The effect of social interactions on calling energetics in the gray treefrog (*Hyla versicolor*). *Behav. Ecol. Sociobiol.* **19**:9–18.
Wickler, W. (1982). Immanuel Kant and the song of the house sparrow. *Auk* **99**:590–591.
Wiepkema, P. R. (1961). An ethological analysis of the reproductive behaviour of the bitterling (*Rhodeus amarus* Bloch). *Arch. Neer. Zool.* **14**:103–199.
Williams, G. C. (1966). *Adaptation and Natural Selection: A Critique of Some Current Evolutionary Thought*, Princeton University Press, Princeton, New Jersey.
Williams, G. C. (1975). *Sex and Evolution*, Princeton University Press, Princeton, New Jersey.
Williams, G. C., and Williams, D. C. (1957). Natural selection of individually harmful social adaptations among sibs with special reference to social insects. *Evolution* **11**:32–39.
Wilson, E. O. (1975). *Sociobiology. The New Synthesis*, Belknap Press of Harvard University Press, Cambridge, Massachusetts.
Wrangham, R. W. (1982). Mutualism, kinship and social evolution. In King's College Sociobiology Group (ed.), *Current Problems in Sociobiology*, Cambridge University Press, Cambridge, pp. 269–289.
Yasukawa, K., and Searcy, W. A. (1985). Song repertoires and density assessment in red-winged blackbirds: Further tests of the Beau Geste Hypothesis. *Behav. Ecol. Sociobiol.* **16**:171–175.
Zahavi, A. (1979). Ritualisation and evolution of movement signals. *Behaviour* **72**:77–81.

Chapter 2

THE FUTURE OF ETHOLOGY: HOW MANY LEGS ARE WE STANDING ON?

Marian Stamp Dawkins

Department of Zoology
University of Oxford
Oxford, England

I. ABSTRACT

The traditional ethological emphasis on asking different sorts of questions about animal behavior (mechanism, function, development, and phylogeny) needs to be reasserted. Overemphasis on questions about function (adaptive significance) has led to a lopsided view of behavior that would be much more valuable (more practical as well as more interesting) if different sorts of questions, particularly those having to do with mechanism, were asked at the same time.

II. INTRODUCTION

There was a time when ethology could fairly claim to be a four-question science. Its great strength lay in the fact that it looked at animal behavior and asked different sorts of questions about it: "How does it work?"; "How does it develop?"; "How did it evolve?"; "What is its survival value?" (Tinbergen, 1963). These questions were not asked in isolation. They fed back upon and cross-fertilized one another so that a question about survival value (e.g., what is the adaptive significance of eggshell removal in the black-headed gull?) would lead to one about mechanism (how does the gull distinguish between an empty eggshell and a full egg?). It was recognized that one needed different sorts of evidence to answer these questions, but thinking about the (adaptive) purpose of the behavior stimulated ideas about what the mechanism underlying it might be and vice versa.

Today, many textbooks still pay lip service to the idea that it is as important to ask about the machinery of behavior as it is to ask why natural selection favored it, how it develops, or how it evolves. A glance at many of the current journals, however, suggests that although the four questions may still be there, one of them—the survival value (adaptation) question—has come to predominate. Ethology may still be a four-legged animal, but it is an animal hopping around on one big leg, with the other three dangling somewhat ineffectively. We seem to have become obsessed with one kind of question, concentrating on inclusive fitness, survival value, and adaptive significance, often almost to the exclusion of others. We ask why animals live in groups of a certain size, meaning (when we think about it that clearly), "Why would their fitness suffer if they were in a group of a different size or on their own?" We rarely mean, "What are the stimuli by which animals recognize each other, how do they know how big a group is getting, how do they regulate its size?" Nor do we ask how the development of the animal affects its responses. And we excuse ourselves from asking questions about phylogeny on the grounds, no doubt, that they are too difficult to answer.

In turning ethology into a one-legged animal (I exaggerate, of course, but not that much), we have made it into a clumsier, less exciting, less progressive one. It is rare today to find questions about mechanism and those about function studied alongside each other. Exceptions such as Maynard Smith and Riechert's (1984) study of spider aggression are refreshing. Measuring inclusive fitness or even just speculating about it often seems to be enough.

I could, of course, be accused of a middle-aged nostalgia for the past. Perhaps I have reached what Donald Griffin has referred to as "the philosopause." I certainly admit to being an ethological chauvinist, and have never been particularly worried about claims that ethology would be "eaten up" by neurobiology or behavioral ecology on the grounds that, being a four-question subject, it was far too dynamic to be subsumed. But it is difficult to be dynamic and to run fast in the one-legged manner that ethology seems often to adopt as its mode of progression.

III. THE QUESTION OF MECHANISM

I would like to suggest that there are three areas where a neglect of questions about mechanism, in particular, has been stifling and damaging. I have nothing against asking questions about adaptation. It is and always has been the lifeblood of ethology and it has been responsible for setting ethology apart from, say, psychology. What I am objecting to is the way it has tended to overshadow other sorts of questions and to become isolated from them. I fix upon mechanism because it is the neglect of this area that seems to have had the most serious consequences.

A. Neuroethology

There should be an exciting new discipline combining ethology with the major advances that have taken place in neurobiology since the advent of breakthrough recording techniques. It is now possible to map the nervous systems of a number of invertebrates. We know which cells are connected to which and what each one does. We ought to be making links between behavior and the circuit diagrams uncovered by neurobiologists. Neurobiologists have told us what many of the underlying mechanisms are. We ethologists should now be telling them what the mechanisms are doing and why. If we really believe in the black box or "whole animal" approach, we should be deducing what is going on inside animal bodies and confirming our models with neurophysiological observations. Neurobiologists, confronted with a bewildering array of nerves and synapses, should be guided in their search for how it all works by the rigorous groundwork—what the animal does with its hardware—that we ethologists have laid for them.

Instead, what do we find? In a recent textbook of neuroethology (Camhi, 1984), the main "contributions from ethology" are listed as *fixed action patterns, sign stimuli,* and *instinctive* or *innate behavior.* Similarly, in an inspiring polemic in favor of neuroethology as a separate discipline in its own right, Hoyle (1984) argues that neuroethologists should be working on the key ethological concepts of fixed action patterns, displacement activities, releasers, and consummatory acts. One reason for the distinctly old-fashioned ring to these authors' understanding of ethology may be the fact that relatively few ethologists work on the mechanism questions of interest to neurobiologists. So in trying to discover what ethologists believe about the mechanisms underlying complex behavior, the neurobiologists find largely the ideas that ethology held 20 or 30 years ago.

It is difficult to understand why this should be so, as those cases where questions have been asked from both the behavioral and neurophysiological angles have often been spectacularly successful. W. J. Davis' work (Davis *et al.,* 1977) on the sea slug *Pleurobranchaea* is an example. The inhibitory connections between different behavior patterns such as feeding, righting, and mating were first deduced by behavioral observations, treating the animal initially as a "black box." Then, the physiological basis for at least some of these connections were established. The work of J.-P. Ewert (1980) on the behavior and neurophysiology of prey-catching toads, the work of Knudsen and Konishi

(1979) on that of barn owls, and the studies of song-learning in birds, also show what can be achieved by uniting ethological and neurophysiological approaches into a true neuroethology.

Part of the problem here is that there is a lack of cooperation between the two disciplines concerning the particular animal and the particular problem to be studied. Neuroethologists tend to choose their animals on the basis of how accessible the nervous system is. They also look for "simple" behavior where they believe there is some hope of finding out about the underlying mechanism. Ethologists study animals that show particularly interesting behavior and are often fond of emphasizing behavior that is not "simple" in the least. But the problem is not just choice of animal or choice of behavior. A large part of the difficulty of linking neuroethology with ethology is that the questions about behavioral mechanisms that would be of most interest to neurobiologists are not the main concern of ethologists. It is as though the neuroethologists are valiantly tunneling through from their side, sincerely believing in a joint discipline, but we are just not doing our bit from our side. Why has there not been more attention paid to the ethology of animals, particularly invertebrates, for which neuroethologists already have a good working knowledge of the nervous system? How many ethologists could really claim, as Tinbergen once did with some justification, that they too are physiologists, but just happen to be physiologists who "do not break the skin"?

B. Ethology, Sociobiology, and Behavioral Ecology

I was never really able to understand why "sociobiology" was considered to be a revolutionary new discipline and why people interested in the way behavior is adaptive now tend to call themselves "behavioral ecologists" rather than ethologists. The biology of social behavior (what else does sociobiology mean?) and the role of behavior in the survival and fitness of animals were part of ethology as perceived by Lorenz and Tinbergen. There was only one difference. The questions about the evolution and adaptive significance of social behavior, which form the backbone of sociobiology and behavioral ecology, were only *part* of ethology, one of its legs, not its be-all and end-all. The "hopping" nature of ethology today is due to the fact that it has become more fashionable to be called a behavioral ecologist and, in following a fashion, the importance of the other three questions has often been played down.

This has left a curious vacuum in the middle of ethology which would not be there at all if all four legs of ethology had been in use at once. When theorists want to know what reasonable assumptions they might make about, say, how animals can assess each other's size, or about how and whether they gain information about each other's territory or parasite load, they do not, in general, find answers.

We do not know the answers to such mechanism questions, which have been of less concern to us than the adaptive significance of the behavior. But how much more impressive models of mimicry and warning coloration are when they are made with, rather than without, knowledge of what a predator can see, what it finds easy or difficult to recognize, and what it can remember. Knowledge of the mechanism by which predators hunt for cryptic prey can affect our view of the adaptive significance of prey appearance (Guilford and Dawkins, 1987). How much better our models of interactions between animals would be if we knew what assumptions (about their memory and perception) were realistic, which were unlikely, and which others were contrary to everything that was known about animal vision and memory. Only recently have those interested in sexual selection (a field that has accumulated a great edifice of adaptive theories) begun to ask questions about the mechanisms by which females might make their choice between males, and only recently has the isolation of evolutionary theory from what animals do begun to break down.

This is all to the good and we need much more of it. Evolutionary theory needs a mechanism just as surely as the study of mechanism needs an evolutionary perspective if it is not to be dry as dust and a catalog of facts.

Our view of animal perception has changed radically since the time when the "sign stimulus" was seen to be the way in which most animal perception worked. It may be the case that some animals, particularly young animals, do pick out simple features of their environment and can be made to respond to quite crude dummies or models. But this is certainly not the rule, and a major feature of behavioral studies over the last few years has been to reveal the complexity of what animals see, hear, and know about their environments. In fact, one of the best ways to annoy someone who works on optimal foraging is to tell them that the major contribution of optimal foraging theory (OFT) studies has been to reveal a great deal about the mechanisms by which animals respond to their environments. (For some reason, they would rather see their studies revealing evolution than mechanism.) No longer can we talk about birds responding to "tree" as though it were a simple sign stimulus. Small birds respond differently to the same tree depending on their expectation of it based on their experience of finding food in other trees. Sparrows recruit others to forage with them, but only if the food is in a form that can be easily divided between different birds and, even then, only if foraging alone would result in some danger from predators (Elgar, 1986a,b). Clearly, there is no simple sign stimulus response to food; on the contrary, sparrows appear to weigh the advantages and disadvantages of having other sparrows feeding near them in a most complex way.

At the same time, the whole of animal communication has undergone something of a silent revolution. Once animals were seen to communicate entirely in conspicuous sign stimuli or releasers, specially evolved by the process of ritualization to be exaggerated. Large signals are clearly important in the assessment of one animal by another (e.g., Clutton-Brock and Albon, 1979), but it is more and more realized that subtlety and complexity (the very opposite of simple sign stimuli) are the characteristics of much animal communication, particularly among animals that know each other (e.g., Cheney and Seyfarth, 1982). The horse called Clever Hans may not have been able to do mathematics, but he was certainly very clever at picking up very minute signs from humans, leaving open the questions of what horses regularly, and perhaps to us imperceptibly, pick up from other horses.

The laboratory studies of Herrnstein *et al.* (1976) have shown that pigeons have the ability to develop complex, humanlike categories of the world that go far beyond simple pattern recognition. Field studies have shown that animals often have a detailed knowledge not just of each other, but of each other's social relationships and kinship bonds as well (Cheney and Seyfarth, 1986). There are clearly immense opportunities for studying the perceptual worlds of animals. A fuller understanding of the mechanisms of social interactions and the mechanisms of animal categorization cannot but help to enrich our knowledge of their behavior. We need ethology with its four questions, not a one-question approach that leaves the study of mechanism out in the cold.

C. Applied Ethology

Applied ethology is now a separate discipline with its own journals, its own conferences, and, more seriously, its own terminology. What may not be commonly realized is that it bears little resemblance to "pure" ethology and many applied ethologists still subscribe to Lorenz's (1950) model as an explanation of behavior (e.g., Vestergaard, 1980; Sambraus, 1984). It uses terms like "ethological need" (Fölsch, 1980) and "social space" that bear little resemblance to any terms in mainstream ethology.

Again, it is ethology's neglect of causal questions that has done much of the damage and has forced applied ethology to go its own way. The questions of interest to applied ethologists—such as "How can an animal be stopped from exhibiting a behavior we consider undesirable?" or "What happens if an animal is deprived of the opportunity of behaving in certain ways?"—are questions about the causal mechanisms of behavior. One of the most critical questions in the field of animal welfare, for instance, is about the consequences that result when animals are kept in zoos or farms and cannot perform much of the behavior typical of free-living members of their species. This is essentially a question about motivation and what happens when animals are deprived of opportunities to do behavior. The answers depend on a knowledge of what happens to factors within the animal as a result of such deprivation. Looking to ethology for such answers, we find little interest in studying the causal basis of motivation. Finding nothing to replace Lorenz's model, applied ethologists have either adopted it or gone on to their own investigations, finding little of relevance to them in modern ethology (Ewbank, 1986). We have been left behind. At a time when funding agencies the world over are increasingly concerned with the practical applications of the research they fund, we have lost out. Our science would be in a much healthier state if it had been more balanced in its outlook and had not thought that evolutionary speculation was more interesting than an interplay of causal and evolutionary questions. It would then be in a far better position to provide answers, or at least know where to start looking for answers, to applied problems and would be seen as far more practical and useful than it is now.

IV. CONCLUSIONS

This has not been just a backward look. It has been a forward look, too. It is important to ask questions about causal mechanism, and the ones asked now will be different, probably more complex, questions than was once the case. They will tell us about complex mechanisms and perceptual worlds probably more sophisticated and possibly more like our own than we have yet attributed to animals. Ethology is stronger—it is more *interesting*—when causal questions go hand in hand with evolutionary ones. And it can still be so, if it continues to emphasize that there is not just one but many kinds of questions to ask about animal behavior. Studies of inclusive fitness and adaptive significance have told us an immense amount about the way in which natural selection has affected behavior. Our understanding of evolution and fitness is greatly advanced over what it was only a few years ago. What we need to do now is to balance this with an equal understanding of the mechanisms that are responsible for behavior. We have a much clearer idea of gene selection down the generations. But genes operate through making bodies do things. These bodies have to develop and they need machinery (sense organs, decision centers, and means of executing action) to be able to pass the genes on to the next generation. To understand this process fully, we need a science that is not only aware of the evolutionary ebb and flow of genotypes over evolutionary time, but can also look at the bridge between generations, at the bodies that grow and move and court and find food and pass their genetic cargo on through time with the frailest and most marvelous of flesh-and-blood machinery. That science, with its eyes on both the long term and the short term, on evolution and mechanism, is ethology. It has been so since its birth and, with a bit of readjustment of its weight onto all four of its legs, it can still be so in the future.

V. REFERENCES

Camhi, J. M. (1984). *Neuroethology,* Sinauer Associates, Sunderland, Massachusetts.
Cheney, D. L., and Seyfarth, R. M. (1982). How vervet monkeys perceive their grunts: field playback experiments. *Anim. Behav.* **30**:737–751.
Cheney, D. L., and Seyfarth, R. M. (1986). The recognition of social alliances by vervet monkeys. *Anim. Behav.* **34**:1722–1731.
Clutton-Brock, T. H., and Albon, S. D. (1979). The roaring of red deer and the evolution of honest advertisement. *Behaviour* **69**:145–170.
Davis, W. J., Mpitsos, G. J., Pinneo, J. M., and Ran, J. L. (1977). Modification of the behavioral hierarchy of *Pleurobranchia*. I: Satiation and feeding mechanisms. *J. Comp. Physiol.* **117**:99–125.
Elgar, M. A. (1986a). House sparrows establish foraging flocks if the resources are divisible. *Anim. Behav.* **35**:169–174.
Elgar, M. A. (1986b). The establishment of foraging flocks in house sparrows: risk of predation and daily temperature. *Behav. Ecol. Sociobiol.* **19**:433–438.
Ewbank, R. (1986). The behavioural needs of farm animals. Paper presented at the Proceedings of the 6th International Conference on Production: Disease in Farm Animals, Belfast.
Ewert, J-P. (1980). *Neuroethology,* Springer-Verlag, Berlin.
Fölsch, D. W. (1980). Essential behavioural needs. In Moss, R. (ed.), *The Laying Hen and Its Environment,* Martinus Nijhoff, The Hague, pp. 121–132.
Guilford, T. C., and Dawkins, M. S. (1987). Search images not proven: a reappraisal of recent evidence. *Anim. Behav.* **35**:1838–1845.
Herrnstein, R. J., Loveland, D. H., and Cable, C. (1976). Natural concepts in pigeons. *J. Exp. Psychol. Anim. Behav. Proc.* **2**:285–302.
Hoyle, G. (1984). The scope of neuroethology. *Behav. Brain Sci.* **7**:367–412.
Knudsen, E. I., and Konishi, M. (1979). Mechanisms of sound localisation in the barn owl (*Tyto alba*). *J. Comp. Physiol.* **133**:13–21.
Lorenz, K. (1950). The comparative method in studying innate behaviour patterns. *Symp. Soc. Exp. Biol.* **4**:221–268.
Maynard Smith, J., and Riechert, S. E. (1984). A conflicting tendency model of spider agonistic behaviour in hybrid–pure line comparison. *Anim. Behav.* **32**:564–578.
Sambraus, H. H. (1984). Accumulation of action specific energy in the eating behaviour of rabbits. In Unselm, J., van Putten, G., and Zeeb, K. (eds.), *Proceedings of the International Congress on the Applied Ethology of Farm Animals, Kiel,* Kuratorium für Technik und Bauwesen in der Landwirtschaft, Darmstadt, pp. 335–338.
Tinbergen, N. (1963). On aims and methods in ethology. *Z. Tierpsychol.* **20**:410–433.
Vestergaard, K. (1980). The regulation of dust-bathing and other behaviour patterns in the laying hen: a Lorenzian approach. In Moss, R. (ed.), *The Laying Hen and Its Environment,* Martinus Nijhoff, The Hague, pp. 101–113.

Chapter 3

THE COMPARATIVE APPROACH IN ETHOLOGY: AIMS AND LIMITATIONS

John L. Gittleman

Department of Zoology and Graduate Programs in Ethology and Ecology
University of Tennessee
Knoxville, Tennessee 37916

I. ABSTRACT

The comparative approach, utilizing cross-species differences in behavior and ecology, is one of the mainstays of contemporary ethology. Increases in the quality, quantity, and quantification of available comparative data, as well as new statistical techniques, have allowed more rigorous applications of the comparative approach. Using examples from the mammalian literature, I review the primary aims in these comparative studies, which include defining new variables, identifying forms in relationships, generating new hypotheses, and testing general or specific evolutionary hypotheses. Along with these uses of the comparative approach, a number of weaknesses or limitations have arisen. Most critical are inherent problems in the comparative data base, biases in the distribution of data, characterizing species with marked intraspecific variation, methodological (statistical) decisions in comparative tests, and establishing causation. I review various ways of partially rectifying some of these problems. Despite such limitations, the comparative approach will increasingly be in the forefront of ethological and evolutionary studies because of (1) the enticing effect of larger data bases becoming available for comparative study, (2) the capacity for pluralistically testing explanations, and (3) this approach being the only direct method of separating evolutionary (adaptive) explanations from phylogenetic effects.

II. INTRODUCTION

The comparative method in its many forms has become one of the cornerstones of modern biology (Zangerl, 1948; Coleman, 1964; Nelson, 1970; Webster, 1976). Practically every specialty has applied the comparative approach in order to identify new and interesting questions, generate testable hypotheses, and test otherwise intractable predictions (Mayr, 1982a). Beginning with Darwin and forming an integral part in the birth of ethology (Lorenz, 1950; Hinde and Tinbergen, 1958; Tinbergen, 1959; for earlier historical papers, see Burghardt, 1985), problems relating to ontogeny (Burghardt, 1978; Bekoff and Byers, 1981; Bekoff, 1988), taxonomy (Greene and Burghardt, 1978; Eberhard, 1982), historical processes (Gittleman, 1981; Ridley, 1983; Lauder, 1986), and evolutionary function (Clutton-Brock and Harvey, 1977, 1979; Harvey et al., 1978; Eisenberg, 1981; Bekoff et al., 1984; Calder, 1984) have become more accessible from cross-species comparisons.

Today, large data banks from long-term field studies coupled with new biometrical techniques have produced powerful means for utilizing a comparative approach. In this chapter, I critically review some of these uses in contemporary ethological studies, emphasizing aims and strengths of recently developed *cross-species comparisons*. As with most rapid scientific developments, a number of limitations or pitfalls have arisen in comparative studies that, when ignored, render comparative work misleading. Thus, I consider various methodological problems in comparative studies. Some of the ideas I present are discussed in more detail, especially with regard to statistical matters, in recent papers by Harvey and Mace (1982), Clutton-Brock and Harvey (1984), and Felsenstein (1985). The examples I use for illustrating aims and limitations in the comparative approach are from mammals, particularly from my work on carnivore behavior, ecology, and morphology. I select these taxa not only because they are more familiar to me but also because I have encountered numerous decision-making problems in my own comparative studies that probably extend to other cross-species comparisons but have yet to be pointed out in the literature.

III. WHEN IS THE COMPARATIVE APPROACH APPROPRIATE?

Clearly, one of the primary contributions of ethological studies has been a steady stream of new methodologies and sophisticated techniques for analyzing behavioral problems. As with decisions required for applying an experimental design, a comparative approach is only useful for particular biological questions (Huey and Bennett, 1986; Huey, 1987). Thus, the critical initial question should be simply whether a comparative approach is appropriate. There are two necessary and sequential parts to this question; the first relates to the data that will be used for comparisons and the second pertains to the question(s) at hand. Ordinarily, this might appear to be moving in a backward fashion, for as scientists most of us are trained to pursue questions in the hypothetic–deductive tradition of sorting out a way to tackle a problem and then gathering data relevant to analyzing it. In utilizing a comparative approach, this is rarely possible (see Jaksíc, 1981); if sufficient comparative data are not available then comparisons obviously cannot be made. For example, problems pertaining to the evolutionary significance of play behavior (Burghardt, 1984, 1988), sex ratios (Clutton-Brock and Albon, 1982), lek mating (Bradbury, 1985), and costs of reproduction (Gittleman and Thompson, 1988) are all in need of comparative treatment, but the sparse comparative data bases are prohibitive. As a tangential point, because availability of data is fundamental to *allowing* a comparative approach, this unusual methodological sequence has drawn some criticism by giving the impression that comparative analyses may only be post hoc. As will be touched on in Sections IV.C and D, by formulating hypotheses from other studies, be they single- or cross-species, comparative tests need not be post hoc.

Prerequisite comparative data. Any comparative study is only as reliable as the data base upon which it is founded. So, it is necessary to ask initially, "When are comparative data appropriate and reliable for cross-species analyses?" First, data should be evenly distributed across taxa. For example, if we are interested in a particular problem across the order Carnivora, which is comprised of seven major taxonomic families, each family must be represented by a substantial representative sample of the species, otherwise explanations for behavioral variation across carnivores may be misleading. Availability and distribution of data must also be matched by quality in terms of (1) number of individuals studied, (2) length of field studies, (3) verification of information on individual species from more than one study site, and (4) quantitative estimates for behavioral characteristics. All of these qualities seem blatantly apparent, but few of them are ever noted in the methods sections of comparative papers. Further discussion of data selection in comparative analyses, particularly in reference to problem sources, is presented in Section V.A.

The need for comparisons. In most ethological studies, the comparative approach is valuable for two kinds of investigation, broad functional (adaptive) or causal questions and phylogenetic (historical) ones. [In this paper, I use the term *adaptive* when referring to traits which allow individuals to compete successfully with other members of the species and provide a solution to extant environmental problems (see Williams, 1966; Lewontin, 1978; Mayr, 1982b); the concept of *function* is closely involved, and hence will be used, in the sense that function may tell us how the trait has contributed to effective competition over evolutionary time (Burian, 1983). I refer to *phylogeny* when trait variation may be ascribed to the geneological history of the species under consideration (see Ghiselin, 1972; Gould, 1977); that is, phylogenetic explanations inherently relate to historical processes.] In the former, explanations for the evolutionary value of a trait in a particular suite of species (e.g., antipredatory strategy of group living in dwarf and banded mongooses) is compared with similar traits in related taxa (e.g., group living in meerkats, Egyptian mongooses, and other viverrids; see Rood, 1986, for details of this example). The degree of (statistical) overlap or divergence among taxa reveals the generality of evolutionary explanation. Other examples of comparative studies, where functional explanations of behavioral variation have been tested across wide taxonomic groups, are available for primates (Clutton-Brock and Harvey, 1977, 1980; Harvey and Clutton-Brock, 1985), marsupials (Lee and Cockburn, 1985), small mammals (Mace, 1979), ungulates (Jarman and Jarman, 1979; Clutton-Brock *et al.*, 1982), and carnivores (Gittleman, 1985a, 1986a,b, 1988b).

The utility of a comparative approach is probably more valuable for testing phylogenetic ideas than consideration of adaptive hypotheses simply because cross-species comparisons are the only heuristic tool that we have for uncovering historical effects. Unfortunately, with the history of comparative biology so grounded in taxonomic questions (Nelson, 1970; Lauder, 1986), applications of species comparisons have not simultaneously focused on functional (adaptive) *and* phylogenetic questions until recently. Greater appreciation for historical effects on behavior (e.g., Gittleman, 1981; Ridley, 1983; Dobson, 1985; Greene, 1986a; Lauder, 1986; Huey, 1987; Huey and Bennett, 1987), the need to disentangle evolutionary origin from evolutionary adaptation (Gould and Vrba, 1982), and recent statistical methods for quantitatively assessing phylogenetic effects (Felsenstein, 1985; Cheverud *et al.*, 1985), suggest that comparative approaches will regain active usage in future studies.

From this brief discussion, it might appear that adopting a comparative approach is an inappropriate method for analyzing many behavioral questions. Indeed, just as we would not use a microscope to monitor home range movements of leopards, the comparative approach should be reserved for appropriate questions of broad-based functional or phylogenetic questions when requisite comparative data are available.

IV. AIMS OF THE COMPARATIVE APPROACH

Comparisons across taxonomic groups can be used in many ways. Generally, five uses are prevalent in ethology; comparisons are used to define or describe new variables, generate new hypotheses, identify forms of relationships, test general hypotheses, and test specific hypotheses (Harvey and

A. Defining and Describing New Variables

In the early stages of an observational science new variables require definition. Such definitions may be formed operationally, as in theoretical models (Maynard Smith, 1978), or as a result of empirical observation. The comparative approach is often used to establish new variables by removing the effects of one variable on another. For example, in studies of mammalian brain size evolution it is standard methodology to remove the effects of body weight (or some other size variable) before examining relationships with brain size. Often referred to as *interspecific allometry* (see Schmidt-Nielson, 1984), this is empirically achieved by drawing a line of best fit across taxa between brain weight and body weight. It has generally been found that the following expression can be used to describe the relationship between a given parameter, P (in this case brain size), and body size W:

$$P = k \times W^\alpha$$

This empirically determined formula can be converted to logarithmic form as follows:

$$\log P = \alpha \log W + \log k$$

Such logarithmic conversion has the advantage of producing a linear expression. A best-fit line drawn through paired logarithmic values of P and W will allow immediate determination of the power function α (the exponent or slope of the line) and the constant k ($\log k$ = the intercept on the ordinate). From this allometric formula we can measure deviations from the best-fit line in relation to taxonomic, behavioral, or ecological variation. In this fashion, the deviations constitute a newly defined variable—the *encephalization quotient* (Jerison, 1973). Encephalization quotients, or other measures of brain size (e.g., Martin and Harvey, 1985; Gittleman, 1986a), have been found to correspond to phylogeny (Jerison, 1973; Gittleman, 1986a), diet (Clutton-Brock and Harvey, 1980), physical stratification (Eisenberg and Wilson, 1978), and metabolic rate (Martin, 1981a). Similar variables relative to body size have also been defined for canine size (Harvey *et al.*, 1978; Leutenegger and Cheverud, 1985), antler length (Clutton-Brock et al., 1980), testes size (Harcourt et al., 1981), home range movements (Harvey and Clutton-Brock, 1981; Gittleman and Harvey, 1982), locomotor behavior (Garland, 1983; Van Valkenburgh, 1987), and a variety of life history characteristics (Western, 1979; Eisenberg, 1981; Gittleman, 1986b).

B. Identifying Forms of Relationships

Defining a new variable by measuring the bivariate ratio between two variables often reveals an underlying form in a relationship. Considering again the concept of encephalization quotient, brain size was until recently thought to scale to body size with an exponent of 0.67, which led to the general assumption that a surface area explanation was required (see Harvey and Bennett, 1983). On closer inspection, it was found that relative brain size repeatedly scales to the power of 0.75 across 11 mammalian orders (Martin, 1981a; Hofman, 1983). Thus, the discovery of an exponent of 0.75 led to accounts that invoked metabolic explanations (Martin, 1981a; Armstrong, 1983), since basal metabolic needs scale to the 0.75 power of body size (Kleiber's law). From observing common allometric scaling in mammalian brain size, we may identify particular factors influencing this morphological trait.

Using a comparative approach, we may also identify the form of broad behavioral or ecological relationships. For example, home range movements in carnivores scale to metabolic needs, defined as the number of individuals living in a home range area times body weight to the 0.75 power (Gittleman and Harvey, 1982). After removing the effects of metabolism, considerable variation is observed around a line of best fit which relates to dietary affinities: Meat-eating carnivores (most felids and mustelids) have relatively larger home ranges than do fruit/vegetation-eaters (ursids, pandas, palm civet) and insectivores (banded mongoose, aardwolf, bat-eared fox). The functional explanation for this interspecific result is that species that feed primarily on relatively abundant and more evenly distributed foods (i.e., insects, fruit) do not require a large area to traverse for securing an adequate food supply. It turns out that the general form of this relationship also applies to another variable, namely population density. If a species is territorial, then the number of individuals utilizing a home range area should be inversely related to the population density or number of individuals residing in a given area. From this assumption we would predict that, given the dietary effects on relative home range size, diet should be an important predictor of population density. Indeed, after removing the effects of body size, we find that insectivores and frugivores live at higher population densities than do meat-eating species (Gittleman, 1984), thus the flip-side of the home range result.

Therefore, the form of the relationships showing dietary effects on home range size and, independently, population density are similar and give more support to the functional explanation for these interspecific trends. This does not exclude the possibility, however, that another factor (e.g., metabolic rate) may be involved; further study may indeed prove this to be the case (see McNab, 1983). The home range example simply emphasizes how repeated empirical trends suggest general relationships and the predictive power of these findings.

C. Generating Hypotheses

Using a comparative approach to define new variables and identify forms of relationships elicits many intriguing questions but usually provides few answers to biological problems. Yet, this does not devalue the importance of comparative study. Novel hypotheses, which often would go unnoticed, are frequently generated from such exercises (Harvey and Mace, 1982; Clutton-Brock and Harvey, 1984). A simple example is presented in comparative analyses of growth rate in carnivores (Gittleman and Oftedal, 1987).

Growth rate, measured as grams-per-day weight increase from birth to weaning, scales to maternal body size with a correlation coefficient of 0.87 (number of species = 46, $P < 0.01$). After we subtracted allometric effects (see Section IV.A) and searched for any potential taxonomic effects, striking deviations around a line of best fit showed that herbivorous/frugivorous species (e.g., ursids) and especially the bamboo-eating red panda (*Ailurus fulgens*) have significantly slower growth rates than do other carnivores. After referring to the physiological and anatomical literature, it was possible to formulate a testable hypothesis based on this comparative result: Pandas grow more slowly because their digestive system (simple stomach; lack of a cecum; relatively short intestines) is inefficient for handling an exclusively herbivorous diet (Roberts and Gittleman, 1984; Schaller *et al.*, 1985) and they have a reduced metabolic capacity (low basal metabolic rate: McNab, 1988). A direct test of this hypothesis would be to experimentally manipulate the diet of pandas and study consequential changes in growth, an experiment which is not feasible given the rarity and temperament of this species. An indirect test, though, would be to compare changes in digestive capacity during the growth or lactation phase between red pandas and other species with more efficient systems. Recent behavioral observations indicate that female red pandas increase their food intake up to 200% during lactation, a substantially higher ingestion rate than that observed in other eutherian mammals (Gittleman, 1988a). Thus, from a general comparative result it was possible to formulate a hypothesis to explain the protracted growth rate in red pandas.

D. Testing General Hypotheses

Hypotheses derived from comparative results may be of a general or specific form, depending on the quality of data analyzed, level of significance in the observed trends, and degree of understanding of the problem at hand. An example of testing a general hypothesis from comparative data may be taken from explanations for sexual dimorphism in carnivores. In species in which the breeding system is such that individuals of one sex mate with more than one individual, thereby leaving some individuals with the majority of matings, sexual selection will favor those characters that increase an individual's chances of matings. In mammals, polygynous breeding systems are most common (Ralls, 1976), thus enhancing intermale competition and selecting for increased size (assuming that size is a deciding factor in conflicts). In monogamous species, if there is an unbiased sex ratio and males and females have similar reproductive life spans, competition between males will be less intense. Thus, if sexual selection is the primary cause of sexual dimorphism in carnivore body size then monogamous species should be less dimorphic than are polygynous ones. By examining the quantitative distribution of body size dimorphism in relation to polygyny and monogamy within the order Carnivora, we can see that, similar to other mammalian groups (primates: Clutton-Brock *et al.*, 1977; ungulates and

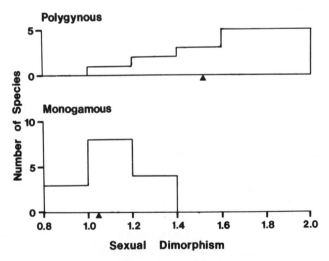

Fig. 1. Distribution of body weight dimorphism across the order Carnivora in relation to mating system. Number of species shown on ordinate. Arrow indicates median value for each type of mating system.

pinnipeds: Alexander *et al.*, 1979), breeding system is associated with degree of dimorphism (Fig. 1). This form of analysis, which is typical for comparative studies that rely on discrete behavioral or ecological categories (e.g., polygyny versus monogamy), is restricted to testing general relationships unless more detailed, quantitative information is unusually available. In this example, to test specifically the breeding system hypothesis, we would require that the degree of breeding competition (e.g., number and intensity of male/male fights during mating) correlates with increased ratio of size dimorphism in carnivores. Although restricting comparative tests to general hypothesis testing is annoying for one trying to establish causation, this does not eliminate their utility. Rather, we may be able to eventually erect causal pathways after numerous general tests have been performed, as seems likely with repetitive observations of size dimorphism and breeding in mammals, and new hypotheses often appear as a result of such general analyses [see, e.g., Harvey and Ralls (1985) with regard to the sexual dimorphism example].

E. Testing Specific Hypotheses

Characteristically, hypothesis testing progresses from a general to a specific kind, and comparative studies are quite illustrative of this. An example may be taken from the development of comparative studies on home range movements of mammals. McNab (1963) found that home range size increases with body weight, and that "hunters" have larger ranges than do "croppers" relative to their body weight. He concluded that home ranges are determined by energetic needs and density of available foods. Thus, McNab was able to examine two *general* explanations for variation in home range size. Yet, these results raise immediate questions of a more specific kind: (1) How does body size reflect energy requirements, i.e., what is the relationship between metabolic rate and home range size? (2) Does home range size vary in proportion to availability in food supply, i.e., what is the correlation between percentage of various foods and home range size? Comparative analyses on mammals and birds indicate that home range size is indeed directly determined by daily energy expenditure, although the allometric exponent may be different for various dietary groups (Mace and Harvey, 1983). Furthermore, studies have examined quantitatively McNab's hypothesis concerning home range size and food availability: Across 37 carnivore species, home ranges are positively correlated ($r = 0.37$, $P < 0.01$) with the proportion of flesh in the diet and negatively correlated ($r = -0.41$, $P < 0.01$) with the proportion of fruit/vegetation and insects (Gittleman and Harvey, 1982). Therefore, the ability to specifically test a hypothesis using a comparative approach is closely tied to the availability of quantitative information and knowledge of the interrelationships among the variables examined.

V. LIMITATIONS OF THE COMPARATIVE APPROACH

Most problems in the comparative approach rest with inherent weaknesses in the data base or with inadequate understanding of the variables, statistical relationships, and causal pathways resulting from interspecific comparisons. Some of these problems have been improved in recent years by a greater appreciation for the constraints involved in not being able to "experiment" or directly manipulate variables with a comparative approach (Huey and Bennett, 1986), and for the assumptions involved with unusually heavy reliance on statistical methodology (Harvey and Mace, 1982; Clutton-Brock and Harvey, 1984). In the following, I reiterate using different examples some of the points made by Harvey and Mace (1982) and Clutton-Brock and Harvey (1984), and discuss further problems prevalent in comparative studies that form the main reasons why comparative analyses are harshly criticized and thus avoided by many behavioral workers. I present these weaknesses, along with some suggestions for rectifying them, in the hope that future comparative studies will proceed on more solid ground. My precautionary words are not meant, however, to be applied to all comparative research.

A. The Data

Perhaps the most important feature in recent comparative analyses is an emphasis on quantitative measurement. Early comparative work in ethology was hampered by subjective descriptions of species characteristics and the statistical weaknesses of using only discrete, noncontinuous characters (Clutton-Brock and Harvey, 1977, 1984; Terborgh and Janson, 1986). Studies within the last decade or so have quantified various morphological characters (body weight, head and body length, brain size, testes size, antler length), life history traits, population characteristics (home range size, population density, day range length), and indices of social interaction (population group size, foraging group size, socionomic sex ratio), as well as a range of ecological variables (diet, habitat, zonation, activity cycle). Although such extensive data are available for a number of taxonomic groups that present opportunity for comparative analyses (see Greene, 1986b), it is imperative that extraction and use of this information be systematic and rigorous. First, careful consideration must be given to measurement accuracy in the original studies. Even with precise anatomical variables, measurements may include considerable error. For example, in a study of allometric and ecological effects of muskrat (*Ondatra zibethicus*) skull characters, the proportion of measurement error varied from 3 to 23% if measuring accuracy was 0.1 mm and from 13 to 45% if accuracy was 0.5 mm, with means of the skull characters ranging from 4 to 63 mm (Pankakoski et al., 1987). Because behavioral and ecological features are usually not measured with the precision of morphometric data, caution must be taken when first culling comparative information (see Bekoff et al., 1984; Bekoff, 1988).

Once particular studies are selected from which to utilize information, species values should be calculated from as many independent samples as possible; if we are considering ecological associations with behavioral variation it may, indeed, be revealing to use information from different geographical localities or to rank-order individual studies in terms of number of individuals studied, length of study, and quality of field methods. Review compilations, encyclopedias, and volumes with generally unidentified reference sources should be used with extreme caution; as a rule, if no information is given in data tables on where the data are derived, then it is best to search for original sources [comparative studies which do not present a data base or cite reference sources are not useful (e.g., Hayssen, 1984; Kruuk, 1986)]. Attention must also be given to environmental conditions of the animals studied, especially when combining data from natural populations and zoo animals. As a general rule, it is best to separate data taken from zoo animals except for variables which might not be influenced by artificial rearing conditions (e.g., in carnivores, some life history traits such as litter size). When multiple field studies are available for a given species, it is advisable to calculate species points from median values across populations, because skewness from aberrant points will least affect species values. Clearly, if particular points deviate by orders of magnitude from the cross-species trend, then these points must be reevaluated for accuracy (see Clutton-Brock and Harvey, 1984). Even though aberrant points are unhelpful for comparative purposes, such outliers may reveal why such intraspecific variation is induced (see Caro and Bateson, 1986).

For information that is not quantifiable (e.g., habitat types: open grassland, thick forest, scrub), it is important to use descriptors for species that are recurrent in the literature, thus acceptable to those researchers who have actually studied them, and probably better reflect the actual biology of the species. Species labels should not be imposed because of a particular orientation or bias of the investigator (Drummond, 1981; Jarman, 1982; van Roosmalen, 1984; McNab, 1987). This problem often surfaces when species are observed to have flexible ecologies, such that one ecological descriptor may not fit. For example, it is widely known that carnivores are very difficult to classify in relation to habitat because they traverse many vegetational zones. The white-tailed mongoose (*Ichneumia albicauda*) is described as living in savannah, woodland, open grassland, and a combination of these vegetational zones (Kingdon, 1977; Waser and Waser, 1985). After rank-ordering available studies of this species in terms of the number of individual animals studied and length of study, I was able to characterize its primary habitat as open grassland and woodland (Gittleman, 1985a, 1986a, b), thus not restricting characterization to one ecological descriptor. Also, by including some flexibility in the ecological classification of the species, inherent biases in the comparative data base are minimized and some intraspecific variation is incorporated into species labels.

Finally, once species are characterized with sufficient accuracy, it may become apparent that the sample sizes of certain taxonomic groups will make cross-species comparisons difficult. For example, in testing for phylogenetic effects (at the family level) of litter weight across carnivores, the following sample sizes of species within each family were used (Gittleman, 1986b): Canidae, 11: Procyonidae, 3; Ursidae, 4; Mustelidae, 15; Viverridae, 8; Hyenidae, 2; Felidae, 16. With such reduced samples, an error value in any family could significantly affect observed differences. Cross-species comparisons within some taxonomic groups will always be constrained by this problem unless analyses are restricted to larger taxa (Jarman, 1982). One method of partly avoiding this problem, however, is to match sample sizes of the larger taxa with the one having the lowest number of species. In testing the hypothesis that duration of parental care and intrinsic rate of natural increase are different between eutherian and metatherian (marsupial) mammals, Thompson (1987) was faced with comparing a small sample of marsupials with a substantially larger data set on eutherians. To control for this difference, he selected a eutherian subset to match female body mass, diet, and, where possible, habitat of the marsupials. Selecting data in this fashion may remedy any biases in taxonomic distribution but, at the same time, introduce subjective biases by using particular subsets of the data. Making further comparisons with randomly generated permutations of subsets in the data is one method for minimizing subjective influences in data usage. Extreme care must be taken when making decisions to include or exclude segments of comparative data.

B. Biases in Distribution of Data

Clearly, if certain taxonomic groups or ecological categories are omitted from consideration because of skewed data availability, then comparative tests will be misleading. Biases sometimes may be more subtle, however (Jarman, 1982; Clutton-Brock and Harvey, 1984). Comparative results often reveal more about the kinds of data collected, the ways in which they are gathered, and actual decision-making factors of field workers than anything concerning cross-species findings. Although probable biases in the data may introduce a fascinating trend previously unnoticed, they more often produce an embarassing result which is of little biological value.

An example of how data distribution may influence cross-species comparisons may be taken from life history analyses. An important variable influenc-

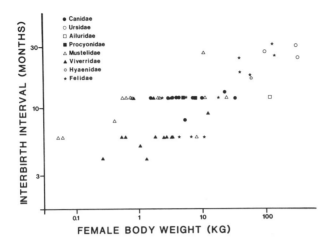

Fig. 2. Relationship between interbirth interval (period between successive births) and female body weight across major taxonomic families in the order Carnivora. Spurious distribution of points is discussed in the text.

ing reproductive rate and generation time in mammals is interbirth interval (Eisenberg, 1981), the period between successive births. Across 54 carnivore genera, interbirth interval is correlated with female body size ($r = 0.67$, $P < 0.01$). After size effects are removed, significant differences relate to taxonomic affiliation (see Fig. 2): Interbirth intervals are comparatively brief in viverrids, long in mustelids, and long in canids (Gittleman, 1986b). Such taxonomic effects are illusory, however, because of the artificial clustering of species points around 9 and 12 months. The distribution of points suggests that the original studies rounded off values from weeks or days to months. Without a quantitative, comparative analysis this trend would not be detected. Obviously, such biases in the sources from which data compilations are derived will dramatically influence comparative results. Therefore, decisions concerning measurement techniques, standardized scales, and selection of particular individual animals for study should be considered in collating comparative data.

C. Intraspecific Variation

Most comparative studies begin with establishing species points or species classifications, even though these may be converted into congeneric, familial, or some other taxonomic level of analysis. Constructing realistic species points is problematic when intraspecific variation is marked (Clutton-Brock and Harvey,

1977; Macdonald, 1983; Jarman, 1982). For example, population group sizes in wolves (*Canis lupus*) vary from a mean of 7 (range 2–16) in Minnesota, to packs with up to 21 individuals in Alaska, to solitaries in various parts of Canada and Europe (Mech, 1970; Zimen, 1981; Scott and Shackleton, 1982; Messier, 1985). Across the Carnivora, similar cases of wide intraspecific variation in population group size are known for coyotes (Kleiman and Brady, 1978; Bekoff and Wells, 1986), silverbacked jackals (Moehlman, 1988), red foxes (Macdonald, 1983), African hunting dog (Frame *et al.*, 1979), African lions (Van Orsdol *et al.*, 1985; Packer, 1986), spotted hyenas (Kruuk, 1972, 1975; Frank, 1986), European badgers (Kruuk and Parish, 1987), and dwarf mongooses (Rood, 1986; Rasa, 1987). Likewise, in a diversity of other taxa intraspecific variation in variables such as home range size, population density, parental care, life history traits, and social organization are legion (Lott, 1984; Gittleman, 1985b). In cross-species comparisons extreme care must be taken with *each variable* in calculating species points so as not to misrepresent such variation. The following are examples of how variation within each variable must be treated separately.

With respect to morphological variables, intraspecific variation is usually not as extensive as is interspecific variation. In a detailed study of eight morphological traits across 184 mammal species, Cherry *et al.* (1982) calculated the mean morphological "distance" (i.e., variation at different taxonomic levels) among subspecies, species, genera, subfamilies, families, superfamilies, suborders, and orders. They found only a slight change in variation between subspecies and species but at least a twofold change between species and all higher taxonomic levels. Similarly, across primates, the coefficient of variation in brain size within species is of the order of 10% whereas there is a tenfold level of variation among species (Harvey and Martin, 1988). Therefore, while it appears that morphological variables vary within species, such variation is trivial in most cases in comparison to distinctions which are made interspecifically. Nevertheless, it should be mentioned that body weight in some taxa, particularly within larger species (e.g., primates, ursids), undergoes marked seasonal fluctuations, and therefore care must be taken not to use measurements of animals studied during extreme environmental conditions (e.g., immediately following hibernation; during the breeding season).

Behavioral and ecological variables are notoriously difficult to interpret or quantify for cross-species study. However, some variables are more difficult than others and hence it is helpful to turn initially to those variables which may be less influenced by intraspecific variation (Martin, 1981b). As an example, population group size across the order Carnivora is actually rather modal, despite the illustration given above for the wolf; grouping outside of the breeding season is only found in approximately 10–15% of all carnivores (Bekoff *et al.*, 1984; Gittleman, 1988b) and, of the grouping species, many have a strong tendency to show some form of grouping behavior rather than live as solitaries (Macdonald and Moehlman, 1982; Macdonald, 1983; Bekoff *et al.*, 1984; Gittleman, 1988).

Therefore, the possibility of misclassifying a carnivore species according to grouping behavior is reduced in this variable.

Other behavioral or ecological variables, however, appear to retain greater intraspecific variation, which presents serious problems for comparative classifications. Home range size and population density in carnivores have extensive intraspecific variation, sometimes showing a hundredfold difference (Macdonald, 1983), and such variation is widespread across the order. Similarly, intraspecific variation in some temporal measures of life histories (e.g., gestation length, interbirth interval, longevity) is considerable. Until there are sufficient data to analyze comparative trends of these variables at the subspecific (e.g., Harvey and Clutton-Brock, 1981) or populational levels (e.g., Laundré and Keller, 1984; Van Orsdol *et al.*, 1985), intraspecific variation will weaken the reliability and generality of cross-species comparisons. At present, the best partial solution to the problem is to consider the extent of intraspecific variation with *each variable* and then calculate medians for intraspecific differences. Also, it should be remembered that even though intraspecific variation may introduce error, there is no reason to suppose that it could be responsible for creating a comparative relationship across a large array of species where none exists. As Martin (1981b, p. 295) forewarned in the context of comparative studies on primates, ". . . overemphasis on the variability of primates societies can . . . obscure the significance of the average condition for each species, and it is oversimplistic to rule out any firm conclusions about the relationship between [primate] behaviour and ecology merely because variability exists."

D. Methodological Decisions in Data Usage

Numerous methodological decisions, often involving statistical problems, arise from cross-species comparisons and many of these have recently been examined (Harvey, 1982; Harvey and Mace, 1982; Clutton-Brock and Harvey, 1984; Smith, 1980, 1984; Felsenstein, 1985; Rayner, 1985; Reist, 1985, 1986). In this brief section, I highlight the main points from this work relevant to establishing appropriate taxonomic levels of analysis and drawing lines of best fit for cross-species comparisons.

When analyzing the comparative distribution of a behavioral or ecological trait, one of the initial decisions to be made is the taxonomic level at which to make comparisons. As Harvey and Mace (1982) point out, two decisions are critical. First, at the lower taxonomic level, choice must be made of a taxon which will represent statistically independent points (e.g., subspecies, species, genus, etc.). Second, at the upper taxonomic level, the range of taxa over which to gather the data must be decided upon.

The main problem at the lower taxonomic level is establishing independence of data points. Frequently, species are not evenly distributed across genera because species within a genus tend to have similar characteristics. For example, with respect to mammalian energetics (basal metabolic rate), home range size, and brain size, values of slope are significantly altered when using species points rather than generic ones because of clustering (see Harvey and Mace, 1982; Elgar and Harvey, 1987). When slopes are influenced from such nonindependence of points, deviations for individual species will significantly alter comparative results. Therefore, if species within genera, or indeed other lower taxonomic groups at which analyses are performed, possess similar characteristics due to phylogenetic constraints, they cannot be considered statistically independent and the analysis will be biased by their presence. A nested analysis of variance may be used to indicate maximum variance at different taxonomic levels to determine the appropriate level for statistical analysis (see Harvey and Clutton-Brock, 1985). This should not, however, be interpreted as an all-encompassing solution for making cross-species comparisons at *one* level: Even within a given order, there may be appropriate level(s) of analysis at genera, families, or subfamilies, depending on degrees of variation at different taxonomic levels.

Choice of upper taxonomic level should include the possibility of differences among taxa before amalgamating them. Most comparative studies have examined species differences across mammalian orders (e.g., Chiroptera: Eisenberg and Wilson, 1978; Primates: Leutenegger and Cheverud, 1982; Pinnipedia: Kovacs and Lavigne, 1986; Innes *et al.*, 1987) or across the entire class Mammalia (e.g., McNab, 1986; Millar, 1977, 1981; Western, 1979; Hayssen, 1984) without testing for variation at lower levels. If statistical tests reveal heterogeneity (either in slope or elevation around a line of best fit) among, say, families within a given order, then such differences must be accounted for.

This point is extremely important not only for establishing statistical rigor, but also for identifying potential phylogenetic effects. As mentioned in Section II, one of the primary strengths of the comparative approach is to control for and discover historical effects of behavior (for further discussion, see Section VI, entitled "Future Directions"). Nevertheless, the comparative literature is replete with studies that fail to consider phylogeny. For example, well-cited investigations of brain size (Jerison, 1973; Sacher and Staffeldt, 1974; Eisenberg and Wilson, 1978), life histories (e.g., Millar, 1977, 1981; Western, 1979; Eisenberg, 1981), and metabolic rate (McNab, 1986, 1987) examined trends across broad taxonomic groups, usually at the ordinal or class level, without searching for variation at lower taxonomic levels. Virtually every comparative study of these variables that has considered variation at the family or subfamily levels shows significant differences (see Clutton-Brock and Harvey, 1980; Elgar and Harvey, 1987; Gittleman, 1986a,b; Harvey and Clutton-Brock, 1985; Harvey and Elgar, 1987; Stearns, 1983). Without considering variation at lower levels, comparative conclusions may be incorrect. For example, Millar (1981) compared relative litter weight (i.e., deviations from a common mammalian slope of litter weight on body weight) among 10 taxonomic orders, and found that litter weight is lightest in carnivores and edentates and heaviest in insectivores, lagomorphs, rodents, hyracoideans, perissodactyls, and artiodactyls. He then went on to argue that these differences are associated with various adaptive patterns, such as size of pelvis canal, placental structure, and available food resources. Millar's comparative conclusions and ensuing explanations assume no heterogeneity *within* the examined taxonomic groups. After accounting for familial differences in the carnivores, we observe that Procyonidae and Viverridae have heavier litter weights than do other carnivore families and, on comparison with other mammalian orders, are comparable with those found in lagomorphs and rodents. Therefore, it is important that comparative studies at least *account for* lower taxonomic effects, otherwise conclusions are likely to be misleading.

Lastly, with studies increasingly showing the importance of size (for reviews, see Calder, 1984; Clutton-Brock and Harvey, 1983; Peters, 1983; Schmidt-Nielson, 1984), the comparative approach is now commonly used to remove allometric effects. Essentially, as mentioned above (see Section IV.A), this requires drawing a line of best fit through some size variable (e.g., body weight, head and body length, skull length) and an independent variable which may be correlated with size (e.g., group size, population density, day range length). Until recently (Harvey and Mace, 1982), a least-squares regression model was used to describe the line of best fit. Such a model may be inappropriate for comparative studies because it assumes that values for the independent variable are measured without error and that a causal relationship is known between the two variables being examined. With many size-related variables in behavior and ecology, such assumptions are rarely met: Measurement error in variables such as interbirth interval or home range size are often considerable (see above) and the causal relationship usually cannot be determined (Clutton-Brock and Harvey, 1984). As a result, regression analysis may underestimate values of slope, especially when correlation coefficients are low (Harvey and Mace, 1982; Rayner, 1985; Seim and Saether, 1983), which will significantly influence comparative results. Although other statistical methods (e.g., reduced major axis, major axis) contain equally restrictive assumptions, major axis analysis is probably the most appropriate model to use when (1) data are converted to logarithmic values prior to analyses, (2) differences in slope among taxonomic or behavioral groups are likely, and (3) the assumptions for measurement error and/or causal interactions between variables are violated.

E. Establishing Causation

Testing for a causal relationship across taxonomic groups is fraught with problems. Unlike experimentation, a comparative analysis cannot manipulate variables in a stepwise fashion without unusually complete and detailed data.

And, even then, while relying on multivariate statistics, interpretation of comparative results is difficult because of uncertain relationships involving three-way interactions, frequent use of dummy variables for missing cells (Snedecor and Cochran, 1967), and heavy reliance on correlation coefficients (see Harvey and Clutton-Brock, 1985).

Part of the problem relates more to a misunderstanding of aims than to a shortcoming of comparative study. Comparative analyses are not always meant to infer causation. As previously discussed, most bivariate relationships with body size (e.g., life histories, brain size, sexual dimorphism in body weight) indicate strong associations or constraints imposed by allometry, with few statements requiring causation (see Schmidt-Nielson, 1984; Barrett, 1988). When causation is inferred, however, the strength of explanation depends on the quality of data, knowledge of the variables analyzed, and, most importantly, causal-type studies derived from single-species accounts. Because I have already touched on the first two qualities in previous discussion (see Section IV, "Aims of the Comparative Approach"), I will elaborate here on the importance of single-species study for making causal statements.

Even though there is no reason to assume that evolutionary mechanisms function in similar ways at different taxonomic levels, quite often there are remarkable parallels. For example, in carnivores, after removing allometric and phylogenetic effects, variation in dietary characteristics is associated with differences in three life history traits (Gittleman, 1986b): Compared to carnivorous species (e.g., wolf, coyote, African wild dog), omnivorous species (gray fox, fennec fox, crab-eating fox) have heavier birth weights, longer gestation lengths, and longer lactation periods. These comparative findings are given even more support by the observation that other ecological variables such as habitat, zonation, and activity cycle are not related to differences in life histories. On referring to single-species studies, we find similar patterns of dietary effects in coyotes (Todd *et al.*, 1981), stoats (Erlinge, 1981), weasels (King, 1983), and black bears (Rogers, 1977). The most common explanation for these findings is that species with a wider food base (i.e., omnivores) devote a greater energetic investment to reproduction, thus selecting for divergent life histories. In this fashion, causal explanations derived from single-species results give credibility to causal statements in comparative studies. Further, when causal hypotheses are formulated from single-species studies *prior to* carrying out comparative tests, criticisms of post hoc methods are not justified. Clearly, though, if causal pathways from single-species studies occur in an opposite direction, then statements of causation in cross-species work must be reappraised.

Causation may also be partly uncovered by finding similarities across comparative studies (Eisenberg, 1981). From independent cross-species studies of primates (Harvey and Clutton-Brock, 1981), carnivores (Gittleman and Harvey, 1982), and rodents (Mace and Harvey, 1983), it may be seen that differences in home range movements relate to whether species feed on readily available and evenly distributed foods, resulting in smaller ranges, or feed on scarcer foods, resulting in larger ranges. Causal statements are more solid and may be advanced a priori when multiple cross-species findings are available that replicate presumed causal interactions, but again these statements hinge on whether single-species findings are also in agreement or at least can specify particular mechanisms that are operating.

Finally, the ability to make causal arguments using the comparative approach often simply depends on common sense. As Simpson *et al.* (1960) cautioned when trying to interpret correlation coefficients, causal patterns must not be forced between variables that obviously have no relationship. Sometimes behavioral or ecological factors interrelate in only one direction, which allows a more certain causality. Using the home range example, it is difficult to imagine causal mechanisms that would elicit home range movements to cause dietary differences across species from independent lineages. Therefore, comparative analyses of relationships that are restricted to causal pathways occurring in one direction often may be more productive in establishing causation (see Ridley, 1986).

VI. FUTURE DIRECTIONS

The preceding discussion outlines general aims and limitations of contemporary ethological studies using the comparative approach. Future behavioral studies should, I suggest, utilize this approach more than they have in the past for two reasons. First, an abundance of standardized, quantitative, and long-term single-species studies (for mammals, at least) will provide more enticing questions to be answered across broad taxonomic groups. Second, and perhaps more importantly, particular problems of the origins and phylogeny of behavior as well as the extent to which parallel evolution occurs at intra- versus interspecific levels are rapidly developing into predominant issues in ethology, all of which demand comparative analyses (Felsenstein, 1984, 1985; Huey, 1987). I wish to conclude by discussing some of these future directions.

One of the primary reasons the comparative approach will continue as a mainstay in future ethological studies is that new comparative methods are being developed to directly test phylogenetic hypotheses of behavior. In the following, I briefly summarize these recent methods (see also Huey, 1987).

Four approaches are currently being used to analyze the phylogeny of behavior. First, comparisons of the distribution of behavioral or ecological traits among species are superimposed on a well-supported phylogeny (for examples, see Greene and Burghardt, 1978; Dobson, 1985; Lauder, 1986). If the distribution of traits is inconsistent with the phylogeny, then several independent evolutionary events must be invoked to explain the distribution of traits. With such a comparative trend, it is unlikely that the distribution of traits is strongly influenced by phylogenetic history. The main weakness of this approach is that variables are usually discrete, and therefore relative phylogenetic influence cannot be partitioned statistically. However, if variables are continuous, application of Cheverud et al.'s (1985) autocorrelational model may assess phylogenetic effects, a method conceptually similar to variance partitioning in quantitative genetics or in analysis of variance.

Second, comparisons of the distribution of a trait may be made across divergently evolved lineages (for examples, see Alexander et al., 1979; Western, 1979; Eisenberg, 1981). For example, Kenagy and Trombulak (1986) found that larger relative testes size is associated with multimale breeding systems (promiscuous or polygynous systems) in three distantly related taxa (carnivores, primates, and artiodactyls). Although phylogeny is not explicitly addressed in these comparisons, the repeated distribution of traits across distantly related lines makes historical effects unlikely. The difficulty with this approach is that, along with polyphyletic origins of divergent taxa, many other factors may be involved with a comparative result, but because of such large-scale taxonomic comparisons these cannot be accounted for.

Third, establishing the independent evolution of a trait may be made through assigning traits to ancestors by successive outgroup comparisons (for examples, see Gittleman, 1981; Ridley, 1983; Huey and Bennett, 1987). Essentially, this method involves finding a phylogeny for the taxonomic group under study and then, by minimizing the number of times that an independent origin must be assumed, inferring those points at which the trait has arisen independently. Although this method is initially useful for explicitly mapping behavioral phylogenies, it contains two problems (Felsenstein, 1985). First, one must presume monophyletic origin, or essentially a well-accepted phylogeny, in the taxa under examination. Second, because of the statistical nature of the direction and frequency of behavioral transitions (changes in behavioral attributes between taxa), determining the actual places in which transitions are placed is subject to considerable error (Felsenstein, 1983). Nevertheless, this is an important advance for at least addressing phylogenetic questions.

Fourth, *if* molecular, gene-frequency, or quantitative morphological characters allow an accurate tree topology to be erected, Felsenstein (1985) has developed a method for correcting the statistical hitch in the previous method: Modeling the characters by Brownian motion on a linear scale, we can specify a set of "contrasts" among species that allow accurate phylogenies to be established which are statistically independent. Admittedly, there are considerable barriers to using this technique (see Felsenstein, 1985), the main one being that the critical information for mapping trees is not available; however, for the first time in using the comparative method to explain underlying causes of behavioral variation, Felsenstein's model conceptually permits us to actually separate phylogenetic and adaptive explanations.

As a whole, these four methods represent an important step toward directly testing historical effects of behavior and, more importantly as a consequence, teasing apart the relative effects of phylogeny versus function. How might we refine our comparative analyses even further? Given the methods and results currently available, three exciting possibilities arise. First, it would be informative to apply different phylogenetic methods to the same data set to establish if various methods produce different results. This might reveal something about historical processes and/or the validity of our methods. Second, if complete comparative data sets are not amenable to testing with multiple methods, it would be helpful to simulate data in order to analyze phylogenetic effects (e.g., see Elgar and Harvey, 1987). Third, with certain problems that have received considerable attention, such that a great deal is known about the functional attributes of particular traits, it would be important to pinpoint those taxa where phylogeny appears more influential than function or vice versa. An example of such a case is a recent study by Huey and Bennett (1987) on coadaptive thermal preferences and thermal dependence of sprinting in some Australian skinks (Lygosominae). After constructing outgroup comparisons of both thermal traits across taxa, they found that coadaptation of traits is tight only for genera with high thermal preferences and that low thermal preferences appear to have evolved even though this has resulted in decreased sprint capacities. Thus, by virtue of carrying out a phylogenetic analysis, we have a possible case where two traits may have evolved in opposite directions, a phenomenon which Huey and Bennett refer to as "antagonistic coadaptation." Despite such successes and new possibilities for uncovering phylogenetic unknowns with cross-species study, it is perhaps more productive at this early stage to recall a view well stated by Simpson (1960, p. 122): "The great drawback of the comparative method and of contemporaneous evidence is that they are not in themselves historical in nature. The drawing of historical conclusions from them is therefore full of pitfalls unless it can be adequately controlled by *directly* historical evidence." These new methods of considering the phylogeny of behavior may bring us closer to direct kinds of evidence.

Finally, from this review, it should be apparent that severe limitations of the comparative approach (e.g., establishing causation; intraspecific variation) lie in what relationships may be inferred from results at different taxonomic levels. To successfully pursue comparative study we must extrapolate information (both original data and causal statements) from single-species studies or, at the very least, refer to observations in other interspecific comparisons. Thus, it is commonplace to observe a comparative result in one taxonomic group being verified with comparative results in other taxonomic groups (e.g., brain size allometry: Clutton-Brock and Harvey, 1980; Martin, 1981a; Gittleman, 1986a; life history traits: Western, 1979; Eisenberg, 1981; Harvey and Clutton-Brock, 1985; Gittleman, 1986b) or, usually in the context of controlling for possible phylogenetic effects, single-species results being compared with those at a comparative level (e.g., group size: Terborgh, 1983; sex differences in growth rate: Clutton-Brock *et al.*, 1982; reproductive rate in marsupials: Thompson and Nicoll, 1986; thermoregulatory benefits in body mass changes: Martin, 1986). These types of comparative study will undoubtedly continue to flourish. However, I would like to suggest that another neglected form of comparative study be adopted. As illustrated in Section IV.C using the comparative approach to generate hypotheses, the logical sequence of moving from a general comparative trend (as in the growth rate allometry example) back down to a single-species characteristic (e.g., protracted growth in the red panda) may give insight into problems which otherwise would be undetected. Moreover, by first analyzing comparative trends and then turning to intraspecific variation, it is possible to stave off criticisms that adaptive explanations of behavior ignore phylogenetic constraints (see Clutton-Brock and Harvey, 1979; Greene, 1986b). In sum, to successfully test functional or causal explanations of behavior, we *must* traverse information at different taxonomic levels: Cross-species comparisons should rely on single-species studies for pinning down causality and providing knowledge of the variables being analyzed; single-species study should use comparative approaches for testing phylogenetic effects and suggesting novel hypotheses from a comparative backdrop. Such reciprocal exchange of information will be important for the growth and development of ethological studies.

VII. ACKNOWLEDGMENTS

I am grateful to Paul Harvey, Hans Kruuk, and Peter Slater for initially commenting on portions of this chapter in its thesis form. I thank Marc Bekoff, Gordon Burghardt, Sandy Echternacht, Paul Harvey, Peter Klopfer, Stuart Pimm, and especially Pat Bateson and Ray Huey for detailed comments on the present manuscript, and Pat Bateson for suggesting that I place it in this volume. Many of these ideas on the comparative approach have developed over the years during conversations and correspondence with Paul Harvey, without whose guidance and support they would have never developed. During preparation of this paper I received financial support from the Graduate Programs in Ethology and Ecology and the Department of Zoology, University of Tennessee, and a NIH Training Grant (T32-HD-07303).

VIII. REFERENCES

Alexander, R. D., Hoogland, J. L., Howard, R. D., Noonan, K. M., and Sherman, P. W. (1979). Sexual dimorphism and breeding systems in pinnipeds, ungulates, primates and humans. In Chagnon, N., and Irons, W. (eds.), *Evolutionary Biology and Human Social Behavior: An Anthropological Perspective*, Duxbery, North Scituate, Massachusetts, pp. 402–435.

Armstrong, E. (1983). Metabolism and relative brain size. *Science* **220**:1302–1304.

Barrett, C. (1988). Causal analysis in behavioural ecology. *Anim. Behav.* **36**:310.

Bekoff, M. (1988). Behavioral development in terrestrial carnivores. In Gittleman, J. L. (ed.), *Carnivore Behavior, Ecology and Evolution*, Cornell University Press, Ithaca, New York, in press.

Bekoff, M., and Byers, J. A. (1981). A critical reanalysis of the ontogeny and phylogeny of mammalian social and locomotor play: An ethological hornet's nest. In Barlow, G. W., Petrinovich, L., and Main, M. (eds.), *Behavioral Development*, Cambridge University Press, Cambridge, pp. 296–337.

Bekoff, M., and Wells, M. C. (1986). Social ecology and behavior of coyotes. In Rosenblatt, J. S., Beer, C., Busnel, M.-C., and Slater, P. J. B. (eds.), *Advances in the Study of Behavior*, Vol. 16, Academic Press, New York, pp. 251–338.

Bekoff, M., Daniels, T. J., and Gittleman, J. L. (1984). Life history patterns and the comparative social ecology of carnivores. *Ann. Rev. Ecol. Syst.* **15**:191–232.

Bradbury, J. W. (1985). Contrasts between insects and vertebrates in the evolution of male display, female choice, and lek mating. In Hollbobler, B., and Lindauer, M. (eds.), *Experimental Behavioral Ecology and Sociobiology*, Sinauer Associates, Sunderland, Massachusetts, pp. 273–289.

Burghardt, G. M. (1978). Behavioral ontogeny in reptiles: whence, whither, and why? In Burghardt, G. M., and Bekoff, M. (eds.), *The Development of Behavior*, Garland STPM Press, New York, pp. 149–174.

Burghardt, G. M. (1984). On the origins of play. In Smith, P. K. (ed.), *Play: In Animals and Humans*, Blackwell, Oxford, pp. 5–41.

Burghardt, G. M. (ed.), (1985). *Foundations in Comparative Ethology*, Van Nostrand Reinhold, New York.

Burghardt, G. M. (1988). Precocial behavior, play, and the ectotherm–endotherm transition: profound reorganization or superficial adaptation? In Blass, E. M. (ed.), *Handbook of Behavioral Neurobiology*, Vol. 9, Plenum Press, New York, pp. 107–148.

Burian, R. M. (1983). "Adaptation." In Grene, M. (ed.), *Dimensions of Darwinism*, Cambridge University Press, Cambridge, pp. 287–314.

Calder, W. A. III (1984). *Size, Function, and Life History*, Harvard University Press, Cambridge, Massachusetts.

Caro, T. M., and Bateson, P. (1986). Organization and ontogeny of alternative tactics. *Anim. Behav.* **34**:1483–1499.

Cherry, L. M., Case, S. M., Kunkel, J. G., Wyles, J. S., and Wilson, A. C. (1982). Body shape metrics and organismal evolution. *Evolution* **36**:914–933.

Cheverud, J. M., Dow, M. M., and Leutenegger, W. (1985). The quantitative assessment of phylogenetic constraints in comparative analyses: sexual dimorphism in body weight among primates. *Evolution* **39**:1335–1351.

Clutton-Brock, T. H., and Albon, S. D. (1982). Parental investment in male and female offspring in mammals. In King's College Sociobiology Group (eds.), *Current Problems in Sociobiology*, Cambridge University Press, Cambridge, Massachusetts, pp. 223–247.

Clutton-Brock, T. H., and Harvey, P. H. (1977). Primate ecology and social organisation. *J. Zool.* **183**:1–39.

Clutton-Brock, T. H., and Harvey, P. H. (1979). Comparison and adaptation. *Proc. R. Soc. Lond.* **205:**547–565.

Clutton-Brock, T. H., and Harvey, P. H. (1980). Primates, brains, and ecology. *J. Zool.* **190:**309–323.

Clutton-Brock, T. H., and Harvey, P. H. (1983). The functional significance of variation in body size among mammals. In Eisenberg, J. F., and Kleiman, D. G. (eds.), *Advances in the Study of Mammalian Behavior*, American Society of Mammalogists, Spec. Publ. No. 7, pp. 532–563.

Clutton-Brock, T. H., and Harvey, P. H. (1984). Comparative approaches to investigating adaptation. In Krebs, J. R., and Davies, N. B. (eds.), *Behavioural Ecology: An Evolutionary Approach*, 2nd ed., Sinauer Associates, Sunderland, Massachusetts, pp. 7–29.

Clutton-Brock, T. H., Harvey, P. H., and Rudder, B. (1977). Sexual dimorphism, socionomic sex ratio and body weight in primates. *Nature* **269:**797–800.

Clutton-Brock, T. H., Albon, S. D., and Harvey, P. H. (1980). Antlers, body size and breeding group size in Cervidae. *Nature* **285:**565–567.

Clutton-Brock, T. H., Guiness, F. E., and Albon, S. D. (1982). *Red Deer: The Ecology of Two Sexes*, University of Chicago Press, Chicago.

Coleman, W. (1964). *Georges Cuvier, Zoologist*, Harvard University Press, Cambridge, Massachusetts.

Dobson, F. S. (1985). The use of phylogeny in behavior and ecology. *Evolution* **39:**1384–1388.

Drummond, H. (1981). The nature and description of behavior patterns. In Bateson, P. P. G., and Klopfer, P. H. (eds.), *Perspectives in Ethology*, Vol. 4, Plenum Press, New York, pp. 1–33.

Eberhard, W. G. (1982). Behavioral characters for the higher classification of orb-weaving spiders. *Evolution* **36:**1067–1095.

Eisenberg, J. F. (1981). *The Mammalian Radiations*, University of Chicago Press, Chicago.

Eisenberg, J. F., and Wilson, D. C. (1978). Relative brain size and feeding strategies in the Chiroptera. *Evolution* **32:**740–751.

Elgar, M. A., and Harvey, P. H. (1987). Basal metabolic rate in mammals: allometry, phylogeny and ecology. *Functional Ecology* **1:**25–36.

Erlinge, S. (1981). Food preference, optimal diet and reproductive output in stoats, *Mustela erminea*, in Sweden. *Oikos* **36:**303–315.

Felsenstein, J. (1983). Parsimony in systematics: biological and statistical issues. *Ann. Rev. Ecol. Syst.* **14:**313–333.

Felsenstein, J. (1984). Comparatively better (review of M. Ridley's *The Explanation of Organic Diversity*). *Nature* **308:**656.

Felsenstein, J. (1985). Phylogenies and the comparative method. *Am. Nat.* **125:**1–15.

Frame, L. H., Malcolm, J. R., Frame, G. W., and van Lawick, H. (1979). Social organization of African wild dogs (*Lycaon pictus*) on the Serengeti Plains, Tanzania (1967–1978). *Z. Tierpsychol.* **50:**225–249.

Frank, L. G. (1986). Social organization of the spotted hyaena (*Crocuta crocuta*). I. Demography. *Anim. Behav.* **34:**1500–1509.

Garland, T., Jr. (1983). The relation between maximum running speed and body mass in terrestrial mammals. *J. Zool.* **199:**157–170.

Ghiselin, M. T. (1972). Models in phylogeny. In Schopf, T. J. M. (ed.), *Models in Paleobiology*, Freeman, Cooper and Co., San Francisco, pp. 130–145.

Gittleman, J. L. (1981). The phylogeny of parental care in fishes. *Anim. Behav.* **29:**936–941.

Gittleman, J. L. (1984). The behavioural ecology of carnivores. Ph.D. dissertation, University of Sussex, Brighton, England.

Gittleman, J. L. (1985a). Carnivore body size: Ecological and taxonomic correlates. *Oecologia* **67:**540–554.

Gittleman, J. L. (1985b). Functions of communal care in mammals. In Greenwood, P. J., Harvey, P.

H., and Slatkin, M. (eds.), *Evolution: Essays in Honour of John Maynard Smith,* Cambridge University Press, Cambridge, pp. 187–205.

Gittleman, J. L. (1986a). Carnivore brain size, behavioral ecology, and phylogeny. *J. Mamm.* **67:**23–36.

Gittleman, J. L. (1986b). Carnivore life history patterns: allometric, phylogenetic, and ecological associations. *Am. Nat.* **127:**744–771.

Gittleman, J. L. (1988a). The behavioral energetics of lactation in a herbivorous carnivore, the red panda (*Ailurus fulgens*). *Ethology,* in press.

Gittleman, J. L. (1988b). Carnivore group living: comparative trends. In Gittleman, J. L. (ed.), *Carnivore Behavior, Ecology, and Evolution,* Cornell University Press, Ithaca, New York, in press.

Gittleman, J. L., and Harvey, P. H. (1982). Carnivore home-range size, metabolic needs, and ecology. *Behav. Ecol. Sociobiol.* **10:**57–63.

Gittleman, J. L., and Oftedal, O. T. (1987). Comparative growth and lactation energetics in carnivores. In Loudon, A., and Racey, P. A. (eds.), *Reproductive Energetics in Mammals,* Oxford University Press, Oxford, pp. 41–77.

Gittleman, J. L., and Thompson, S. D. (1988). Energy allocation in mammalian reproduction. *Am. Zool.,* in press.

Gould, S. J. (1977). *Ontogeny and Phylogeny,* Harvard University Press, Cambridge, Massachusetts.

Gould, S. J., and Vrba, E. (1982). Exaptation—a missing term in the science of form. *Paleobiology* **8:**4–15.

Greene, H. W. (1986a). Diet and arboreality in the emerald monitor, *Varanus prasinus,* with comments on the study of adaptation. *Fieldiana* **31:**1–12.

Greene, H. W. (1986b). Natural history and evolutionary biology. In Feder, M. E., and Lauder, G. V. (eds.), *Predator–Prey Relationships,* University of Chicago Press, Chicago, pp. 99–108.

Greene, H. W., and Burghardt, G. M. (1978). Behavior and phylogeny: constriction in ancient and modern snakes. *Science* **200:**74–77.

Harcourt, A. H., Harvey, P. H., Larson, S. G., and Short, R. V. (1981). Testis weight, body weight and breeding system in primates. *Nature* **293:**55–57.

Harvey, P. H. (1982). On rethinking allometry. *J. Theor. Biol.* **95:**37–41.

Harvey, P. H., and Bennett, P. M. (1983). Brain size, energetics, ecology and life history patterns. *Nature* **306:**314–315.

Harvey, P. H., and Clutton-Brock, T. H. (1981). Primate home-range size and metabolic needs. *Behav. Ecol. Sociobiol.* **8:**151–155.

Harvey, P. H., and Clutton-Brock, T. H. (1985). Life history variation in primates. *Evolution* **39:**559–581.

Harvey, P. H., and Elgar, M. A. (1987). In defence of the comparative method. *Functional Ecology* **1:**160–161.

Harvey, P. H., and Mace, G. M. (1982). Comparisons between taxa and adaptive trends: problems of methodology. In King's College Sociobiology Group (eds.), *Current Problems in Sociobiology,* Cambridge University Press, Cambridge, pp. 343–362.

Harvey, P. H., and Martin, R. D. (1988). The comparative method in evolutionary biology. In DeVore, I. B., and Tooby, J. (eds.), *Evolution, Adaptation, and Behavioral Science,* Aldine, Chicago, in press.

Harvey, P. H., and Ralls, K. (1985). Homage to the null weasel. In Greenwood, P. J., Harvey, P. H., and Slatkin, M. (eds.), *Evolution: Essays in Honour of John Maynard Smith,* Cambridge University Press, Cambridge, pp. 155–171.

Harvey, P. H., Kavanagh, M., and Clutton-Brock, T. H. (1978). Sexual dimorphism in primate teeth. *J. Zool.* **186:**475–486.

Hayssen, V. D. (1984). Mammalian reproduction: constraints on the evolution of infanticide. In Hausfater, G., and Hrdy, S. B. (eds.), *Infanticide: Comparative and Evolutionary Perspectives*, Aldine, New York, pp. 105–123.

Hinde, R. A., and Tinbergen, N. (1958). The comparative study of species-specific behavior. In Roe, A., and Simpson, G. G. (eds.), *Behavior and Evolution*, Yale University Press, New Haven, Connecticut, pp. 251–268.

Hofman, M. A. (1983). Energy metabolism, brain size and longevity in mammals. *Quart. Rev. Biol.* **58**:495–512.

Huey, R. B. (1987). Phylogeny, history, and the comparative method. In Feder, M. E., Bennett, A. F., Burggren, W., and Huey, R. B. (eds.), *New Directions in Ecological Physiology*, Cambridge University Press, Cambridge, pp. 76–98.

Huey, R. B., and Bennett, A. F. (1986). A comparative approach to field and laboratory studies in evolutionary biology. In Feder, M. E., and Lauder, G. V. (eds.), *Predator–Prey Relationships*, University of Chicago Press, Chicago, pp. 82–98.

Huey, R. B., and Bennett, A. F. (1987). Phylogenetic studies of coadaptation: preferred temperatures versus optimal performance temperatures of lizards. *Evolution* **41**:1098–1115.

Innes, S., Lavigne, D. M., Earle, W. M., and Kovacs, K. M. (1987). Feeding rates of seals and whales. *J. Anim. Ecol.* **56**:115–130.

Jakšić, F. M. (1981). Recognition of morphological adaptations in animals: the hypothetico-deductive method. *Bioscience* **31**:667–670.

Jarman, P. J. (1982). Prospects for interspecific comparison in sociobiology. In King's College Sociobiology Group (eds.), *Current Problems in Sociobiology*, Cambridge University Press, Cambridge, pp. 323–342.

Jarman, P. J., and Jarman, M. V. (1979). The dynamics of ungulate social organization. In Sinclair, A. R. E., and Norton-Griffiths, M. (eds.), *Serengeti: Dynamics of an Ecosystem*, University of Chicago Press, Chicago, pp. 185–220.

Jerison, H. J. (1973). *Evolution of the Brain and Intelligence*, Academic Press, New York.

Kenagy, G. J., and Trombulak, S. C. (1986). Size and function of mammalian testes in relation to body size. *J. Mamm.* **67**:1–22.

King, C. M. (1983). Factors regulating mustelid populations. *Acta Zool. Fenn.* **174**:217–220.

Kingdon, J. (1977). *East African Mammals*, Vol. 3A, Academic Press, New York.

Kleiman, D. G., and Brady, C. (1978). Coyote behavior in the context of recent canid research: problems and perspectives. In Bekoff, M. (ed.), *Coyotes: Biology, Behavior, and Management*, Academic Press, New York, pp. 163–186.

Kovacs, K. M., and Lavigne, D. M. (1986). Maternal investment and neonatal growth in phocid seals. *J. Anim. Ecol.* **55**:1035–1051.

Kruuk, H. (1972). *The Spotted Hyena: A Study of Predation and Social Behavior*, University of Chicago Press, Chicago.

Kruuk, H. (1975). Functional aspects of social hunting in carnivores. In Baerends, G., Beer, C., and Manning, A. (eds.), *Function and Evolution of Behavior*, Oxford University Press, Oxford, pp. 119–141.

Kruuk, H. (1986). Interactions between Felidae and their prey species: a review. In Miller, S. D., and Everett, D. D. (eds.), *Cats of the World: Biology, Conservation, and Management*, National Wildlife Federation, Washington, D.C., pp. 353–374.

Kruuk, H. H., and Parish, T. (1987). Changes in the size of groups and ranges of the European badger (*Meles meles* L.) in an area of Scotland. *J. Anim. Ecol.* **56**:351–364.

Lauder, G. V. (1986). Homology, analogy, and the evolution of behavior. In Nitecki, M. H., and Kitchell, J. A. (eds.), *Evolution of Animal Behavior: Paleontological and Field Approaches*, Oxford University Press, New York, pp. 9–40.

Laundré, J. W., and Keller, B. L. (1984). Home range of coyotes: a critical review. *J. Wild. Mgt.* **48:**127-139.

Lee, A. K., and Cockburn, A. (1985). *Evolutionary Ecology of Marsupials,* Cambridge University Press, Cambridge.

Leutenegger, W., and Cheverud, J. M. (1982). Correlates of sexual dimorphism in primates: ecological and size variables. *Int. J. Primatol.* **3:**387-402.

Leutenegger, W., and Cheverud, J. M. (1985). Sexual dimorphism in primates: the effects of size. In Jungers, W. L. (ed.), *Size and Scaling in Primate Biology,* Plenum Press, New York, pp. 33-50.

Lewontin, R. C. (1978). Adaptation. *Sci. Am.* **239:**157-169.

Lorenz, K. (1950). The comparative method in studying innate behavior patterns. *Symp. Soc. Exp. Biol.* **4:**221-268.

Lott, D. F. (1984). Intraspecific variation in the social systems of wild vertebrates. *Behaviour* **88:**266-325.

Macdonald, D. W. (1983). The ecology of carnivore social behaviour. *Nature* **301:**379-384.

Macdonald, D. W., and Moehlman, P. D. (1982). Cooperation, altruism, and restraint in the reproduction of carnivores. In Bateson, P. P. G., and Klopfer, P. H. (eds.), *Perspectives in Ethology,* Vol. 5, Plenum Press, New York, pp. 433-467.

Mace, G. M. (1979). The evolutionary ecology of small mammals. D. Phil. thesis, University of Sussex, Brighton, England.

Mace, G. M., and Harvey, P. H. (1983). Energetic constraints on home range size. *Am. Nat.* **121:**120-132.

Martin, R. A. (1986). Energy, ecology, and cotton rat evolution. *Paleobiology* **12:**370-382.

Martin, R. D. (1981a). Relative brain size and basal metabolic rate in terrestrial vertebrates. *Nature* **293:**57-60.

Martin, R. D. (1981b). Field studies of primate behaviour. *Symp. Zool. Soc. Lond.* **46:**287-331.

Martin, R. D., and Harvey, P. H. (1985). Brain size allometry: ontogeny and phylogeny. In Jungers, W. L. (ed.), *Size and Scaling in Primate Biology,* Plenum Press, New York, pp. 147-173.

Maynard Smith, J. (1978). In defence of models. *Anim. Behav.* **26:**632-633.

Mayr, E. (1982a). *The Growth of Biological Thought,* Harvard University Press, Cambridge, Massachusetts.

Mayr, E. (1982b). Adaptation and selection. *Biol. Zbl.* **101:**161-174.

McNab, B. K. (1963). Bioenergetics and the determination of home range size. *Am. Nat.* **97:**133-140.

McNab, B. K. (1983). Ecological and behavioral consequences of adaptation to various food resources. In Eisenberg, J. F., and Kleinan, D. G. (eds.), *Advances in the Study of Mammalian Behavior,* American Society of Mammalogists, Lawrence, Kansas, pp. 664-697.

McNab, B. K. (1986). The influence of food habits on the energetics of eutherian mammals. *Ecol. Mon.* **56:**1-19.

McNab, B. K. (1987). Basal rate and phylogeny. *Functional Ecology* **1:**159-160.

McNab, B. K. (1988). Rate of metabolism, body size and food habits in the order Carnivora. In Gittleman, J. L. (ed.), *Carnivore Behavior, Ecology, and Evolution,* Cornell University Press, Ithaca, New York, in press.

Mech, L. D. (1970). *The Wolf,* Natural History Press, New York.

Messier, F. (1985). Solitary living and extraterritorial movements of wolves in relation to social status and prey abundance. *Can. J. Zool.* **63:**239-245.

Millar, J. S. (1977). Adaptive features of mammalian reproduction. *Evolution* **31:** 370-386.

Millar, J. S. (1981). Pre-partum reproductive characteristics of eutherian mammals. *Evolution* **35:**1149-1163.

Moehlman, P. D. (1988). Intraspecific variation in Canidae. In Gittleman, J. L. (ed.), *Carnivore Behavior, Ecology, and Evolution,* Cornell University Press, Ithaca, New York, in press.

Nelson, G. J. (1970). Outline of a theory of comparative biology. *Syst. Zool.* **19**:373–384.

Packer, C. (1986). The ecology of sociality in felids. In Rubenstein, D. I., and Wrangham, R. W. (eds.), *Ecological Aspects of Social Evolution,* Princeton University Press, Princeton, New Jersey, pp. 429–451.

Pankakoski, E., Vaisanen, R. A., and Nurmi, K. (1987). Variability of muskrat skulls: measurement error, environmental modification and size allometry. *Syst. Zool.* **36**:35–51.

Peters, R. H. (1983). *The Ecological Implications of Body Size,* Cambridge University Press, Cambridge.

Ralls, K. (1976). Mammals in which females are larger than males. *Quart. Rev. Biol.* **51**:245–276.

Rasa, O. A. E. (1987). The dwarf mongoose: a study of behavior and social structure in relation to ecology in a small, social carnivore. In Rosenblatt, J. S., Beer, C., Busnel, M.-C., and Slater, P. J. B. (eds.), *Advances in the Study of Behavior,* Vol. 17, Academic Press, New York, pp. 121–163.

Rayner, J. M. V. (1985). Linear relations in biomechanics: the statistics of scaling functions. *J. Zool.* **206**:415–439.

Reist, J. D. (1985). An empirical evaluation of several univariate methods that adjust for size variation in morphometric data. *Can. J. Zool.* **63**:1429–1439.

Reist, J. D. (1986). An empirical evaluation of coefficients used in residual and allometric adjustment of size covariation. *Can. J. Zool.* **64**:1363–1368.

Ridley, M. (1983). *The Explanation of Organic Diversity: The Comparative Method and Adaptations for Mating,* Clarendon Press, Oxford.

Ridley, M. (1986). The number of males in a primate troop. *Anim. Behav.* **34**:1848–1858.

Roberts, M. S., and Gittleman, J. L. (1984). *Ailurus fulgens,* Mammalian Species Account #222, pp. 1–8.

Rogers, L. L. (1977). Social relationships, movements and population dynamics of black bears in northeastern Minnesota, Ph.D. dissertation, University of Minnesota.

Rood, J. P. (1986). Ecology and social evolution in the mongooses. In Rubenstein, D. l., and Wrangham, R. W. (eds.), *Ecological Aspects of Social Evolution,* Princeton University Press, Princeton, New Jersey, pp. 131–152.

Sacher, G. A., and Staffeldt, E. E. (1974). Relation of gestation time and brain weight for placental mammals: implications for the theory of vertebrate growth. *Am. Nat.* **105**:593–615.

Schaller, G. B., Jinchu, H., Wenshi, P., and Jing, Z. (1985). *The Giant Pandas of Wolong,* University of Chicago Press, Chicago.

Schmidt-Nielson, K. (1984). *Scaling: Why Is Animal Size So Important?* Cambridge University Press, Cambridge.

Scott, B. M. V., and Shackleton, D. M. (1982). A preliminary study of the social organization of the Vancouver Island wolf (*Canis lupus crassodon;* Hall, 1932). In Harrington, F. H., and Paquet, P. C. (eds.), *Wolves of the World,* Noyes Press, Park Ridge, New Jersey, pp. 12–25.

Seim, E., and Saether, B.-E. (1983). On rethinking allometry: which regression model to use? *J. Theor. Biol.* **104**:161–168.

Simpson, G. G. (1960). The history of life. In Tax, S., (ed.), *The Evolution of Life,* Vol. 1, University of Chicago Press, Chicago, pp. 117–180.

Simpson, G. G., Roe, A., and Lewontin, R. C. (1960). *Quantitative Zoology,* Harcourt, Brace, New York.

Smith, R. J. (1980). Rethinking allometry. *J. Theor. Biol.* **87**:97–111.

Smith, R. J. (1984). Allometric scaling in comparative biology: problems of concept and method. *Am. J. Physiol.* **246**:R152–R160.

Snedecor, G. W., and Cochran, W. G. (1967). *Statistical Methods*, University of Iowa Press, Ames, Iowa.
Stearns, S. C. (1983). The influence of size and phylogeny on patterns of covariation among life-history traits in the mammals. *Oikos* **41**:173–187.
Terborgh, J. (1983). *Five New World Primates: A Study in Comparative Ecology*, Princeton University Press, Princeton, New Jersey.
Terborgh, J., and Janson, C. H. (1986). The socioecology of primate groups. *Ann. Rev. Ecol. Syst.* **17**:111–135.
Thompson, S. D. (1987). Body size, duration of parental care, and the intrinsic rate of natural increase in eutherian and metatherian mammals. *Oecologia* **71**:201–209.
Thompson, S. D., and Nicoll, M. E. (1986). Basal metabolic rate and energetics of reproduction in eutherian mammals. *Nature* **321**:690–693.
Tinbergen, N. (1959). Comparative studies of the behavior of gulls (Laridae): a progress report. *Behaviour* **15**:1–70.
Todd, A. W., Keith, L. B., and Fischer, C. A. (1981). Population ecology of coyotes during a fluctuation of snowshoe hares. *J. Wild. Mgt.* **45**:629–640.
Van Orsdol, K. G., Hanby, J. P., and Bygott, J. D. (1985). Ecological correlates of lion social organization (*Panthera leo*). *J. Zool.* **205**:97–112.
van Roosmalen, M. G. M. (1984). Subcategorizing foods in primates. In Chivers, D. J., Wood, B. A., and Bilsborough, B. (eds.), *Food Acquisition and Processing in Primates*, Plenum Press, New York, pp. 167–175.
Van Valkenburgh, B. (1987). Skeletal indicators of locomotor behavior in living and extinct carnivores. *J. Vert. Paleontol.* **7**:162–182.
Waser, P. M., and Waser, M. S. (1985). *Ichneumia albicauda* and the evolution of viverrid gregariousness. *Z. Tierpsychol.* **68**:137–151.
Webster, D. B. (1976). On the comparative method of investigation. In Masterton, R. B., Hodos, W., and Jerison, H. (eds.), *Evolution, Brain, and Behavior*, Lawrence Erlbaum, Hillsdale, New Jersey, pp. 1–11.
Western, D. (1979). Size, life history and ecology in mammals. *Afr. J. Ecol.* **17**:185–204.
Williams, G. C. (1966). *Adaptation and Natural Selection*, Princeton University Press, Princeton, New Jersey.
Zangerl, R. (1948). The methods of comparative anatomy and its contribution to the study of evolution. *Evolution* **2**:351–374.
Zimen, E. (1981). *The Wolf*, Delacorte, New York.

Chapter 4

A BRIEF HISTORY OF THE STUDY OF ANIMAL BEHAVIOR IN NORTH AMERICA

Donald A. Dewsbury

Department of Psychology
University of Florida
Gainesville, Florida 32611

I. ABSTRACT

North America has a rich tradition in the study of animal behavior. This is traced and analyzed through four phases. It begins with pioneering studies from such workers as J. J. Audubon and L. H. Morgan. The developmental phase was the product of biologists, such as W. Craig, H. S. Jennings, J. Loeb, the Peckhams, W. M. Wheeler, and C. O. Whitman, and psychologists, such as J. M. Baldwin, T. W. Mills, J. B. Watson, and R. M. Yerkes. In its third phase, activity was sustained by such workers as W. C. Allee, C. R. Carpenter, S. J. Holmes, K. S. Lashley, M. M. Nice, G. K. Noble, T. C. Schneirla, and C. P. Stone. The post-World War II phase reflects a continuation of these traditions, interactions with European ethology, and the development of behavioral ecology. Throughout its history and today, the study of animal behavior in North America has represented an interaction of workers from different disciplines, especially psychology and zoology, interested in a broad range of problems concerning the immediate causation, development, evolutionary history, and adaptive significance of behavior.

II. INTRODUCTION

The structure of any interdisciplinary intellectual endeavor, such as the study of animal behavior, must often appear to be a web of complex and puzzling interactions. Often, these become simplified if one can place events and developments in historical context. Many peculiarities can be understood as resulting from unique historical events and encounters. To some extent this has been attempted for the study of European ethology (Thorpe, 1979) and comparative psychology (Dewsbury, 1984). However, there has been less attention devoted to the overall development of the comprehensive, interdisciplinary study of animal behavior in North America.

In writing of the study of animal behavior in North America, I may appear chauvinistic in neglecting activities in Europe and elsewhere. This is not intended. Developments outside of North America were obviously of great importance. However, they lie beyond the scope of the present paper.

The subject clearly begs for a book-length treatment; such is precluded by the length of this paper. This paper may be regarded as an outline for such a full-length treatment. I shall focus on some of the prominent contributors to the study of animal behavior in North America, stressing some of their backgrounds and approaches and some of the things they did. A broader approach would do more to place these "great men" within the context of their times. Rather, I hope to identify some of the people who shaped the study of animal behavior in North America and to give some brief indication of why they may be viewed as important today.

My approach in this paper will place much emphasis on those aspects of earlier work that are particularly relevant to current activity in the field. Thus, my approach can correctly be regarded as somewhat "presentist" or "Whiggish" (e.g., Stocking, 1965). I believe that it is legitimate to seek and credit work antecedent to the concerns of the present; the approach has its defenders in historical writing (e.g., Buss, 1977). However, an in-depth historical treatment, in which the work of these scientists is placed and evaluated within the context of their times, is also needed.

My perception of this history differs in some respects from that normally promulgated; the reader should approach all accounts with an open mind. Judged against contemporary standards, virtually all older work has both strengths and weaknesses. As in my previous book (Dewsbury, 1984), I prefer to focus on the strengths. Perhaps more work on the history of the study of animal behavior can be generated from honest disagreements in perceptions of, and approaches to, the history of the field.

III. PIONEERS

White Europeans were not the first careful observers of animal behavior in North America; native Americans preceded them. American Indians lived close to nature, and legends and myths are replete with references to animals. For example, pronghorn antelope were said to be adversaries of rattlesnakes, decoying them into a striking attitude and then killing them by leaping on them with all four hooves. The Snake–Antelope ceremony of the Pueblo Indians was based on such interactions and included both fertility motifs and war elements. The Pueblo Indians also freely used imagery based on rabbits in fertility rituals (Tyler, 1975). Hopi stories are especially rich in coyote-related imagery (Malotki, 1985). According to one story passed on by oral tradition, the saying that "there is a coyote mating somewhere" refers to the occurrence of rain while the sun is shining and relates to mating by coyotes in the open, undeterred even by rain.

Naturally, the early settlers were attuned to animal life as well. Gray (1987) calls Thomas Morton (1579–1647) "America's first behavioral observer" (p. 69). Morton was a gentlemen, lawyer, and sportsman who settled in Massachusetts, where he catalogued and studied animals. In his *New English Canaan or New Canaan,* Morton (1637) described many native species and some behavioral patterns, such as those used by beavers in cutting down trees and transporting them to the dam site, sometimes cooperatively.

The scientific study of animal behavior matured near the end of the 19th century largely owing to the influence of Charles Darwin. Prior to Darwin, the literature concerning the study of the behavior of animals and its interpretation was largely the province of naturalists and philosophers. Surely the best known of the naturalists was John James Audubon (1785–1851). Audubon is best known for his drawings of birds and other animals. Whereas others portrayed birds in unnatural postures and out of ecological context, Audubon placed the animal in its appropriate orientation and context. His journals (Audubon, 1897) are full of observations on the habits of North American fauna. We learn of the flight of Sprague's lark, the behavior of opossums, and the feeding patterns of bears. Knowledge of behavior aided Audubon in drawing the animals. Audubon posed his dead specimens using wires to stiffen the bodies in the desired postures.

Among the most prolific of North American naturalists before and after Darwin was John Burroughs. In Burroughs we find a mixture of objective observations on behavior and excessive sentiment. In Burroughs' essay on "The Idyl of the Honeybee" (see Burroughs, 1875), he describes scouts seeking a future home, and the spiral flights of foragers noting the location of a food source. Burroughs noted that "the bee does not tell its fellows what it has found, but they smell out the secret" (p. 53). In one experiment, Burroughs trapped a group of foraging bees in a box containing the substances that attracted them and displaced the box. He found that, when released, the bees continued in the original direction of travel, not compensating for the displacement. Similar studies are discussed by Lindauer (1961). In contrast to these observations, however, one also reads anthropomorphic descriptions, such as that of robins crying "Thief, thief" on the approach of a jay (Burroughs, 1875, p. 7).

The writings of Ernest Thompson Seton (e.g., Seton, 1913) inspired many young, aspiring naturalists.

Although he apparently resided in the United States just four or five years, David Fredrich Weinland (1829–1915), a German zoologist working with Louis Agassiz at Harvard, was active in the American Association for the Advancement of Science (A.A.A.S.) and other organizations (Cadwallader, 1984). His paper in the A.A.A.S. proceedings, "A Method of Comparative Animal Psychology" (Weinland, 1858), is especially notable. It antedates the book by Pierre Fluorens, which Jaynes (1969) credits with founding comparative psychology in name. Weinland traces the history of the study of animal psychology from the Greeks to the 19th century. He decries "the anecdotal character with which this science unluckily has been stamped" (p. 260), and cautions against comparison before a species is well understood:

> ... the object of animal psychology is precisely this, to trace that whole circle, that whole sphere of psychical life of each species of animals, without respect to the psychical life of any other animal, and particularly of man. Then only when the whole circle is traced is it time to compare. (Weinland, 1858, p. 259)

A naturalist whose observations were sophisticated enough to be treated as a true forerunner of ethology was Lewis Henry Morgan (1818–1881). Morgan was a lawyer with strong interests in anthropology and animal psychology. His major achievement, *The American Beaver and His Works* (1868), originated with his work as a director of a company building a railroad through northern Michigan. Morgan became fascinated with the beavers and gathered material from many diverse sources, including his own observations, to produce a comprehensive treatise on the anatomy, ecology, and habits of beavers. The work was cited by Darwin and Romanes. Morgan, too, was an early advocate of the ethogram:

> A monograph upon each of the principal animals seems, therefore, to be desirable, if not absolutely necessary. . . . These should contain a minute exposition of their artificial works, where such are constructed; of their habits, their modes of life, and their mutual relations. (Morgan, 1868, p. vi)

The final chapter of Morgan's book is devoted to animal psychology. Morgan strongly supports the notion of mental continuity and attributes much behavior to the action of a free intelligence, while advocating abandonment of the term "instinct" to explain the intelligent acts of animals.

There was much for the naturalist with a pioneering spirit to explore. C. Hart Merriam (1855–1942) is best remembered for his work in mammalogy. He discovered and described approximately 660 new mammals (Osgood, 1945). However, he also had strong interests in animal behavior, writing on topics such as hibernation in mammals (Merriam, 1884a), and spearheading a cooperative investigation of migratory patterns in birds (Merriam, 1884b).

Representative of the contributions from philosopher–theologians was the work of John Bascom (1827–1911), who was much concerned with harmonizing theology and evolution. Bascom, who taught at Williams College and was president of the University of Wisconsin, favored intuitional over empirical philosophy. Nevertheless, he wrote the first book on comparative psychology in North America, *Comparative Psychology or The Growth and Grades of Intelligence* (Bascom, 1878). Bascom cited work from Darwin, Wallace, Lubbock, Galton, and Spalding and provided an account of many varieties of behavior. Included is a section on comparative neuroanatomy, with a discussion of brain size and relative brain/body weight ratios.

He noted that "as a rule, animals are made more dull by domestication, civilization" (p. 206). Bascom was not always on target; he accepted John Lubbock's work concluding that "bees and ants communicate but little with each other, far less than has been supposed" (p. 162).

Bascom left the University of Wisconsin in 1887; in 1888 the first North American chair in Experimental and Comparative Psychology was established there. It was occupied by Joseph Jastrow, who was never really an active researcher in the field.

The first learned society devoted to comparative psychology was the Association for the Study of Comparative Psychology, founded by T. Wesley Mills (1847–1915), of the Veterinary College at McGill University. Regrettably, the society appears to have had little influence outside of Montreal.

IV. 1890–1917

Although important antecedents to the modern study of behavior appeared prior to 1890, it was in the period from about 1890 until World War I that the developments came to fruition. I shall discuss developments within zoology and psychology, as well as their interaction, each in turn.

A. Zoology

Although it is a mistake to overstate the extent of the differentiation, there were two general trends in the study of animal behavior in zoology during this time, one stemming from field naturalists and one from laboratory experimentalists.

1. The Tradition of Natural History

The tradition of natural history lived on in the work of George W. Peckham (1845–1914) and Elizabeth G. Peckham (b. 1854). Although others had studied arthropods in the field, the studies of the Peckhams carried the depth of analysis, with respect to detail of both description and interpretation, to extents not previously approached. In works such as *On the Instincts and Habits of the Solitary Wasps* (Peckham and Peckham, 1898) and *Wasps Social and Solitary* (Peckham and Peckham, 1905) (with an introduction by John Burroughs), the Peckhams provided behavioral descriptions that would be models for future workers. For example, there is a detailed description of the reproductive behavior of the wasp *Ammophila*. The Peckhams worked on the protective resemblance in mimicry and were advocates of the mechanism of sexual selection, particularly female choice, in a time when such advocacy was unpopular.

Charles Otis Whitman (1842–1910) was a very influential biologist as the founder of the *Journal of Morphology*, the first director of the Marine Biological Laboratory at Woods Hole, the long-term autocratic chair of the influential Department of Zoology of the University of Chicago, and the major professor of a string of outstanding students. Whitman's influence on the study of animal behavior was substantial, although in his lifetime he really published just two papers in the field. One was a paper on myths in animal psychology, in which he tried to debunk a set of cases wherein authors had attributed animal behavior to complex processes without fully understanding the normal behavioral repertoire of the species (Whitman, 1899). The other was the Woods Hole lecture (Whitman, 1898) in which it is said he "wrote the sentence that initiated the birth of modern ethology" (Hess, 1962, p. 168): "Instincts and organs are to be studied from the common viewpoint of phyletic descent" (Whitman, 1898, p. 328). The paper is a wide-ranging treatise on instinct that should be read by all ethologists. Whitman viewed instincts as less rigid than did many of his contemporaries and stressed the comparative approach. His detailed research on the behavior of pigeons was published posthumously by psychologist Harvey Carr (1919).

Among Whitman's students were William Morton Wheeler (1865–1937) and Wallace Craig (1876–1954). Wheeler had the good fortune to grow up in Milwaukee at a time that permitted him to work with the Peckhams, as George Peckham taught in and supervised the Milwaukee city schools during this period, and with Whitman, who worked at Milwaukee's Lake Laboratory during the years 1886 to 1889. Later, in contrast to his mentor, Wheeler could boast of a long list of publications dealing with animal behavior, especially the behavior of arthropods. In his paper "On Instincts," Wheeler contrasts the experimental, historical, and psychopathic methods of studying instinctive behavior. He favored the historical method and illustrated its utility by analyzing the phylogeny of nuptial gift presentation in the balloon fly, *Hilara sartor* (Wheeler, 1921). Wheeler cautioned against too mechanistic a view of instincts. In a 1902 paper, Wheeler became the first North American to advocate use of the term "ethology" in the modern sense:

> The only term hitherto suggested which will adequately express the study of animals, with a view to elucidating their true character as expressed in their physical and psychical behavior towards their living and inorganic environment, is *ethology*. (Wheeler, 1902, p. 974)

Wheeler's pen contained both wit and bite, as in his review of J. B. Watson's *Behaviorism:* "Psychology is universally admitted to be a queer science . . . due . . in no small measure to the mental peculiarities of its devotees" (Wheeler, 1926, p. 439).

Alfred North Whitehead regarded Wheeler as the only man he had known who would have been worthy and able to carry on a conversation with Aristotle (Evans and Evans, 1970).

Although there are various sources of good material on Whitman and Wheeler, much less has been written of Craig. He received his Ph.D. under Whitman at Chicago in 1908 and then taught at the University of Maine and at Harvard. He is best known for his paper "Appetites and Aversions as Constituents of Instincts," one version of which was presented at the American Psychological Association Convention in 1914 and was later published in full (Craig, 1918). Craig's distinction between appetitive behavior and consummatory action has been most influential; indeed, Wheeler used it in analyzing the behavior of ant lions. Most of Craig's empirical work was on pigeons. His work on social influences on the stimulation and inhibition of ovulation in ring doves (e.g., Craig, 1913) was later elaborated on in the work of Daniel Lehrman and his colleagues.

2. The Experimentalist Tradition

Perhaps the type specimen of the laboratory zoologist was Jacques Loeb (1859–1924). Loeb envisioned biology as a deterministic physical science and was a strong opponent of teleological thinking. He is best known for his tropistic view of behavior, treating behavior as the product of physical forces resulting from physicochemical actions in the body (e.g., Loeb, 1918). Loeb's great opponent in this view was Herbert Spencer Jennings (1868–1947). Jennings, too, was a devotee of the experimental, objectivist approach, but thought the tropism theory too limited to account for the range of behavior proposed by Loeb. Jennings found behavior, even in lower organisms, to be much more variable and affected by internal factors than was implied by the tropism theory (e.g., Jennings, 1906). Jennings was squarely in the historical line favoring a view of continuity in the mental processes of all organisms and had strong evolutionary leanings. Jennings' (1927) discussion of the factors leading to monogamous and polygamous mating systems addressed many issues currently under debate. The work on the behavior of invertebrates by Raymond Pearl (1879–1940) and Samuel O. Mast (1871–1947) is also notable.

Although he, too, was a student of tropisms, Samuel Jackson Holmes (1868–1964) emphasized the variability inherent in behavior and analyzed tropisms primarily as they related to the evolution of intelligence, his primary area of interest (e.g., Holmes, 1911, 1916). Holmes considered the full range of the evolution of intelligence to encompass tropisms, instincts, and intelligent behavior in progressively higher animals. Late in his career, Holmes became quite controversial because of his extreme views on eugenics. In places, Holmes' writing has a contemporary ring, coming close to the concept of the innate schoolmarm, selection at the level of the individual, and reciprocal altruism: "It is inheritance that affords the means by which inheritance is improved." (Holmes, 1916, p. 164); "Most creatures care not a fig for the welfare of any but their own immediate kin; and where we find any consideration bestowed upon alien forms. . . it is done for the sake of something to be gained in return." (Holmes, 1916, 36).

Another group of biologists focused on the physiological factors involved in the control of vertebrate behavior, with a strong neurological approach. H. H. Donaldson (1857–1938) studied the role of the nervous system in behavior. He was one of the first to study laboratory rats and is notable for having studied the brains of three prominent scholars: E. S. Morse, Sir William Osler, and G. Stanley Hall (Donaldson, 1928). The career of G. H. Parker (1864–1955) was largely devoted to an analysis of sensory factors in the control of the behavior of both vertebrates and invertebrates (Romer, 1967). C. Judson Herrick (1866–1960) had a long career that focused on the analysis of brain and behavior in amphibians, culminating in the publication of *The Brain of the Tiger Salamander, Amblystoma tigrinum* (Herrick, 1948). Herrick (1907) proposed that "The first task of comparative psychology. . . is to define as accurately as we may with the imperfect means at our command the sensorimotor life of the whole range of lower organisms" (p. 78). Herrick's older but less well-known brother, Clarence Luther Herrick (1858–1904), had similar interests in the brain and behavior (e.g., Herrick, 1893). Like his brother, C. L. Herrick had a strong interest in comparative psychology: Cadwallader (1984) located just one course in Comparative Psychology offered prior to that of the elder Herrick at Denison in 1885, the exception having been given by John Dewey at the University of Michigan in 1884. Perhaps C. L. Herrick's most lasting contribution was his founding of the *Journal of Comparative Neurology* in 1891. Gottlieb (1987) regards C. L. Herrick as the "founder of developmental psychobiology" (p. 1).

Although both Herricks had a strong predilection for developmental analyses, it was George E. Coghill (1872–1941) who brought developmental analysis to fruition. Coghill opposed the prevalent view that complex behavior was built up from simple reflexes, proposing instead that the first movements were whole-body reactions and that partial patterns emerge from the whole (see Oppenheim,

1982). Tinbergen (1951) devoted four pages of *The Study of Instinct* to a discussion of this work.

One of C. L. Herrick's students at the University of Cincinnati was Charles H. Turner (1867–1923), who received his M.S. from Cincinnati in 1892 and his Ph.D. from the University of Chicago in 1906. Like others, he straddled the border between psychology and biology, spending most of his career in biology departments and writing on comparative psychology. Cadwallader (1984) calls Turner "the first black American psychologist" (p. 33). Turner's work on web construction in gallery spiders (Turner, 1892), and his work on the behavior of ants (Turner, 1907), deserve more attention than they have received. T. C. Schneirla (1929) wrote that Turner's "place in the literature should not be a minor one" (p. 24).

B. Comparative Psychology

Some consideration of my definition of comparative psychology is necessary at this point. I use the term "animal psychology" to encompass all studies in psychology with nonhuman animals. Within animal psychology one can differentiate three areas, corresponding roughly to the three journals in the field published by the American Psychological Association—physiological psychology, process-oriented learning and motivation, and comparative psychology. Comparative psychology is the area in psychology concerned with the evolution, function, immediate causation, and development of behavior in a variety of species. Although I believe these three areas have been and are somewhat distinct, I do not advocate separation in contemporary research; rather, increased interaction is desirable. I do believe that differentiation is useful for historical analysis. Classically, individuals regarded as psychologists did graduate work and were employed in psychology departments. However, happily, there are many intermediate cases that make assignment difficult.

Although the evolution of mind and intelligence were undoubtedly prominent themes in comparative psychology as it developed around the turn of the twentieth century, there were also strong emphases on the study of evolution, the natural lives of animals, the comparative method, and behavioral ontogeny.

1. *John Broadus Watson (1878–1958)*

Before tracing these themes I would like to consider a representative comparative psychologist who I believe deserves credit as an early protoethologist—the unlikely person of the young John B. Watson. In the ethological literature Watson is perhaps the most reviled of all psychologists. He was the developer of behaviorism, the view that the science of behavior should be based on objective behavioral observations. The young John Watson had a broad view of behavior and its determinants. The older Watson adopted an extreme environmentalist position, a position that is not a necessary correlate of behaviorism (e.g., Watson, 1930). Beginning in about 1916, Watson both worked almost exclusively with humans and adopted the radical environmentalist position not present in his earlier writings (Logue, 1978). His oft-quoted extreme statements were generally never intended to apply to nonhuman animals. Though guilty of disbelief in the continuity of species, with respect to the regulation of behavior, even the older Watson may not have been guilty of denying instinctive behavior to nonhuman animals.

The young Watson received his Ph.D. from the University of Chicago in 1903, with a dissertation, "Animal Education" (Watson, 1903), that was a developmental study of the ontogeny of behavior and the nervous system in laboratory rats. The work includes what may have been the first laboratory demonstration of communication via chemical signals, as Watson demonstrated that rats showed a preference for paths containing the odor of the opposite sex. His most ethological work was that done during three summers in the Dry Tortugas Islands off the coast of Florida, the work of the last summer being collaborative with Karl S. Lashley. Watson surveyed the populations of noddy and sooty terns and studied a wide range of behavioral patterns, including eating and drinking, the nesting cycle, mating behavior, mate and egg recognition, development, orientation, and even maze learning (e.g., Watson, 1908). Here we have Watson doing comparative field studies of two closely related species of shore birds. Surely, Watson did not knowingly or deliberately set out to anticipate ethology; the main impact of the corpus of his work led to a behavioristic approach to human behavior. Nevertheless, the resemblance between Watson's early work and later core ethology is notable and important. It is difficult not to see this work as a precursor of modern ethology.

Consider also the following quotations from Watson (1912):

> . . . "true instincts" [are] inherited fixed modes of responding to definite objects or classes of objects which arise independently of tuition. . . . (p. 376)

> We must bring up certain members of a given species in isolation from their kind in order to watch the development of activity without tuition, and compare the results with those . . . in which the animals are brought up in social contact with fellows of their own age and with adults of the same species. (p. 376)

> Experiment shows that young animals without previous tuition from parents or from their mates and without assistance from the human observer can and do perform the correct act the very first time they are in a situation which calls for such an act. (p. 377)

The resemblance to later statements by Konrad Lorenz should be apparent. Watson described courtship feeding in noddy terns:

> Suddenly, one of the birds (male) began nodding and bowing to a bird standing nearby (female). The female gave immediate attention and began efforts to extract fish from the throat of the male. The male would first make efforts to disgorge, then put the tip of the beak almost to the ground and incline it to the angle most suitable to admit her beak. . . . (Watson, 1908, p. 196)

There is a resemblance between the methods of Watson ["An isolated noddy nest . . . was moved 3 feet farther out on the limb . . . nest locality exerts the stimulus for nest orientation" (Watson, 1908, p. 224)] and Tinbergen ["The pine-cones were thereupon taken away and deposited a foot away . . . homing ability . . . depends on an amazing learning capacity" (Tinbergen, 1951, pp. 147–148)].

In the manifesto that established behaviorism as a viable movement, Watson (1914) adopted some positions that have not withstood the test of time. However, in his earlier work and in much in that volume he demonstrates sensitivity to and concern for many of the concerns that would later characterize core European ethology.

2. The Study of Learning

The comparative study of learning has always been a major interest within comparative psychology. The leader in this area was Edward Lee Thorndike (1874–1949). He was a leader in establishing the use of objective methods in psychology and is best known for his studies of animals learning to escape confinement in puzzle boxes (e.g., Thorndike, 1898). His conclusion that the process of learning is essentially the same in all vertebrate species led to a decrease of comparative study in the area of process-oriented learning and may have increased its separation from true comparative psychology.

The study of learning in rats was initiated by Linus Kline (born 1866), a graduate student in "zoological psychology" at Hall's Clark University. Kline devised some boxes from which rats had to learn to secure food. Whereas Thorndike's animals had to get out of boxes for rewards, Kline's rats had to get in. Kline's fellow graduate student, Willard S. Small (1870–1943) was the first to study rats in mazes (see Miles, 1930). There were many comparative studies of learning published during this period, as such universities as Harvard, Clark, and Chicago became centers for such research.

3. Evolution in Comparative Psychology

Darwin's theory of evolution had profound implications for psychology, for it established the likelihood of mental continuity between human and nonhuman animals. Darwin (1859) believed that psychology would become securely based on the foundation of an evolutionary perspective. Darwin's protege, G. J. Romanes, developed the field in his *Animal Intelligence* (1882) and subsequent works with the object "of considering the facts of animal intelligence in their relation to the theory of Descent" (Romanes, 1882, p. vi). The impact of the theory of evolution on North American comparative psychologists was considerable. The young John B. Watson called for the establishment "of an experimental station for the study of the evolution of the mind" (Watson, 1906, p. 156).

One of the most influential psychologists of this period was G. Stanley Hall (1844–1924). Hall was the first president of the American Psychological Association and it was Hall who brought Whitman, Wheeler, Donaldson, and others to Clark University as the president of that institution (see Ross, 1972). He was a strong advocate of the study of evolution in psychology:

> I must have been almost hypnotized by the word "evolution," which was music to my ear and seemed to fit my mouth better than any other. (Hall, 1927, p. 357)

Hall called for "an evolutionary synthesis in the psychological domain" (1909, p. 267). Beginning in 1893 he regularly taught a course in "Psychogenesis," which dealt with a variety of topics related to behavior and evolution.

The sixth A.P.A. president was James Mark Baldwin (1861–1934). He was another psychologist with a strong interest in evolution. Baldwin (1896), along with C. Lloyd Morgan (1896) and H. F. Osborn (1896), proposed an evolutionary process potentially capable of explaining some phenomena attributed to the inheritance of acquired traits. It is still known today as the "Baldwin effect." Baldwin was forced from his academic career at the Johns Hopkins University when, in the words of Cohen (1979, p. 53), "The lively philosopher committed a heinous academic crime; he was caught in a negro brothel in a position that could not be described as philosophical." Watson, too, would be forced from Johns Hopkins, upon revelation of an extramarital affair.

Baldwin (1909), Hall (1909), and James Rowland Angell (1909) marked the fiftieth anniversary of the publication of the *Origin of Species*, remarking on how profoundly the theory had affected psychology; Howard (1927) proposed that we speak of psychology as pre-Darwinian and post-Darwinian.

Thorndike (1899) called attention to the similarity with which frogs, lizards, chicks, and cats react to irritation of the head region by scratching. He used this as an example of the way in which instinctive activities could persist, with some modification, in a wide range of forms.

4. The Natural Lives of Animals

Concern for the study of the natural lives of animals was prevalent during this period. Hall called for research "with reference to the life history and habits of the species in the state of nature" (1909, p. 253). Kline wrote that "a careful study of the instincts, dominant traits and habits of an animal as expressed in its free life—in brief its natural history should precede as far as possible any experimental study" (1899, p. 399). According to Thorndike "the first task of comparative psychology is to find out the instinctive equipment of any animal studied" (1899, p. 58).

This became a major issue in the study of learning. The study of rats and mazes was initiated at Clark not as an arbitrary task, but because the mazes were designed to mimic the winding passages found dug by feral rats under an old cabin belonging to Kline's father. Mills criticized Thorndike's use of puzzle boxes for a failure to consider the natural behavior of the animal: "As well enclose a living man in a coffin, lower him, against his will, into the earth and attempt to deduce normal psychology from his conduct" (Mills, 1899, p. 266).

Watson's field study of noddy and sooty terns (e.g., Watson, 1908) provides an excellent example of how this advice can be put into practice. Robert Yerkes (1903) also did field work, on the reactions of frogs to sounds.

5. Study of a Range of Species

Although laboratory rats became the most popular species for study, even early in the field, comparative psychologists of this era were also interested in the study of a wide range of species. In making suggestions for a laboratory course in comparative psychology, Kline (1899) recommended work on ameba, vorticella, paramecia, hydra, earthworms, slugs, fish (including sticklebacks), chicks, white rats, and cats. The prototypical comparative psychologist for the century, Robert Mearns Yerkes (1875–1956), studied earthworms, crabs, frogs, turtles, mice, rats, and many species of primates during his long career. John F. Shepard (1881–1965) studied the behavior of ants, an enterprise to be carried on by his student, T. C. Schneirla. There are many other examples of the range of species studied by comparative psychologists (see Dewsbury, 1984).

6. Behavioral Ontogeny

From the beginning, psychologists were especially concerned with the problem of how behavior develops. Development entails the interaction of genetic and environmental factors. In Mills' (1898) *The Nature and Development of Animal Intelligence* he reported developmental data and distinguished stages in behavioral development suggestive of what would later be termed "critical periods" (e.g., Scott, 1962). Watson's (1903) dissertation was a developmental study. Small (1899) completed a normative study of the ontogeny of behavior in rats.

An early study of genetic influences on behavior can be found in Yerkes' (1907) *The Dancing Mouse,* a study of the sensory capacities, learning abilities, and genetic inheritance of a "dancing" mutation in house mice. Jennings (1908) called it "doubtless the fullest, most satisfactory, and most suggestive experimental account that we have on the behavior of any higher animal" (pp. 93–94).

7. The "Ethological Attitude"

Burghardt (1973) considered five attributes as constituting the "ethological attitude": (1) studying ecologically meaningful behavioral patterns, (2) beginning analysis with descriptive studies, (3) studying a wide range of species and behaviors, (4) comparing similar behaviors in closely related species, and (5) disparaging the exclusive use of domesticated animals. Early comparative psychologists clearly were not ethologists and did not develop this attitude with the completeness that characterized core European ethology. However, the essence of this attitude clearly was present in comparative psychology.

C. Interdisciplinary Interactions

Psychologists and zoologists interacted in building a science of animal behavior in North America. There were many bases for such interaction. For one, they often were located on the same campuses. Shown in Figs. 1–4 are diagrams of the tenure of some important figures in the study of animal behavior

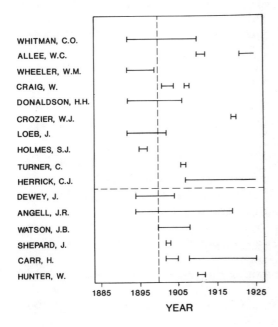

Fig. 1. Portrayal of the overlapping tenures of individuals interested in the study of animal behavior at the University of Chicago, 1885–1925.

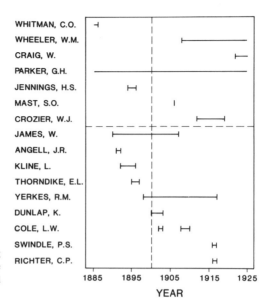

Fig. 2. Portrayal of the overlapping tenures of individuals interested in the study of animal behavior at Harvard University, 1885–1925.

at four of the major universities of the time—Chicago, Harvard, Johns Hopkins, and Clark, respectively,—during the years 1885 to 1925. The rapid dissipation of the early promise shown by Clark while under Hall is apparent in the tenure of its promising faculty (Fig. 4). Chicago was founded, in part, with departed Clark faculty and had strength in both zoology and psychology (Fig. 1). All four universities had prominent faculty in the study of animal behavior in both departments. There appear to have been real interactions between these departments.

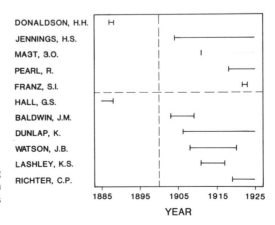

Fig. 3. Portrayal of the overlapping tenures of individuals interested in the study of animal behavior at Johns Hopkins University, 1885–1925.

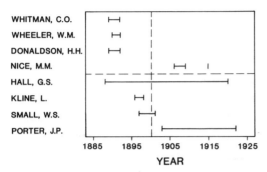

Fig. 4. Portrayal of the overlapping tenures of individuals interested in the study of animal behavior at Clark University, 1885–1925.

For example, Watson was influenced by Angell, Loeb, and Donaldson at Chicago; he began his research with rats by caring for Donaldson's colony. Watson took courses from Jennings while on the faculty at Johns Hopkins. Carr (1919) edited the posthumous publication of Whitman's work on pigeons. Lashley was influenced by Watson, Jennings, and Franz.

Whitman was not the only animal behaviorist to lecture at Woods Hole; Thorndike (1899) delivered a lecture on "Instinct." What is notable is the similarity of the two papers; Thorndike supported Whitman in almost all respects.

There was also collaboration in editing journals. During the time when the *Journal of Comparative Neurology* became the *Journal of Comparative Neurology and Psychology* (1904–1910), psychologists Watson and Yerkes joined Donaldson, Jennings, and the Herricks on the editorial board. When the *Journal of Animal Behavior* was founded in 1911, its editorial board included psychologists Madison Bentley, Harvey Carr, Edward Thorndike, Margaret Floy Washburn, John B. Watson, and Robert Yerkes, along with zoologists S. J. Holmes, H. S. Jennings, and William Morton Wheeler. When *Psychobiology* was founded in 1920, it was edited by psychologist Knight Dunlap in association with Walter B. Cannon, Franz, Jennings, and Parker.

Craig, Donaldson, Holmes, and Wheeler were members of the American Psychological Association. The 1907 meeting of the A.P.A. was held in conjunction with the American Society of Naturalists.

V. 1917–1945

Many of the trends in the study of animal behavior established prior to World War I continued during the period between the two world wars.

A. Zoology

The activities of some of the zoologists who were prominent prior to World War I continued after the war; the careers of scientists refuse to respect arbitrary boundaries of time. Wheeler, Loeb, Jennings, Parker, C. J. Herrick, and Coghill all continued well into the period now under consideration. However, new names also came to the fore. The differentiation between the naturalist and experimentalist traditions also became progressively blurred, as in the work of Allee and Noble.

1. The Naturalist Tradition

A major figure during this important period was Warder Clyde Allee (1885–1955). Although he conducted experimental research, the conceptual framework of Allee's work was generally ecological and thus he fits best within the naturalistic tradition, in some sense carrying on the Whitman–Craig tradition at the University of Chicago (see Banks, 1985; Schmidt, 1957). After tussling with Jennings and Mast over invertebrate behavior early in his career, Allee's later career was concentrated on the study of vertebrate social behavior. He placed behavior in an ecological framework and is best known for stressing the role of cooperation in the regulation of social interactions, sometimes with clear-cut group selectionist considerations. According to *Allee's principle,* undercrowding of organisms can be as bad for survival as overcrowding; aquatic species, especially, condition their environments to make living conditions more hospitable. From the mid-1930s on, Allee's research was concentrated on social organization, especially dominance–subordinance relationships, leadership, and territoriality (e.g., Allee, 1938). He and his students conducted a long series of studies of dominance relationships in domestic chickens. The 36 Ph.D. students Allee supervised, and others he influenced, went on to have major impact on the study of animal behavior in North America—an influence that is still being felt. For example, it was individuals influenced by Allee who formed the nucleus of the Animal Behavior Society (see Guhl and Schein, 1976).

Margaret Morse Nice (1883–1974) made major contributions to the understanding of avian behavior (see Nice, 1979; Trautman, 1977). I place her in the naturalist tradition in zoology, where I think she best fits. However, her only doctorate was an honorary degree and her interest in child psychology led her to take an A.M. in psychology; psychologists would be proud to claim her as one of our own. Margaret Nice fully understood the importance of long-term observation of animals under natural conditions, and that an understanding of social structure required that the observer be able to recognize individuals. Her study of the role of territoriality in bird life (Nice, 1941) expanded on the work of Altum and Howard and is a classic in the field. Her best-known research was a long-term study of the life history of song sparrows. Allee (1937) wrote of "the evident love of the subject matter in all its details" (p. 540) in Nice's work. Margaret Nice was one of the first North Americans to appreciate the importance of European ethology, as in her review of Tinbergen's work on territoriality (Nice, 1935) and her application of Lorenz's *Kumpan* notion to the behavior of her song sparrows as early as 1939 (Nice, 1939).

Another researcher difficult to pigeonhole is Margaret Altmann (1900–1984), who fits at the interface of comparative psychology, ethology, and animal husbandry (Chiszar and Wertheimer, 1988). Her Ph.D. was from a "psychobiology" program at Cornell, but there was a strong agricultural component to the program. Margaret Altmann understood the importance of ethological concerns for animal husbandry and did much research on the topic (e.g., Altmann, 1941). She also conducted important field studies on large mammals, especially moose and elk (e.g., Altmann, 1963).

2. The Experimentalist Tradition

William J. Crozier (1892–1955) was another prototype of the experimentalist. He applied the concept of tropisms, developed by Loeb, to vertebrate behavior using parametric analyses in which he attempted to fit complex equations to animal behavior (e.g., Crozier, 1928) (see Hoagland and Mitchell, 1956). Crozier studied such behavior as the path taken by an animal moving up an inclined plane and the effect of various stimuli on the path taken. He found a strong genetic influence on such behavior.

A major force in the field during this period was G. Kingsley Noble (1894–1940), an experimentalist especially interested in such topics as social behavior, sex recognition, brain function, development, and hormones in fish, amphibians, reptiles, and birds. Noble's experimental work was based on an understanding of the natural behavior of the species under study (see R. C. Noble, 1946), as, for example, in his detailed study of the social behavior of American chameleons (Greenberg and Noble, 1944). This behavior was the basis for experimental analysis, as in his work with the effects of hormones (e.g., Noble and Greenberg, 1941). Noble worked to dissociate effects of female choice and male–male competition (e.g., Noble, 1938). At the American Museum of Natural History, Noble created the Department of Experimental Biology, a unit devoted to the study of animal behavior. It later became the Department of Animal Behavior and was influential in shaping the study of animal behavior through the work of such people as F. A. Beach, T. C. Schneirla, and D. S. Lehrman.

William C. Young (1899–1965) approached the study of hormone–behavior interactions from the endocrinological perspective, where behavior had been viewed as too ephemeral for serious study (see Goy, 1967). However, he persevered and made a major contribution in establishing behavioral endocrinology as a viable field with his research on hormone–behavior interactions in guinea pigs and primates and by editing the influential *Sex and Internal Secretions* (Young, 1961).

3. Recruits to the Population

The survival of a discipline depends on the quality of the young scientists attracted to the field. The years 1917 through 1945 saw the addition of important animal behaviorists, most of whom would make their major contributions after World War II. They included: Lester Aronson (New York University, 1945); Theodore Bullock (University of California at Berkeley, 1940); John Calhoun (Northwestern University, 1943); Archie Carr (University of Florida, 1937); Nicholas Collias (University of Chicago, 1942); David Davis (Harvard University, 1939); Vincent Dethier (Harvard University, 1939); John Emlen (Cornell University, 1934); William Etkin (University of Chicago, 1934); Hubert Frings (University of Minnesota, 1940); John Fuller (Massachusetts Institute of Technology, 1935); Benson Ginsburg (University of Chicago, 1943); Donald Griffin (Harvard University, 1942); Alphaeus Guhl (University of Chicago, 1943); Arthur Hasler (University of Wisconsin, 1937); Charles Michener (University of California at Berkeley, 1941); John Paul Scott (University of Chicago, 1935); and Herman Spieth (University of Indiana, 1931). Sherwood Washburn (Harvard University, 1940) became an important anthropologist, who would play a major role in establishing the study of primate behavior in that discipline. This group provided the impetus for massive development of the study of animal behavior in North America after World War II.

B. Comparative Psychology

The accepted view of the history of comparative psychology is that it was in a state of decline during the period 1917–1945 (e.g., Beach, 1950). I have argued that there was an appreciable body of work in the field during this time and that there is thus historical continuity in comparative psychology from the turn of the century to the present. The view of a comparative psychology in decline during this period is based, in part, on analyses of data on the numbers of papers of certain types published in a limited number of journals. I believe these analyses to be misleading (Dewsbury, 1984, pp. 18–26).

My argument has been interpreted as meaning that I believe comparative psychology to have been dominant during this period. This is clearly not the case. The dominant trend in animal psychology was the development of global learning theories such as those of Clark Hull, Edwin Guthrie, and Edward Tolman. As the learning theory approach grew and prospered, comparative psychology, as I am defining it, became smaller in relation to it, but not necessarily in absolute terms. Some confusion resulted because the *Journal of Comparative Psychology* served all three areas of animal psychology—physiological psychology, process-oriented learning studies, and comparative psychology. I believe the argument that animal psychology became dominated by the study of learning is essentially correct. It is incorrect, however, to neglect the important work done by true comparative psychologists during this period.

The contrast between pre- and post-World War II activity is somewhat greater in psychology than in zoology. Only Yerkes and Lashley, of the prewar cohort, exerted significant influence on comparative psychology after the war. However, the contributions of postwar Ph.D. recipients were considerable.

1. The Natural Lives of Animals

Although the study of learning was a very important part of comparative psychology, as it was elsewhere in animal psychology, the study of the natural lives of animals was far from neglected. Calvin P. Stone (1892–1954) stated the theme for the study of the natural lives of animals in comparative psychology. He detected a gradual change in the field away from the study of conscious and organic processes toward the study of "congenital" behavior. He stated the need for "comprehensive studies of fundamental types of native behavior" (Stone, 1922, p. 95). Stone presented data on the ontogeny of copulatory behavior in laboratory rats. He would be a leader in the study of congenital behavior for the next 30 years.

Perhaps the most significant contribution of comparative psychology during this period was the beginning of the modern era of primate field studies—an approach that grew out of comparative psychology. The "grandfather" of the approach was Robert Yerkes. After returning from government service following World War I, Yerkes turned his attentions increasingly to the study of primate behavior. He studied primates in his laboratories at Yale University, at his summer home in New Hampshire, at the summer home of the Ringling Brothers circus in Florida, in the collection of a Madam Abreu in Cuba, and at G. V. Hamilton's private primate facility in California. Finally, in 1930, the facility that would be named the Yerkes Laboratory of Primate Biology was opened in Orange Park, Florida. It would become a training ground for many leading animal behaviorists for the next quarter of a century. The studies at the facility encompassed a wide range of behavioral patterns, including social and sexual behavior, in addition to the study of learning (e.g., Yerkes, 1943).

Yerkes arranged for three significant field studies of primate behavior in the early 1930s. Henry W. Nissen (1901–1958) studied chimpanzees and provided descriptions of social organization, nests, feeding, and social behavior (see Nissen, 1931). H. C. Bingham (1888–1964) studied a similar range of behavioral patterns in gorillas (see Bingham, 1932). Although these studies appear primitive by today's standards, of the latter study Allee wrote "at last such field studies have been put on a sound basis which should result in the hunting of information rather than of specimens" (1933, p. 320).

Even more significant, however, was Yerkes' arranging to send C. Ray Carpenter (1905–1975) to study howler monkeys on Barro Colorado Island in Panama (see Carpenter, 1934). After receiving his doctorate under Stone in 1932, Carpenter became associated with the Yale Laboratories and began a career that would see him complete several significant field studies of primates and facilitate many more (see Price, 1968). He established the colony of rhesus macaques on Cayo Santiago in Puerto Rico. Carpenter is properly regarded as the "father" of the modern field study of nonhuman primates.

A significant program of field research on various species of ants was initiated during this period by Shepard's student, Theodore C. Schneirla (1902–1968) (see Aronson et al., 1972). Schneirla made at least 15 major field trips to Barro Colorado Island, the Philippines, Thailand, Mexico, and field sites in the United States to study the behavior of ants under field conditions. His was probably the most extensive field research program in comparative psychology. Schneirla is best known for his studies of the control of the alternation in army ant colonies between migratory and nonmigratory phases; the stimulus is provided by the larval brood. Schneirla wrote "Only in consequence of detailed observational and experimental study may we hope to work out the causal factors which underlie the characteristic behavior of a given animal; a self-evident rule which is frequently ignored by impatient theorists" (Schneirla, 1938, p. 53).

Frank A. Beach (1911–1988) was a prominent force in the study of naturally occurring behavior for half a century; his first publication was on the maternal behavior of rats (Beach, 1937). He concentrated his studies on reproductive behavior, especially its neural and hormonal determinants. Beach was a major force in attracting younger workers to study such behavioral patterns.

Harry F. Harlow (1905–1981) divided his career between the study of learning and of naturally occurring behavior. Early in his career, Harlow studied learning in a variety of species, including goldfish, rats, and primates. Even at this stage, he had an interest in the social influences on behavior, as in his work on social facilitation (Harlow, 1932). Later, he became world famous for his studies of primate social interactions and the role of developmental factors on adult social behavior in rhesus macaques (e.g., Harlow et al., 1971).

Karl S. Lashley (1890–1958) is best known for his research on the neural bases of learning in rats. However, in his long career he studied sexuality in paramecia (with Jennings), development in monkeys (with Watson), homing in birds in the field (with Watson), sound acquisition in parrots, reproduction in *Hydra,* color vision in birds, handedness in monkeys, and neuroanatomy.

Although between the wars Yerkes, Carpenter, Schneirla, Beach, Harlow, and Lashley were the most prominent students of naturally occurring behavior, others made contributions. Carl Murchison (1887–1961), best known for his work as a journal producer, also produced a series of papers on dominance relationships in domestic chickens (e.g., Murchison, 1935), a topic also studied by Allee and his associates. Perry F. Swindle (1889–1972) wrote on the integration of relatively simple reflexes into complex movement patterns, as in nest building in birds (see Swindle, 1919). Crawford (1939) reviewed available information on the social behavior of vertebrates, discussing Lorenz's work on imprinting and the *Kumpan.*

A sampling of other research in comparative psychology during this period shows studies of homing in birds (Warner, 1931), homing in pigeons (Grundlach, 1932), bird flight (Warner, 1928), nest building in rats (Kinder, 1927), dominance in primates (Maslow, 1936), the development of feeding in chicks (Bird, 1926), and song learning in birds (Sanborn, 1932).

2. *Evolution*

The study of evolution was somewhat less prevalent during this period than before or after. Nevertheless, some relevant work continued. Swindle (1917) considered the adaptive significance of various eye appendages, such as the presence of white whiskers in nocturnal species and dark whiskers in diurnal species, concluding that they aid in visual fixation. Yerkes (1933) was interested in the evolution of grooming patterns in primates and wrote that "if we are to trace the evolution of primate social service the world must be our laboratory and all students of life investigators" (p. 23).

Schneirla's long research program on doryline ants provides a model program. In Scheirla's work we see a comparative psychologist studying a taxon of invertebrates under field conditions. He reviewed his work late in his career. Schneirla compared New World and Old World dorylines and speculated that in the New World "Far more extensive adaptive radiations may have arisen, advancing them much further into specializations. . . . We therefore find closely related species of *Eciton,* as for example *E. hamatum* and *E. burchelli,* strikingly different. . . , whereas closely related species of *Aenictus,* as for example *A. laeviceps* and *A. gracilis,* are strikingly similar" (Schneirla, 1965, p. 883). Schneirla wrote "Adaptation in terms of a nomad–statary functional cycle may have evolved homologously in epigaeic and hypogaeic doryline species of both Old and New Worlds, with secondary differences as noted above" (1957, p. 849).

Stone (1943) used the occasion of his A.P.A. presidential address to review for his fellow psychologists the current status of evolutionary theory. His conclusion included one of the first uses of a term that was to come into wide use:

> I can think of no better attitude with which to indoctrinate our colleagues of tomorrow who would make animal psychology their speciality than one of constant vigilance for opportunities to study the instincts as they are related to the subject of *behavioral ecology.*" (Stone, 1943, p. 24, italics mine)

3. Development

Studies of behavioral development continued to be prominent in comparative psychology. The developmental approach was especially prominent in the writings of Schneirla, but Beach, Stone, and Yerkes all made substantial contributions in this area. Perhaps the best known work is that of Leonard Carmichael (1898–1973), who studied the development of behavior in frogs and salamanders by rearing them under conditions that prevented movement. He found that genetic factors were more important in development than anticipated. However, like many psychologists, Carmichael stressed the interaction of genes and environment: "heredity and environment are *interdependently* involved in the perfection of behavior. . . there may be great variations in the necessary amount of environmental stimulation required to develop typical behavior patterns" (Carmichael, 1928, p. 259).

Behavior genetics has developed as an area of considerable interest in psychology. Tolman's (1924) study of selection based on maze performance and the more substantial work of his student, R. C. Tryon (e.g., Tryon, 1940), are early examples of the application of genetic methodology to the study of behavior. Stone considered the methods available for the study of behavior genetics (Stone, 1947) and analyzed wildness and savageness in different strains of rats (Stone, 1932).

The proper use of the concept of instinct was debated throughout this time period. The central figures in the controversy were Watson, Zing-Yang Kuo (1898–1970), and William McDougall (1872–1938). The Watson–McDougall debate was one of the great media events of psychology (Watson and McDougall, 1929). After initially taking a more radical position, Kuo settled on a position according to which he recognized the existence of behavior that is not acquired postnatally, citing the work of Bird, Carmichael, and Stone as demonstrating this. Rather, he stressed that heredity and environment cannot be properly separated (Kuo, 1929). The concept of instinct had its defenders during this time, among them Stone (1943) and Lashley (1938). In his presidential address to the Eastern Psychological Association, Lashley approached such ethological concepts as the *Sollwert* and the sign stimulus. He believed that "there is good evidence that animals without previous experience may give specific reactions to biologically significant objects and that the recognition or discrimination of these objects may be quite precise" (Lashley, 1938, p. 452). Thorpe (1979, p. 47) wrote that this paper "independently of Lorenz, but almost exactly at the same time, expressed almost every point of importance which came to characterize the ethological view of instinct."

4. Study of a Wide Range of Species

Perusal of the above sections on comparative psychology between the wars will reveal that, in addition to the prevalent use of laboratory rats and other domesticated species, a wide range of nondomesticated species were also studied. There are studies of paramecia, hydra, ants, frogs, salamanders, primates, and various species of birds; additional species could have been cited.

5. Integration

It is surely fair to state that comparative psychologists did not achieve the degree of integration that characterized European ethology. Nevertheless, important steps were made. Comparative psychologists published comprehensive textbooks in which they brought together many of the significant facts of the time. Notable are the three-volume treatise of Warden *et al.* (1935, 1936, 1940), the edited volume from Moss (1934) (which would go through several editions), and the text of Maier and Schneirla (1935).

The remarkable theoretical paper of Lashley (1938) was discussed in Section 3, on behavioral development. The theoretical papers of T. C. Schneirla (e.g., Schneirla, 1939) still provide the basis for some work in comparative psychology.

6. Recruits to the Population

Among those joining the ranks of comparative psychologists and not discussed previously were: Norman Maier (University of Michigan, 1928); Josephine Ball (University of California at Berkeley, 1929); Winthrop Kellogg (Columbia University, 1929); Donald Hebb (Harvard University, 1936); Austin Riesen (Yale University, 1939); William Verplanck (Brown University, 1941); and M. E. Bitterman (Cornell University, 1945).

C. The Status of the Study of Animal Behavior in North America circa 1945

In the period following 1945, European ethology would have great influence on the study of animal behavior in North America. I believe that the above discussion demonstrates that, prior to this, a considerable number of zoologists and psychologists conducted some very important studies of animal behavior in North America. Many of these projects reflect the essence of the "ethological attitude," in that they entailed biologically meaningful behavior, description, a wide range of species, comparison of similar behaviors, and nonexclusive use of domesticated species.

Comparative psychology has been treated as a very small area at this time. In comparison to the number of process-oriented learning theorists at the time, comparative psychology was small. When compared to the number of ethologists at the time or to the number of comparative psychologists in earlier times, however, the numbers of true comparative psychologists were respectable.

It is true, however, that no major, synthetic theory was developed during this time. In the words of Thorpe (1979):

> The more one studies the American situation at that time the more extraordinary it seems that the American group did not become the modern founders of ethology. They came so near it and were well in advance of workers anywhere else in the world. (Thorpe, 1979, p. 50)

There are probably many answers to the questions about the limitations of the accomplishments of North American animal behaviorists during this period. Comparative psychologists, such as Schneirla and Carpenter, were not widely respected in psychology. They received minimal encouragement and support. Further, there was little unity of approach or mission. The various contributors were somewhat isolated, each mining an area somewhat independently of the others. There was no unifying theory to tie together the important, yet disparate, studies that were being conducted. Although some comparative analyses of homologous movements in closely related species were conducted, these were not the focus of attention. Finally, there was no charismatic figure to publicize and promote the field more broadly. The result was that while significant contributions were made by psychologists, and zoologists, throughout the century, no identifiable school developed. That developed in European ethology.

Professor Lorenz has written of my previous book (Dewsbury, 1984) "I cannot agree with Dr. Dewsbury's history book that seems to contend that comparative psychologists have really been doing ethological work all along" (Lorenz, 1985, p. xiii). I think that this is both true and not true, depending on which facets of ethology one emphasizes. Comparative psychology and core ethology were much closer than is generally recognized with respect to interest in naturally occurring behavior, evolution, genetics, and the development of behavior. Comparative psychologists compared species and did some field studies. These facets of comparative psychology have generally been underestimated. At this level, the most substantial difference lies in the extent to which core ethologists analyzed homologous behavioral patterns in closely related species. If one takes this as the essence of ethology, the two disciplines were separated by more than it appears if one adopts the broader "ethological attitude" (Burghardt, 1973) as characterizing ethology. There were differences. Nevertheless, comparative psychologists were doing studies that would normally have been treated as ethology, had they not been conducted by psychologists. They were, in this sense, "doing ethological work all along." Comparative psychologists neither concentrated on the study of homologous movements in related species nor developed a comprehensive theory. In this sense, they were not "doing ethology."

This in no way detracts from the enormous contributions of the core ethological approach. The accomplishments of Lorenz, Tinbergen, Baerends, von Frisch, and others in reorienting the study of animal behavior are deserving of the credit they have received. Their impact was greater than that of the comparative psychologists of their time. Although they were not the only ones doing important work on animal behavior, the appeal of the work of European ethologists to North American workers was considerable.

VI. 1945–PRESENT

I shall not attempt a comprehensive treatment of the period from 1945 until the present. The study of animal behavior in North America has grown astronomically during this period and could be the topic of a paper or book in itself. Also, some time is needed for achievement of a proper perspective. I shall confine myself to brief mention of some of the new animal behaviorists who arrived in the decade following World War II and who effected much of this growth, and a consideration of the impact of European ethology on the study of animal behavior in North America.

A. Recruits to the Population, 1945–1955

With the postwar opportunities for education, the number of animal behaviorists grew considerably and helped secure the foundation for the expansion that has led to the current vigorous state of the field. Among zoologists added to the population at this time were: Kenneth Armitage (University of Wisconsin, 1954); Edwin Banks (University of Florida, 1955); George Bartholomew (Harvard University, 1947); Charles Carpenter (University of Michigan, 1947); John Christian (Johns Hopkins University, 1954); Elsie Collias (University of Wisconsin, 1948); William Dilger (Cornell University, 1955); Thomas Eisner (Harvard University, 1955); Howard Evans (Cornell University, 1949); Edgar Hale (University of Chicago, 1950); John King (University of Michigan, 1951); Sol Kramer (University of Illinois, 1948); Donald Maynard (U.C.L.A., 1955); Martin Schein (Johns Hopkins University, 1954); Allen Stokes (University of Wisconsin, 1952); Charles Southwick (University of Wisconsin, 1953); Edward Wilson (Harvard University, 1955); Roslyn Warren (New York University, 1954); and Howard Winn (University of Michigan, 1955).

Within psychology, there were: George Collier (University of Indiana, 1951); Allen Gardner (Northwestern University, 1954); Robert Goy (University of Chicago, 1953); Jerry Hirsch (University of California at Berkeley, 1955); Eckhard Hess (Johns Hopkins University, 1948); Daniel Lehrman (New York University, 1954); William Mason (Stanford University, 1954); Gerald McClearn (University of Wisconsin, 1954); Howard Moltz (New York University, 1953); Stanley Ratner (University of Indiana, 1954); Arthur Riopelle (University of Wisconsin, 1950); Jay Rosenblatt (New York University, 1953); Duane Rumbaugh (University of Colorado, 1955); William Thompson (University of Chicago, 1951); J. Michael Warren (University of Wisconsin, 1953); and Everett Wyers (University of California at Berkeley, 1955).

B. The Influence of European Ethology on North America

Core ethology evolved during the twentieth century, primarily in Europe (see Thorpe, 1979), and came to fruition in the work of Konrad Lorenz and Niko Tinbergen. In the years following World War II, especially in the 1950s, the face of the study of animal behavior in North America was transformed with the full force of the arrival of European ethology. In the process of the interaction of ethology and American animal behavior studies, both were permanently transformed. Whatever the strengths of animal behavior in North America at the time, young Americans found great inspiration in core European ethology.

The beginnings of the arrival of ethology were apparent in the 1930s, as with the citations in the work of Nice (1935, 1939), Crawford (1939), and Maier and Schneirla (1935). Tinbergen first came to the United States in 1938 and visited such places as the American Museum of Natural History and Yerkes' laboratories in New Haven and Orange Park (Tinbergen, 1985). Lorenz was more familiar with the American scene, having studied briefly at Columbia University in New York in 1922. Although some contacts were made at this time, progress in American ethology was slowed by the war, and recovered slowly after it.

1. The Early 1950s

When Tinbergen, Baerends, and Thorpe founded the journal *Behaviour* in 1948, they reached out to the United States by placing two comparative psychologists, Beach and Carpenter, on the editorial board. Contact was facilitated by the appearance in English of a number of readily available publications; Americans are not noted for their fluency in languages. The publication of a symposium of the Society for Experimental Biology on *Physiological Mechanisms of Animal Behaviour* (Danielli and Brown, 1950) included important chapters by Lorenz, Tinbergen, Baerends, and Thorpe. (Lashley also contributed a chapter.) Tinbergen's *The Study of Instinct* appeared in 1951 and Lorenz's *King Solomon's Ring* in 1952. These publications encouraged several workers to seek out and read the less accessible work. In a pivotal chapter, Beach (1951) discussed much ethological work.

Perhaps the most significant event of this time was the publication of Lehrman's (1953) "A Critique of Konrad Lorenz's Theory of Instinctive Behavior." Although toned down from its initial version, the paper included a stinging attack on some of the fundamental principles of core ethology. It was a "shot heard round the world," as can be seen by the various reactions of ethologists to the publication of this paper (in Dewsbury, 1985). Although the paper initially had negative impact, it ultimately had the effect of clarifying points of difference and opening, rather than closing, discussion. Thus, it was of major importance in bringing comparative psychologists and ethologists together, and ultimately to the influence of ethology on North America.

The first interaction of comparative psychologists and ethologists after the Lehrman paper occurred when Gerard Baerends and Jan van Iersel met with Beach, Lehrman, Hess, and Verplanck in Montreal at the 14th Congress of Psychology in 1954 (Beer, 1975; Baerends, 1985). After initial uneasiness, the sources of misunderstanding were clarified and lasting friendships were built. It was realized that both groups had a serious and valid interest in the study of animal behavior and the differences between them were smaller than they had appeared. The process continued at another meeting in Boston a month later (Baerends, 1985; Verplanck, 1955). Americans Aronson, Beach, Bullock, Estes, Griffin, Harlow, Pribram, and Verplanck, among others, met with Baerends and van Iersel, with the significant addition of Robert Hinde. Hinde, together with others attending these conferences, would be a major force in bringing about a reconciliation. Lehrman and Lorenz first met a few weeks later at a conference in Paris. Beach, Lehrman, Lorenz, and Tinbergen participated in a conference on "Group Processes" at Cornell University in 1954 and contributed major chapters. Among others at the meeting were J. S. Bruner, E. Mayr, T. C. Schneirla, E. H. Erikson, J. Crane, N. E. Collias, M. Mead, E. H. Hess, K. H. Pribram, and H. S. Liddell.

The most prolonged interactions occurred when Beach hosted a meeting lasting several weeks at the Center for Advanced Study in the Behavioral Sciences in Palo Alto. The group included Beach, Harlow, Hebb, Hess, Lehrman, and Rosenblatt from comparative psychology and Baerends, Hinde, van Iersel, and Vowles from ethology (Baerends, 1985; Hinde, 1978). There was no formal schedule, no publication, and no pressure—just a lot of time for deep and informal interaction.

This critical period in the early 1950s effected the interaction between core ethology and North American animal behaviorists. The interactions continued and grew in various ways in the years that followed.

2. Publications

Ethological work became progressively more accessible to English-speaking audiences. The publication of the *Group Processes* volume (Schaffner, 1955) allowed the reader to follow dialogue among the participants as it unfolded. Claire H. Schiller (1957) published a volume of translations of ethological papers, making some of the seminal works readily available in English for the first time; Lashley wrote the introduction. Tinbergen's *Curious Naturalists* appeared in 1958 and Eckhard Hess (1962) wrote a comprehensive summary of the fundamentals of ethological theory. Verplanck (1957) published a glossary designed to improve mutual comprehension of the terms from different fields with common interests. Hinde (1959) summarized recent developments in ethology. W. H. Thorpe's (1956) *Learning and Instinct in Animals* and Robert Hinde's (1966) *Animal Behaviour: A Synthesis of Ethology and Comparative Psychology* were notable for their blending of work from core ethology and comparative psychology, and played an important role in effecting synthesis.

3. International Ethological Conferences

The first official International Ethological Conference (I.E.C.) was held in Buldern, West Germany, in 1952, the second at Oxford in 1953; there was just one American registrant at each—Bernard Greenberg at Buldern and Russell DeValois at Oxford. Ethological conferences have been held in alternate years ever since. In 1955, at Groningen, four North Americans—Beach, Demorest Davenport, Hess, and Lehrman—appeared on the program. Six North Americans were on the program in 1957 at Freiburg, with 10 attending. The growth of North American participation can be seen in Fig. 5 and Table I. By 1967, when the I.E.C. was held in Stockholm, 80 North Americans attended and 27 appeared on the program. The I.E.C. finally came to North America for meetings in Washington, D.C., in 1973 and Vancouver, British Columbia, in 1979.

4. Americans in Europe

It was natural that bright and energetic students maturing in this era should be attracted to the scene of the action in core ethology and study in Europe. Some

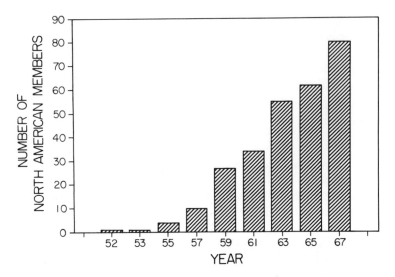

Fig. 5. Portrayal of the growth in the number of North Americans attending the International Ethological Conferences, 1952–1967.

of these are listed in Table II. Among those completing their doctorates in European ethological laboratories were Martin Moynihan, Miles Keenleyside, Beatrice Tugendhat Gardner, and John Fentress. An even larger number of researchers went to Europe after completing their doctorates in North America; they, too, are listed in Table II.

5. *Europeans in North America*

The reciprocal process saw American universities bringing trained European ethologists to serve on their faculties. Frank McKinney emigrated to Canada in 1954 and moved to the University of Minnesota in 1963; he did outstanding work on the behavior of water fowl. Peter Marler moved to the University of California at Berkeley in 1957 and to Rockefeller University in 1966; he would be a major force in cementing the ties between Europe and North America. Colin Beer first studied at and then moved to the Institute of Animal Behavior at Rutgers University in 1962; Beer wrote significant papers on the history and philosophy of ethology. N. Robin Liley moved to Canada in 1963 and to the University of British Columbia in 1965; he hosted the Vancouver conference. Walter Heiligenberg, a Lorenz Ph.D., came to the University of California at San Diego in 1972. Thelma Rowell, of the University of California at Berkeley, was another prominent addition from Europe. Among those who came to the United States to teach, but later returned to Europe, were Richard Andrew, Monica Impekoven, E. G. Franz Sauer, and Wolfgang Schleidt.

6. Effects of Interaction

Both ethology and the study of animal behavior in North America were markedly changed by this extensive interaction. There are still some on both sides of the Atlantic who resist the change that has been effected. Most workers, however, have welcomed the change and development of their disciplines. Most of the old hostilities are gone. Although there are some disturbing signs of new fractures, the study of animal behavior is generally remarkably unified—in zoology and psychology, on both sides of the Atlantic, and around the world.

There have been many developments in the study of animal behavior since the reconciliation of ethology and comparative psychology. The most prominent of these is the emergence of the behavioral ecological–sociobiological approach.

Table I. North American Animal Behaviorists Participating in the International Ethological Conferences between the Years 1955 and 1965[a]

Host city: Year held:	Groningen (1955)	Freiburg (1957)	Cambridge (1959)	Starnberg (1961)	The Hague (1963)	Zurich (1965)
	Beach	Beach	Aronson	Beach	Aronson	Aronson
	Davenport	Beckwith	Bullock	Beckwith	Banks	Brown
	Hess	Dethier	Crane	Dethier	Barlow	Cooper
	Lehrman	Hess	Davenport	Dilger	Beach	Denenberg
		Lehrman	Dethier	Fuller	Bitterman	Eisenberg
		Roeder	Dilger	Hale	Brody	Fentress
			Emlen	Harlow	Collias	Ficken
			Hess	Hess	Dane	Flynn
			Lehrman	Klinghammer	Eisenberg	Freedman
			Moynihan	Klopfer	Emlen	Gottlieb
			Pittendrigh	Kramer	Ficken	Griffin
			Schein	Lehrman	Gardner	Kleiman
			Scott	Levine	Ginsburg	Klopfer
			Stokes	Moltz	Hess	Lehrman
			Verplanck	McCleary	Komisaruk	Levine
				Reese	Konishi	Mason
				Roeder	Lehrman	Maynard
				Schein	Maynard	Myrberg
				Shaw	McConnell	Payne
				Winn	Nelson	Phillips
					Payne	Reese
					Roeder	Rosenblatt
					Rosenblatt	Scott
					Ward	Shaw
					Wilson	Warren
						Wilson

[a]Based upon conference program listings for these years.

Table II. Listing of North Americans Working in European Laboratories of Ethology, c. 1950–Present

Doctoral work

Cambridge	Oxford	Groningen
Klopman, R. B. (1960)	Moynihan, M. H. (1953)	Keenleyside, M. (1955)
Fentress, J. (1965)	Gardner (*Tugendhat*), B. (1959)	
		Leiden
		Symons, P. E. K. (1965)

Postdoctoral work

Cambridge	Leiden	Groningen
Klopfer, P. H. (1957–1958)	Collias, N. E. (1949)	Greenberg, B. (1951–1952)
Warren, R. P. (1957–1959)	Ulmer, N. (1959–1960)	Tucker, A. (1954–1955)
Stokes, A. W. (1958–1959)	Nelson, K. (1963–1965)	Plecha, N. (1958?)
Kaufman, I. C. (1958–1959)	Wyers, E. J. (1964–1965)	Jensen, D. D. (1958–1960)
Zeigler, H. P. (1958–1960)	Dunham, D. W. (1964–1966)	Reese, E. S. (1961–1962)
Ratner, S. C. (1959–1960)	Jenni, D. A. (1964–1966)	Miller, R. (1961–1962)
Griffin, D. R. (1960–1961)	Borowski, R. (1969–1970)	Drent, R. (1961–1967; 1972)
Nottebohm, F. (1964–1965)	Vandervoort, F. (1969–1970)	Hogan, J. A. (1961–1964; 1985–1986)
Baum, W. (1965–1966)	Seewiesen	
Miller White, S. (1965–1971)	Kramer, S. (1955–1957)	Erickson, C. (1965–1966)
Reese, E. P. (1966–1967)	Barlow, G. W. (1958–1960)	Marshall, J. A. (1966–1967)
Reese, T. W. (1966–1967)	Burchard, J. (1959–1965)	Rowland, W. J. (1970–1971)
Thompson, T. (1967–1968)	Myrberg, A. A. (1961–1964)	Oxford
Kling, J. (1968–1969)	Gruber, S. (1971–1972)	Verplanck, W. S. (1953)
Honig, W. (1970–1971)	Beckwith, W.	Barlow, G. W. (1974–1975)
Copenhagen	Edinburgh	Tubingen
Oring, L. (1966–1967)	Noakes, D. (1970–1972)	Hailman, J. P. (1964)

Although this approach is sometimes billed as swallowing up the older approaches (e.g., Wilson, 1975), indications are that just as a synthesis of ethology and comparative psychology was effected, so sociobiological principles are being integrated into the mainstream approach to animal behavior. Diverse viewpoints are being modified and reconciled and new general principles developed. The chronicle of these developments must await future writers.

VII. CONCLUSION

A scientific study of animal behavior developed in North America, as it did in Europe, throughout the 19th century and matured around the turn of the century. Prominent animal behaviorists worked in the traditions within both zoology and psychology. Both comparative psychology and zoology produced some outstanding animal behaviorists. With some minor interruptions, the tradition of studying behavior in relation to the natural lives of animals lived throughout the century in both disciplines. The similarities between the two and to European ethology are closer than has often been recognized. The disciplines have changed with time as new methods and ideas have developed. However, through two world wars, various debates over instinct, the mushrooming of American science in the 1950s, the arrival of ethology, and the challenges of sociobiology, the study of animal behavior has continued to grow. If the past truly provides the best basis for predictions of the future, the prospects for continued development appear bright.

VIII. ACKNOWLEDGMENTS

I thank the large number of workers who answered my letters and provided autobiographical and other information helpful in writing this paper. Regrettably, not all of this information could be used in this paper. Gerard Baerends, Robert Hinde, Arthur Myrberg, and Ernst Reese provided especially extensive help. Patrick Bateson and Wilse B. Webb provided helpful feedback on an earlier draft of this paper.

IX. REFERENCES

Allee, W. C. (1933). Gorillas in a native habitat. *Ecology* **14**:319–320.
Allee, W. C. (1937). Life history of the song sparrow. *Ecology* **18**:540–541.
Allee, W. C. (1938). *The Social Life of Animals,* Norton, New York.
Altmann, M. (1941). Interrelationships of the sex cycle and the behavior of the sow. *J. Comp. Psychol.* **31**:481–498.
Altmann, M. (1963). Naturalistic studies of maternal care in moose and elk. In Rheingold, H. L. (ed.), *Maternal Behavior in Mammals,* Wiley, New York, pp. 233–253.
Angell, J. R. (1909). The influence of Darwin on psychology. *Psychol. Rev.* **16**:152–169.
Aronson, L. R., Tobach, E., Rosenblatt, J. S., and Lehrman, D. S. (eds.) (1972). *Selected Writings of T. C. Schneirla,* Freeman, San Francisco.
Audubon, M. R. (ed.) (1897). *Audubon and His Journals,* Scribner's, New York.
Baerends, G. P. (1985). Two pillars of wisdom. In Dewsbury, D. A. (ed.), *Leaders in the Study of Animal Behavior: Autobiographical Perspectives,* Bucknell University Press, Lewisburg, Pennsylvania, pp. 12–40.
Baldwin, J. M. (1896). A new factor in evolution. *Am. Nat.* **30**:441–451, 536–553.
Baldwin, J. M. (1909). The influence of Darwin on theory of knowledge and philosophy. *Psychol. Rev.* **16**:207–218.

Banks, E. M. (1985). Warder Clyde Allee and the Chicago school of animal behavior. *J. Hist. Behav. Sci.* **21**:345–353.
Bascom, J. (1878). *Comparative Psychology or the Growth and Grades of Intelligence,* Putnam's, New York.
Beach, F. A. (1937). The neural basis of innate behavior. I. Effects of cortical lesions upon the maternal behavior pattern in the rat. *J. Comp. Psychol.* **24**:393–436.
Beach, F. A. (1950). The snark was a boojum. *Am. Psychol.* **5**:115–124.
Beach, F. A. (1951). Instinctive behavior: Reproductive behavior. In Stevens, S. S. (ed.), *Handbook of Experimental Psychology,* Wiley, New York, pp. 387–434.
Beer, C. G. (1975). Was Professor Lehrman an ethologist? *Anim. Behav.* **23**:957–964.
Bingham, H. C. (1932). Gorillas in a native habitat. *Publ. Carnegie Inst.* **426**:1–65.
Bird, C. (1926). The effect of maturation upon the pecking instinct of chicks. *J. Genet. Psychol.* **33**:212–234.
Burghardt, G. M. (1973). Instinct and innate behavior: Toward an ethological psychology. In Nevin, J. A. (ed.), *The Study of Behavior,* Scott Foresman, Glenview, Illinois, pp. 321–400.
Burroughs, J. (1875). *Birds and Bees Sharp Eyes and Other Papers,* Houghton-Mifflin, Boston.
Buss, A. R. (1977). In defense of a critical–presentist historiography: The fact–theory relationship and Marx's epistemology. *J. Hist. Behav. Sci.* **13**:252–260.
Cadwallader, T. C. (1984). Neglected aspects of the evolution of American comparative and animal psychology. In Greenberg, G., and Tobach, E. (eds.), *Behavioral Evolution and Integrative Levels,* Erlbaum, Hillsdale, New Jersey, pp. 15–48.
Carmichael, L. (1928). A further experimental study of the development of behavior. *Psychol. Rev.* **35**:253–260.
Carpenter, C. R. (1934). A field study of the behavioral and social relations of howling monkeys (*Alouatta palliata*). *Comp. Psychol. Monogr.* **10**(48):1–168.
Carr, H. A. (ed.) (1919). *The Behavior of Pigeons. Posthumous Works of Charles Otis Whitman,* Carnegie Institution, Washington, D.C.
Chiszar, D., and Wertheimer, M. (1988). Margaret Altmann: A rugged pioneer in rugged fields. *J. Hist. Behav. Sci.* **24**:102–106.
Cohen, D. (1979). *J. B. Watson: The Founder of Behaviorism,* Routledge & Keegan Paul, London.
Craig, W. (1913). The stimulation and the inhibition of ovulation in birds and mammals. *J. Anim. Behav.* **3**:215–221.
Craig, W. (1918). Appetites and aversions as constituents of instincts. *Biological Bulletin* **34**:91–107.
Crawford, M. P. (1939). The social psychology of the vertebrates. *Psychol. Bull.* **36**:407–446.
Crozier, W. J. (1928). Tropisms. *J. Gen. Psychol.* **1**:213–238.
Danielli, J. F., and Brown, R. (eds.) (1950). *Physiological Mechanisms of Animal Behaviour,* Academic Press, New York.
Darwin, C. (1859). *On the Origin of Species by Means of Natural Selection, or the Preservation of Favoured Races in the Struggle for Life,* John Murray, London (Modern Library Edition, n.d.).
Dewsbury, D. A. (1984). *Comparative Psychology in the Twentieth Century,* Hutchinson Ross, Stroudsburg, Pennsylvania.
Dewsbury, D. A. (ed.) (1985). *Leaders in the Study of Animal Behavior: Autobiographical Perspectives,* Bucknell University Press, Lewisburg, Pennsylvania.
Donaldson, H. H. (1928). A study of the brains of three scholars. *J. Comp. Neurol.* **46**:1–95.
Evans, M. A., and Evans, H. E. (1970). *William Morton Wheeler, Biologist,* Harvard University Press, Cambridge, Massachusetts.
Gottlieb, G. (1987). A tribute to Clarence Luther Herrick (1858–1904): Founder of developmental psychobiology. *Devel. Psychobiol.* **20**:1–5.
Goy, R. W. (1967). William Caldwell Young. *Anat. Rec.* **157**:3–11.

Gray, P. H. (1987). Thomas Morton as America's first behavioral observer (in New England 1624–1646). *Bull. Psychon. Sci.* **25**:69–72.
Greenberg, B., and Noble, G. K. (1944). Social behavior of the American chameleon (*Anolis carolinensis* Voigt). *Physiol. Zool.* **17**:392–439.
Guhl, A. M., and Schein, M. W. (1976). *The Animal Behavior Society: Its Early History and Activities.* Animal Behavior Society, Morgantown, West Virginia.
Grundlach, R. H. (1932). A field study of homing in pigeons. *J. Comp. Psychol.* **13**:397–402.
Hall, G. S. (1909). Evolution and psychology. In Chamberlin, T. C. (ed.), *Fifty Years of Darwinism: Modern Aspects of Evolution,* Holt, New York, pp. 251–267.
Hall, G. S. (1927). *Life and Confessions of a Psychologist,* Appleton, New York.
Harlow, H. F. (1932). Social facilitation of feeding in the albino rat. *J. Genet. Psychol.* **41**:211–221.
Harlow, H. F., Harlow, M. K., and Suomi, S. J. (1971). From thought to therapy: Lessons learned from a primate laboratory. *Am. Sci.* **59**:538–549.
Herrick, C. J. (1907). Comparative psychology. *Pop. Sci. Month.* **70**:76–78.
Herrick, C. J. (1948). *The Brain of the Tiger Salamander, Amblystoma tigrinum,* University of Chicago Press, Chicago.
Herrick, C. L. (1893). Methods and scope of comparative psychology. *Denison Q.* **1**:1–10, 134–141, 179–187.
Hess, E. H. (1962). Ethology: An approach toward the complete analysis of behavior. In Brown, R., Galanter, E., Hess, E. H., and Mandler, G. (eds.), *New Directions in Psychology,* Holt, Rinehart and Winston, New York.
Hinde, R. A. (1959). Some recent trends in ethology. In Koch, S. (ed.), *Psychology: A Study of a Science,* Vol. 2, McGraw-Hill, New York, pp. 561–610.
Hinde, R. A. (1966). *Animal Behaviour: A Synthesis of Ethology and Comparative Psychology,* McGraw-Hill, New York.
Hinde, R. A. (1978). Foreword. In McGill, T. E., Dewsbury, D. A., and Sachs, B. D. (eds.), *Sex and Behavior: Status and Prospectus,* Plenum Press, New York.
Hoagland, H., and Mitchell, R. T. (1956). William John Crozier: 1892–1955. *Am. J. Psychol.* **69**:135–138.
Holmes, S. J. (1911). *The Evolution of Animal Intelligence,* Henry Holt, New York.
Holmes, S. J. (1916). *Studies of Animal Behavior,* Richard G. Badger, Boston.
Howard, D. T. (1927). The influence of evolutionary doctrine on psychology. *Psychol. Rev.* **34**:305–312.
Jaynes, J. (1969). The historical origins of 'ethology' and 'comparative psychology.' *Anim. Behav.* **17**:601–606.
Jennings, H. S. (1906). *Behavior of Lower Organisms,* Columbia University Press, New York.
Jennings, H. S. (1908). Review of *The Dancing Mouse. Psychol. Bull.* **5**:92–94.
Jennings, H. S. (1927). From amoeba up, the biological basis of the family. *Survey* **59**:272–276, 341.
Kinder, E. F. (1927). A study of the nest-building activity of the albino rat. *J. Exp. Zool.* **47**:117–161.
Kline, L. W. (1899). Suggestions toward a laboratory course in comparative psychology. *Am. J. Psychol.* **10**:399–430.
Kuo, Z.-Y. (1929). The net result of the anti-heredity movement in psychology. *Psychol. Rev.* **36**:181–199.
Lashley, K. S. (1938). Experimental analysis of instinctive behavior. *Psychol. Rev.* **45**:445–471.
Lehrman, D. S. (1953). A critique of Konrad Lorenz's theory of instinctive behavior. *Q. Rev. Biol.* **28**:337–363.
Lindauer, M. (1961). *Communication among Social Bees,* Harvard University Press, Cambridge, Massachusetts.

Loeb, J. (1918). *Forced Movements, Tropisms, and Animal Conduct,* J. B. Lippincott, Philadelphia.
Logue, A. W. (1978). Behaviorist John B. Watson and the continuity of species. *Behaviorism* **6:**71–79.
Lorenz, K. Z. (1952). *King Solomon's Ring,* Crowell, New York.
Lorenz, K. Z. (1985). Foreword. In Burghardt, G. (ed.), *Foundations of Comparative Ethology,* Van Nostrand Reinhold, New York.
Maier, N. R. F., and Schneirla, T. C. (1935). *Principles of Animal Psychology,* McGraw-Hill, New York.
Malotki, E. (1985). *Gullible Coyote Unaihu,* University of Arizona Press, Tucson.
Maslow, A. H. (1936). The role of dominance in the social and sexual behavior of infrahuman primates: I. Observations at the Vilas Park Zoo. *J. Genet. Psychol.* **48:**261–277.
Merriam, C. H. (1884a). Hibernating mammals. *Science* **3:**616.
Merriam, C. H. (1884b). Migration of North American birds. *Am. Nat.* **18:**310–313.
Miles, W. R. (1930). On the history of research with rats and mazes: A collection of notes. *J. Gen. Psychol.* **3:**324–337.
Mills, W. (1898). *The Nature and Development of Animal Intelligence,* Macmillan, New York.
Mills, W. (1899). The nature of animal intelligence and the methods of investigating it. *Psychol. Rev.* **6:**262–274.
Morgan, C. L. (1896). *Habit and Instinct,* Arnold, London.
Morgan, L. H. (1868). *The American Beaver and His Works,* J. B. Lippincott, Philadelphia.
Morton, T. (1637). *New English Canaan or New Canaan,* Jacob Frederick Stam, Amsterdam.
Moss, F. A. (ed.) (1934). *Comparative Psychology,* Prentice-Hall, New York.
Murchison, C. (1935). The experimental measurement of a social hierarchy in *Gallus domesticus:* I. The direct identification and direct measurement of social reflex no. 1 and social reflex no. 2. *J. Gen. Psychol.* **12:**3–39.
Nice, M. M. (1935). On the meaning of territory in the life of birds. *Bird-Banding* **6:**110.
Nice, M. M. (1939). The social *Kumpan* and the song sparrow. *Auk* **56:**255–262.
Nice, M. M. (1941). The role of territory in bird life. *Am. Midl. Nat.* **26:**441–487.
Nice, M. M. (1979). *Research Is a Passion with Me: The Autobiography of Margaret Morse Nice,* Consolidated Amethyst Communications, Toronto.
Nissen, H. W. (1931). A field study of the chimpanzee: Observations of chimpanzee behavior and environment in western French Guinea. *Comp. Psychol. Monogr.* **8**(36)**:**1–122.
Noble, G. K. (1938). Sexual selection among fishes. *Biol. Rev.* **13:**133–158.
Noble, G. K., and Greenberg, B. (1941). Effects of seasons, castration and crystalline sex hormones upon the urogenital system and sexual behavior of the lizard (*Anolis carolinensis*). I. The adult female. *J. Exp. Zool.* **88:**451–479.
Noble, R. C. (1946). *The Nature of the Beast,* Doubleday, Garden City, New York.
Oppenheim, R. W. (1982). The neuroembryological study of behavior: Progress, problems, perspectives. *Current Topics Devel. Biol.* **17:**257–309.
Osborn, H. F. (1896). A mode of evolution requiring neither natural selection nor the inheritance of acquired characteristics. *Trans. NY Acad. Sci.* **15:**141–142, 148.
Osgood, W. H. (1945). Clinton Hart Merriam, 1855–1942. *Biogr. Mem. Nat. Acad. Sci. USA* **24:**1–57.
Peckham, G. W., and Peckham, E. G. (1898). On the instincts and habits of the solitary wasps. *Bull. Wisc. Geol. Nat. Hist. Surv.* **2:**1–245.
Peckham, G. W., and Peckham, E. G. (1905). *Wasps Social and Solitary,* Houghton, Mifflin, Boston.
Price, B. (1968). *Into the Unknown,* Platt and Munk, New York.
Romanes, G. J. (1882). *Animal Intelligence,* Appleton, New York.
Romer, A. S. (1967). George Howard Parker. *Biog. Mem. Nat. Acad. Sci. USA* **39:**359–390.

Ross, D. (1972). *G. Stanley Hall. The Psychologist as Prophet,* University of Chicago Press, Chicago.
Sanborn, H. C. (1932). The inheritance of song in birds. *J. Comp. Psychol.* **13**:345–364.
Schaffner, B. (ed.), (1955). *Group Processes,* Macy Foundation, New York.
Schiller, C. H. (ed.), (1957). *Instinctive Behavior: The Development of a Modern Concept,* International Universities Press, New York.
Schmidt, K. P. (1957). Warder Clyde Allee, 1885–1955. *Biog. Mem. Nat. Acad. Sci.* **30**:2–40.
Schneirla, T. C. (1929). Learning and orientation in ants. *Comp. Psychol. Monogr.* **6**(4):1–143.
Schneirla, T. C. (1938). A theory of army-ant behavior based upon the analysis of activities in a representative species. *J. Comp. Psychol.* **25**:51–90.
Schneirla, T. C. (1939). A theoretical consideration of the basis for approach–withdrawal adjustments in behavior. *Psychol. Bull.* **37**:501–502.
Schneirla, T. C. (1957). Theoretical considerations of cyclic processes in doryline ants. *Proc. Am. Phil. Soc.* **191**:106–133.
Schneirla, T. C. (1965). Dorylines: Raiding and in bivouac. *Nat. Hist.* **74**(8):44–51 and **74**(9):40–47.
Scott, J. P. (1962). Critical periods in behavioral development. *Science* **138**:949–958.
Seton, E. T. (1913). *Wild Animals at Home,* Grosset & Dunlap, New York.
Small, W. S. (1899). Notes on the psychic development of the young white rat. *Am. J. Psychol.* **11**:80–100.
Stocking, G. W. Jr. (1965). On the limits of 'presentism' and 'historicism' in the historiography of the behavioral sciences. *J. Hist. Behav. Sci.* **1**:211–218.
Stone, C. P. (1922). The congenital sexual behavior of the young male albino rat. *J. Comp. Psychol.* **2**:95–153.
Stone, C. P. (1932). Wildness and savageness in rats of different strains. In Lashley, K. S. (ed.), *Studies in the Dynamics of Behavior,* University of Chicago Press, Chicago, pp. 1–55.
Stone, C. P. (1943). Multiply, vary, let the strongest live and the weakest die—Charles Darwin. *Psychol. Bull.* **40**:1–24.
Stone, C. P. (1947). Methodological resources for the experimental study of innate behavior as related to environmental factors. *Psychol. Rev.* **54**:342–347.
Swindle, P. F. (1917). The biological significance of the eye appendages of organisms. *Am. J. Psychol.* **28**:486–496.
Swindle, P. F. (1919). Analysis of nesting activities. *Am. J. Psychol.* **30**:173–186.
Thorndike, E. L. (1898). Animal intelligence: An experimental study of the associative processes in animals. *Psychol. Rev. Monogr. Suppl.* **2**(4):1–109.
Thorndike, E. L. (1899). Instinct. *Biol. Lect. Marine Biol. Lab. Woods Hole* **7**:57–67.
Thorpe, W. H. (1956). *Learning and Instinct in Animals,* Methuen, London.
Thorpe, W. (1979). *The Origins and Rise of Ethology,* Praeger, New York.
Tinbergen, N. (1951). *The Study of Instinct,* Oxford University Press, Oxford.
Tinbergen, N. (1958). *Curious Naturalists,* Basic, New York.
Tinbergen, N. (1985). Watching and wondering. In Dewsbury, D. A. (ed.), *Leaders in the Study of Animal Behavior: Autobiographical Perspectives,* Bucknell University Press, Lewisburg, Pennsylvania, pp. 440–463.
Tolman, E. C. (1924). The inheritance of maze-learning ability in rats. *J. Comp. Psychol.* **4**:1–18.
Trautman, M. B. (1977). In memorium: Margaret Morse Nice. *Auk* **94**:430–441.
Tryon, R. C. (1940). Genetic differences in maze-learning ability in rats. In Whipple, G. M. (ed.), *The Yearbook of the National Society of Education,* Vol. 39, Public School Publishing Company, Bloomington, Illinois, pp. 111–118.
Turner, C. H. (1892). Psychological notes upon the gallery spider—illustrations of intelligent variations in the construction of the web. *J. Comp. Neurol.* **2**:95–110.

Turner, C. H. (1907). The homing of ants: An experimental study of ant behavior. *J. Comp. Neurol. Psychol.* **17**:367–434.

Tyler, H. A. (1975). *Pueblo Animals and Myths,* University of Oklahoma Press, Norman.

Verplanck, W. S. (1955). Problems of comparative behavior. *Science* **121**:189–190.

Verplanck, W. S. (1957). A glossary of some terms used in the objective science of behavior. *Psychol. Rev.* **64**:(6,2):1–42.

Warden, C. J., Jenkins, T. N., and Warner, L. H. (1935, 1936, 1940). *Comparative Psychology: A Comprehensive Treatise,* 3 vols., Ronald, New York.

Warner, L. H. (1928). Facts and theories of bird flight. *Q. Rev. Biol.* **3**:84–98.

Warner, L. H. (1931). The present status of the problems of orientation and homing birds. *Q. Rev. Biol.* **6**:208–214.

Watson, J. B. (1903). *Animal Education,* University of Chicago Press, Chicago.

Warson, J. B. (1906). The need of an experimental station for the study of certain problems in animal behavior. *Psychol. Bull.* **3**:149–156.

Watson, J. B. (1908). The behavior of noddy and sooty terns. *Publ. Carnegie Inst.* **2**(103):187–255.

Watson, J. B. (1912). Instinctive activity in animals. *Harper's Magazine* **124**:376–382.

Watson, J. B. (1914). *Behavior: An Introduction to Comparative Psychology,* Holt, New York.

Watson, J. B. (1930). *Behaviorism,* 2nd. ed., University of Chicago Press, Chicago.

Watson, J. B., and McDougall, W. (1929). *The Battle of Behaviorism,* Norton, New York.

Weinland, D. F. (1858). A method of comparative animal psychology. *Proc. Am. Assn. Adv. Sci.* **12**:256–266.

Wheeler, W. M. (1902). 'Natural history,' 'ecology' or 'ethology'? *Science* **15**:971–976.

Wheeler, W. M. (1921). On instincts. *J. Abnorm. Psychol.* **15**:295–318.

Wheeler, W. M. (1926). A new word for an old thing. *Q. Rev. Biol.* **1**:439–443.

Whitman, C. O. (1898). Animal behavior. *Biol. Lect. Marine Biol. Lab Wood's Hole,* **6**:285–338.

Whitman, C. O. (1899). Myths in animal psychology. *Monist* **9**:524–537.

Wilson, E. O. (1975). *Sociobiology: The New Synthesis,* Harvard University Press, Cambridge, Massachusetts.

Yerkes, R. M. (1903). The instincts, habits, and reactions of the frog. *Psychol. Rev. Monogr. Suppl.* **17**:579–638.

Yerkes, R. M. (1907). *The Dancing Mouse: A Study in Animal Behavior,* Macmillan, New York.

Yerkes, R. M. (1933). Genetic aspects of grooming, a socially important primate behavior pattern. *J. Soc. Psychol.* **4**: 3–25.

Yerkes, R. M. (1943). *Chimpanzees: A Laboratory Colony,* Yale University Press, New Haven, Connecticut.

Young, W. C. (ed.) (1961). *Sex and Internal Secretions,* Williams and Wilkins, Baltimore.

Chapter 5

ANIMAL PSYCHOLOGY: THE TYRANNY OF ANTHROPOCENTRISM

J. E. R. Staddon

Departments of Psychology and Zoology
Duke University
Durham, North Carolina 27706

I. ABSTRACT

Psychological work with animals has an equivocal status. At one time predominant, now diminished by defections to neuroscience and overshadowed by cognitive psychology, animal psychologists have retreated to two main niches: the study of animal cognition, and the use of animals as models of people. Neither retreat is without its problems. I argue that these flow from psychology's inability to free itself from a fixation on human pathologies and abilities, to the detriment of general scientific issues. I suggest that psychology as a basic science should be about intelligent and adaptive behavior, wherever it is to be found, so that animals can be studied in their own right, for what they can teach us about the nature and evolution of intelligence, and not as surrogate people or tools for the solution of human problems.

II. INTRODUCTION

> No one successfully investigates the nature of a thing in the thing itself; the inquiry must be enlarged, so to become more general.
>
> Francis Bacon
> (*Novum Organum* I, Aph. LXX)

To write on the role of animals in contemporary psychology is a thankless, but necessary, task. It is thankless because the topic puts one on the defensive—

after all, no one has ever been asked to write a chapter on the role of *human beings* in psychology! It is necessary for the same reason: No matter what is said, animal psychology has a somewhat equivocal role. I will say something about the historical and philosophical reasons for this in a moment.

The topic of animal psychology is also timely. Most researchers have at one time or another served on the research review committees set up by governmental granting agencies. In recent years, the feeling has grown among the behavioral scientists who seek support from these agencies (especially the mission-oriented agencies) that research needs to be justified in terms of *direct human benefit*. When the proposed research involves humans, the question of human welfare is somehow less urgent; clearly (researchers seem to think, or, more accurately perhaps, think their political patrons will think), anything we learn from behavioral research with people is likely to have some benefit for people. But when the research is with animals, the major justification for the work is more and more often that the animal is in some undefined way a *model* for the human. The increasing popularity of what is often a questionable justification for work on animal behavior provides a second reason for writing about animal psychology.

III. ANIMALS IN PSYCHOLOGY: A BRIEF HISTORY[1]

How did animals ever get into psychology in the first place? And what then displaced them? After all, the appropriate motto for psychology might seem to be Pope's overworked line about "the proper study of mankind. . . ." It would be tedious and needlessly controversial to go into historical details, but the broad outlines are clear enough. Scientific psychology in the United States was at first *experimental* psychology. And experimental psychology, with its strong interest in sensory and perceptual systems, is very close to physiology. So the stage was set for something even more physiological and biological. Early psychology (remember Dewey and James) was also very much interested in education. The *meliorism* of America—the wish to make things better, the confidence that "better" and "worse" are self evident, and the faith that improvement is always possible—has also always given education, and other ways of changing behavior, a special status. Education implies (if it does not always produce) *learning*, so methods of objectively studying learning had great appeal. These two strands, the physiological and the educational, came together naturally in the form of a fixation on the *reflex*, first as reflex *qua* reflex, but then in the 1920s, with John B. Watson's belated discovery of Pavlov, on the ill-named *conditioned reflex*. It

---

[1]Michael Domjan (1987) has recently published an excellent brief summary of the historical origins of the psychology of animal learning. See also Chapter 4, this volume.

looked to everyone as if experimental psychology had found its atom, its gene, its cell: Everything was reflexes. And reflexes could be studied best, most objectively, in animals.

Darwin and evolution, the true biological basis for a relationship between human and infrahuman animals, was at all times a rather minor theme. The influential and energetic Edward Thorndike, inventor of the law of effect (based on his studies of the instrumental learning of cats and chickens in William James' basement at the turn of the century) knew about Darwin, of course, and clearly thought well of him. But his heart was elsewhere, and in his later career he followed it into education. And Clark L. Hull (1943), erector of edifices of learning theory measureless to man, also wrote approvingly of biology and evolution, as did radical behaviorist B. F. Skinner (1966), inventor of operant conditioning, many years later (although a clear-eyed critic may detect in Skinner's writings more a wish to assimilate evolutionary biology to behaviorism than the reverse). But evolution was always a side issue. Historically, the interest of experimental psychologists in work with animals derived from a basically reflexological, atomistic (i.e., chemical) view of the subject, in which animals were to be studied before people for the same reason that chemists study acids and bases before they study proteins and DNA—because the former are simpler, though built of the same atomic ingredients.

What's wrong with that, the reader may well ask? This strategy, of beginning with elements and simple compounds before tackling stereoisomers, has worked very well in other parts of science, so why not in psychology? The short answer, of course, is that Dalton discovered real atoms—but Pavlov did not. Behavior is not built up out of reflexes, at least not reflexes of the sort envisaged by the early behaviorists. And after countless rats had run numberless kilometers down thousands of runways, this finally became apparent. Hull's overextended stimulus–response theories passed almost into oblivion, leaving behind a legacy of highly articulated experimental method and a faith in associationistic ideas maintained today by a relative handful of ingenious conditioning researchers. "Rat" psychology was oversold and, like bell-bottom jeans, so dominated the scientific marketplace for so many years that now no one will be seen dead wearing it.

The hegemony of the rat-runners was in any case doomed to transience, I think, because the fundamental impulse of psychology is not, at bottom, biological. Over 25 years ago, George Miller wrote a little book called *Psychology: The Science of Mental Life* (1962). Miller had made his name in psychoacoustics, the "hardest" of the hard-science parts of psychology, and his title was something of a challenge to the behaviorists, a call to return to the true religion of psychology: the study of the mind. This book was one of the first shots in a battle which was soon to end in the rout of the behaviorists and victory for the cognitive scientists. The advent of mechanical models for thought, in the form of digital computers, soon allowed cognitive psychologists to do what they had wanted to do all along, namely to wallow in "mental life" free of the pangs of guilt induced by the behaviorist critique.

Thus, psychology returned to concern for fundamentally human problems. Even within that arena, all was not tranquil, of course. Cognitivists, the new "experimental psychologists," upheld the hard-science tradition, aided by their mechanomorphic helpers in the world of artificial intelligence. The limitations and biases of technique edged them away from motivational, emotional, and learning problems, toward sensory, perceptual—and linguistic—problems. The digital computer as a symbol-manipulating device had a decisive effect in turning the cognitivists toward the symbolic activities of people, and thus away from the behavior of animals. Social science psychology cozied up a little to the new cognitive movement, but was most comfortable with an emphasis on methodology; perfectly balanced experimental design and the latest statistics sufficed to reassure its practitioners that they were doing "real science." And both camps sparred with the clinicians—but the less said about that, the better. None of these groups had much place for work with animals. Tradition permits a place for learning studies, and a bit of ethology, within the introductory psychology curriculum, and most cognitivists will pay lip service to the past value of animal studies in teaching us something about motivation—the effects of reward and punishment—and simple learning. Not a few, particularly those whose interests are linguistic, will deny any real relevance of work with animals to our understanding of humans, however. Few human psychologists these days have any strong allegiance to work with animals.

Meanwhile, the evolution of technique began to change the face of what used to be called "physiological psychology." Developments in electrophysiology, in pharmacology, in neuroimmunochemistry, in techniques for staining nerve tracts and in "fixing" physiological activity (e.g., the 2-deoxyglucose method) began to create a new world of neurobiology that soon outgrew its origins in psychology, physiology, and anatomy departments. Psychologist practitioners of this new "little big science" were soon lost to psychology as they left to join the swelling armies of what has collectively become known as *neuroscience*.

The remaining animal psychologists have shown two main reactions to these adverse developments, both sharply modulated by human concerns: animal-as-thinker, and animal-as-human-model. The former, which has come to prominence most recently, is of the "if you can't beat 'em, join 'em" variety: so-called *animal cognition* (see, for example, Roitblat et al., 1984). A few hardy souls have for many years pursued an interest in the limited symbolic, particularly linguistic, activities of animals—the language abilities of apes and parrots, for example (Premack, 1976; Gardner and Gardner, 1978; Rumbaugh, 1977; Pepperberg, 1981). It is impossible to summarize this work in a few sentences, but it is probably fair to say that it is impelled in large measure by the desire to show scientifically (i.e., in a way not open to the charges of anecdotalism leveled at early animal psychologists such as Romanes) that infrahuman animals "have" language—or do not have it (e.g., Terrace, 1979). The presumption is that language—human language—has certain essential features (biologists may cringe at this, remembering Ernst Mayr's lifelong criticisms of "essentialism"), the presence of which can be ascertained in the behavior of animals by sufficiently ingenious and careful experimentation.

To the modest number of ape-language people have been added in recent years a much larger number who are interested in the "cognitive abilities" of a wider range of species. Studies now abound on topics such as spatial learning (e.g., Menzel, 1978; Olton, 1978), the learning of sequences, problem solving, and the mastering of visual and numerical "concepts" and abstractions such as symmetry and transitivity, as studied in rats, pigeons, parrots, and a few other species (cf. Delius, 1985; Herrnstein et al., 1976; Pepperberg, 1987; Riley and Roitblat, 1978). Even the ethologists are in the act, beginning with a resurgence a few years ago of interest in "animal consciousness" (e.g., Griffin, 1976; Pepperberg, 1981).

It is too early to evaluate the animal cognition movement. On the positive side, it has greatly increased psychologists' appreciation for the versatility of animals, which were once thought of as nothing more than simple stimulus–response automatons. But on the negative side, it has often tied research with animals to an agenda drawn from a view of human nature that may be ill-suited to scientific inquiry. And it has also led to a needless "complexification" whereby perfectly good, mechanistic theories are termed "cognitive," simply because it is the mode. [For example, Gibbon et al. (1984) have developed a simple theory of the timing behavior of rats, according to which there is an internal clock, which can go at different rates, and some kind of accumulator, that counts the clock "ticks." Various treatments have been shown in ingenious experiments to affect the clock rate or the accumulator value at which the animal shows behavioral changes. This theory, which has all the conceptual complexity of an egg timer, is for some reason termed "cognitive."]

The anthropocentric problem can be illustrated as follows. What is the proper task for you if you are a psychologist interested in human language (or human verbal behavior, if you are a terminological purist)? Obviously you are not concerned with whether or not people "have" language; your main objective is to discover what language *is*, i.e., what it is that people "have," and how they develop it. What is it about people (and their brains and the way these brains develop) that causes them to learn this rich pattern of behavior? What are the properties of what they have learned? In other words, the language problem is to be studied in people in just the same way as an ethologist might study birdsong or any other species-typical pattern, by looking at environmental effects on development—and by looking at the properties of what had developed. With humans as subjects, of course, the second task will be much more feasible than the first—we cannot emulate King Psammeticus of times past and rear human infants in a speechless environment to see the effect on their language development. Hence, human language research is almost exclusively on what has developed, rather than on how it developed.

Unfortunately, most of the ape-language research, and more than a little research on other aspects of animal cognition, has not followed this strategy. Rather than investigating what the animals have learned, and how they came to learn it, much research has simply proceeded according to the following implicit syllogism: People have language; language has certain essential features (although there is argument about exactly what these are); apes can be taught something like language; Does their language have the same essential features as human language? If the answer is "yes," they "have" language. This is a question not devoid of interest, but it is perhaps not the most fruitful question to ask. One might as well study human mathematical ability by deciding on the essential features of, say, calculus—which would certainly be easier than defining language: Identify a subgroup of humans who have difficulties in learning calculus (not a restricted set, probably) and then, after extensive training, ask if these people "have" calculus. I believe that such an exercise would be unlikely to teach us much about people, calculus, or mathematical ability. It may well be that the ape-language exercise will be similarly unfruitful—not because the problem of language and animals is uninteresting, but because it has often been approached in the wrong way.

Other areas of animal cognition suffer from the same anthropocentrism. Rather than simply studying pigeon perception or problem-solving, for example, there is concern with whether or not pigeons "have" transitivity, self-awareness, or the perception of symmetry. The interest of these questions arises not from what we know of the behavior of birds, but from what philosophers tell us of the understanding of human beings. Past history can give us little encouragement that hypotheses derived in this way will be of much help in understanding pigeons or people. The philosopher's stone turned out to be a mare's nest. I find it hard to believe that research that aims to discover whether chimpanzees are capable of deceit or self-consciousness will teach us very much about chimpanzees—or human morality. The proper point of animal psychology is not to see if animals "have" language or any other human ability, but to find out how they do what they do—which is the same as the objective of human psychology. (I recognize that before one can find out *how* animals do something, one must first find out *if* they can do it: In order to study how they do what they do, one must first define just what it is that they do. This is indeed a legitimate reason to try and teach animals things that humans can learn. My point is that finding out if they can do something is only the first step—and there is no reason to restrict our explorations to tasks that are specifically human. Bees can see ultraviolet light and bats can hear ultrasound; these abilities were not discovered by asking if bats and bees can do human-type things.)

IV. ANIMAL MODELS

What is the proper definition of psychology? Is it "the study of thinking, feeling, and wishing," as one critic of behaviorism has argued (Bakan, 1977)? Is it "the study of behavior," as the behaviorists assert? Is it the study of "information processing"? Or "mental life"? Or simply, "the human mind"? Opinions differ, but an overwhelming majority would agree that it has *humanity* at its center. Animal psychology is for most psychologists the handmaiden of human psychology, not an activity carried on in its own right. The only question is how best to use research with animals as a guide to the psychology of people.

As we have seen, the early animal-learning theorists followed an impeccable scientific strategy: They assumed that people and animals were related in much the same way as complex chemical compounds were related to simple ones. The complex and the simple entities follow the same rules (they believed), but the rules should be easier to discover in the simpler case. Unfortunately, the rules the pioneers proposed turned out to be too simple to accurately describe even the behavior of rats. This failure was sufficient to cast doubt on the whole "pure science" strategy. Supplicant researchers, anxious to avoid the unthinking slanders of Senator Proxmire[2] and other politicians happy to pander to the anti-intellectualism of the electorate at the expense of science, cast about for ways to convince the ignorant that their research was as unworthy of a Golden Fleece Award as was a dam in Wisconsin. Aware of the apparent failure of the "pure science" approach, they sought a more persuasive rationale. Clearly, "human relevance" was the key. Medical tradition provided a vehicle in the form of the "animal model." [See Davey (1983) for a more comprehensive discussion of many of the issues discussed in this section.]

In its original form, the animal-model idea is straightforward and valid. In order to study infectious disease, it is more than helpful to be able to study its effects on a nonhuman subject. An animal model, in the form of an animal susceptible to the human virus or bacterium, enormously speeds the search for a cure. Since the causal agent of the human and animal infection is the same, the relevance of the animal results to the human case can hardly be doubted.

---

[2]Senator William Proxmire of Wisconsin has long been influential in science funding and policy. As the inventor of the "Golden Fleece Award," which is periodically given (with much publicity) to inexpensive basic science projects with funny-sounding titles (including work on ape language), he has had a major effect on the titles of funded research projects.

At the level of cells or genes, infrahuman organisms are also valid models for people, although historically much of this work was done for its own sake, rather than from expectation of imminent human benefit. No one questions the applicability to people of genetic ideas derived originally from experiments with sweet peas and fruit flies, because microanatomical and physiological studies have shown so many similarities of cellular and genetic structure and function across species, and evolutionary and developmental principles provide ample proof of continuity. The giant axon of the squid has provided useful information about nerve function in general because the laws of the action potential are essentially identical for all neurons.

When we move from the cellular to the systemic level, however, the picture clouds considerably. In the visual system, for example, there are close similarities between the receptive field structures of all primates, but cats, dogs, and pigeons differ and frogs are different from all of these. Generalizations at this level are of two kinds: evolutionary, and in terms of *principles of operation*. Evolutionary relationships allow us to make some sense of differences in neuroanatomy between closely or distantly related species, although we still know too little about function to be able to tell much of an evolutionary story there. The chief hope, therefore, is that we will be able to generalize the principles of operation we derive from comparative studies. There is some suggestion, for example, that the visual cortexes of all species are organized as a set of intercalated *maps*—maps of particular line orientations, movement directions, and other visual properties (e.g., Schwartz, 1980). This search for common principles is of course nothing but the "pure science" strategy prematurely abandoned by the early learning theorists. If principles can be generalized from species X to humans, then species X is indeed a "model" for man—but to put it in this way is really to misrepresent and diminish the endeavor. The value of species X is not just its relation to man, but its relation to all organic life.

So we have come full circle. The animal-model idea either compares two systems that are essentially identical (the animal infected by a human disease organism), or it amounts to a search for general principles—in which case the "model" notion is hardly applicable at all. If common principles are not demonstrably involved, generalizations from model to original become highly suspect. This conclusion is underlined when we move from the level of individual systems to that of whole organisms.

Take as an example one of the best-studied and most widely promoted behavioral animal models: the learned helplessness model for human depression. Martin Seligman and his associates found that dogs given severe, unavoidable electric shocks on one day in a pen failed the next day to learn to jump over a low barrier to a "safe" side—something that unshocked dogs learn at once. In a series of experimental and theoretical articles, Seligman and his associates have argued that this phenomenon is a valid model for certain kinds of human depression—for instance, the loss of affect and initiative that follows a traumatic experience such as the loss of a job or a spouse (Seligman, 1975; Peterson and Seligman, 1984). If the similarity is more than a surface resemblance, the implications may be quite important. For example, several studies (e.g., Visintainer *et al.*, 1982) have shown that uncontrollable electric shock—the cause of learned helplessness—impairs immune function (see also Chapter 7, this volume). This has lent support to correlational studies showing, for example, that lung lesions, detected by x ray, are more likely to be malignant in patients that have recently suffered depression-inducing events such as the loss of a spouse, loss of prestige, or job loss (Horne and Picard, 1979). If we are prepared to assess the benefit of a model by the Pascalian calculus of probable validity multiplied by expected benefit, then Seligman's model may well come out high, even if its probable validity is very low indeed. (Such a calculus ignores *opportunity cost,* of course—the potential benefits of opportunities foregone by supporting this line of work rather than another.)

There are clearly similarities between learned helplessness and depression: The loss of affect and initiative is similar in both, and both respond similarly to some antidepressant drugs—although there are substantial individual differences in the susceptibility of human depressives to different drugs. Yet the differences between them are almost as striking. Learned helplessness is a relative phenomenon. For example, it is hard to demonstrate in rats, unless the avoidance response is quite difficult. (Dogs show the effect even with a very easy response.) It is also transient: If the avoidance training is delayed more than 24 hr. after the unavoidable shock, most dogs have no difficulty in learning. Few human depressions, I need hardly add, are so rapidly cured by the "tincture of time." It seems clear that learned helplessness is in large measure the outcome of a positive feedback process: If the dogs fail to respond early in the avoidance period, then the repeated shocks they receive cause and maintain continued passivity. In other words, it is the history of severe and apparently inescapable—or, at least, unescaped—shocks that causes the effect. There is nothing comparable to this history of repeated severe punishment in most of the human depressives for whom helplessness is used as a model. And the helplessness model is of no help in understanding why certain events, such as loss of a spouse, cause crippling depression in one person but not in another.

The enormous role of individual differences in human behavior is something that should make us especially hesitant about using animal models. For example, every introductory textbook repeats B. F. Skinner's explanation of casino gambling as an example of "variable-ratio reinforcement." While a one-armed bandit is undoubtedly formally equivalent to a variable-ratio schedule (both dispense rewards on a probabilistic basis: The more you pull, the more you get, in strict proportion), Skinner's explanation leaves out of account the most interesting aspect of compulsive gambling, namely the fact that only a minority of people are susceptible to it. Everyone knows that *for these people* the torrent of coins when they win is a very powerful reinforcer. The interesting question is why this should be so: In what way are these people different from the rest? The schedule analogy is far from helpful on this point.

Human individual differences are often so large that they may be compared with species differences in behavior. Consider, for example, Harlow's famous experiments [first reported in a paper modestly entitled "The Nature of Love" (Harlow, 1958)—such are the ingredients of fame in the social sciences] on the development of attachment and social behavior in macaque monkeys. In these experiments, baby macaque monkeys were reared apart from their natural mothers. Some of the monkey infants were provided with artificial "mothers" made of terry cloth. For others, the artificial mothers were made of something less cuddly. Some of the artificial mothers dispensed milk, others did not. Harlow found that despite the food reward offered by the milk-giving mothers, the cuddly mothers were preferred—the infant monkeys rushed to these when faced with something novel or frightening, just as they would to their real mothers. In other experiments, Harlow and his associates found that monkey infants reared apart from their mothers grew up with severely distorted social behavior (Harlow, 1958). This work was taken to show the importance of physical contact to normal human upbringing. The extension of these monkey studies to humans was inherently plausible—who can doubt the value of a hug?—but as Daniel Lehrman once pointed out, the validity of the extension depends much more on its inherent plausibility than on the method used. If Harlow had studied any one of a number of other monkey species, for example, he would have found very different results.

One may suspect large individual differences in human susceptibility to the kind of isolation Harlow imposed on his monkeys. I always remember Bertrand Russell's vivid account of his loveless upbringing. His parents having died when he was only two, he was brought up in the house of his cold and distant grandmother. The general flavor of the regimen to which he was exposed is conveyed by the procedure used to dispel young Bertie's natural fear of drowning: The fledgling philosopher was held by a footman upside down with his head underwater in a barrel, until he stopped struggling. Russell recalls that the treatment worked—he was not thereafter afraid of the water. This method would probably not commend itself to modern theories of child rearing. While no one could call Russell "normal," in a normative sense, it is not clear that he was severely impaired by the unfeeling treatment he received as a child—although others might have been.

V. CONCLUSION

Much of animal psychology is good science as well, but there is also much that appears unconvincing to scientists in other disciplines. Harlow's experiments, widely honored in the psychological community, are questioned by biologists; Seligman's work on depression, equally lauded and fascinating to psychologists and medical people, is not part of most ethology courses; the work of B. F. Skinner, the one animal psychologist of whom almost everyone has heard, finds no place in most biology curricula. Psychologists of eminence are gifted at whipping up rich confections of experiment and theory that fascinate as much by their allusions to human experience as by the power of their principles or their firm base in evolutionary biology.

I suggest that the fundamental culprit in the equivocal status of animal psychology is the myopically *anthropocentric* nature of psychology. Every other science has had to overcome man's (and I do not exclude woman's) tendency to see everything in terms of himself. Astronomy advanced only when the earth ceased to be the center of the universe; it advanced further when the sun was similarly displaced. Physics had to get over the idea that the microworld looks just like the macroworld, that everything can be explained in terms that apply to bricks and tables and rocks, things that we can touch and see. And biology, of course, could not proceed until humanity was understood on all fours with subhuman animals—and plants—in the all-embracing worldview of Darwinian evolution. Modern psychology, especially of the behaviorist variety, has been very sensitive about anthropomorphism—the explanation of animal behavior in human terms. But it has accepted unthinkingly the Popesian credo that *man* is its proper object. It is the implicit subordination of psychological research with animals to limited human ends that has led to the parade of half truths and weak analogies of which I have given a tiny sample.

The solution, I believe, is to abandon the idea that psychology (at least in its nonapplied aspects) is the study of the human mind, or human behavior, or human thinking, feelings, etc., indeed of specifically *human* anything. Psychology should be the study of *intelligence,* of adaptive and complex behavior, wherever it is to be found—in animals, people, or even machines—informed throughout by the principles of evolution of which people, as much as animals, are the fruit. I believe we shall find (as several others have recently proposed) that there are principles of intelligence that reflect properties of nature that transcend any particular species, but reflect properties of the world common to all ecological niches. We seem to be finding that the properties of perception reflect the geometry of space, and the solidity of objects, that the properties of associative learning reflect the Humean principles of causality, and that properties of human problem-solving reflect the costs and benefits of simple social contracts (see, e.g., Cosmides and Tooby, 1987; Shepard, 1987; Staddon, 1987). We understand these natural constraints all the better because our computation machines are not subject to them. Thus, by "enlarging the inquiry," as Bacon suggests, we may truly comprehend all things, human and nonhuman.

VI. ACKNOWLEDGMENTS

I thank the editors of this volume and Peter Holland, Kimberly Kirby, and Clive Wynne for comments on an earlier version. I am also grateful to the National Science Foundation for research support over many years.

VII. REFERENCES

Bakan, D. (1977). Political factors in the development of American psychology. *Ann. NY Acad. Sci.* **291**:222–232.
Cosmides, L. and Tooby, J. (1987). From evolution to behavior: Evolutionary psychology as the missing link. In Dupré, J. (ed.), *The Latest on the Best: Essays on Evolution and Optimality,* Bradford Books, Cambridge, Massachusetts, pp. 277–306.
Davey, G. C. L. (ed.) (1983). *Animal Models of Human Behaviour,* Wiley, Chichester.
Delius, J. D. (1985). Cognitive processes in pigeons. In d'Ydewalle, G. (ed.), *Cognitive Information Processing and Motivation,* Elsevier, Amsterdam, pp. 3–18.
Domjan, M. (1987). Animal learning comes of age. *Am. Psychol.* **42**:556–564.
Gardner, R. A., and Gardner, B. T. (1978). Comparative psychology and language acquisition. *Ann. NY Acad. Sci.* **309**: 37–76.
Gibbon, J., Church, R. M., & Meck, W. H. (1984). Scalar timing in memory. *Ann. NY Acad. Sci.* **423**:52–77.
Griffin, D. R. (1976). *The Question of Animal Awareness,* Rockefeller University Press, New York.
Harlow, H. F. (1958). The nature of love. *Am. Psychol.* **13**:673–685.
Herrnstein, R. J., Loveland, D. H., and Cable, C. (1976). Natural concepts in pigeons. *J. Exp. Psychol.: Anim. Behav. Proc.* **2**:285–302.
Horne, R. L., and Picard, R. S. (1979). Psychosocial risk factors for lung cancer. *Psych. Med.* **41**:503–514.
Hull, C. L. (1943). *Principles of Behavior,* Appleton-Century, New York.
Menzel, E. (1978). Cognitive mapping in chimpanzees. In Hulse, S. H., Fowler, H., and Honig, W. K. (eds.), *Cognitive Processes in Animal Behavior,* Erlbaum, Hillsdale, New Jersey, pp. 375–422.
Miller, G. A. (1962). *Psychology: The Science of Mental Life,* Harper & Row, New York.
Olton, D. S. (1978). Characteristics of spatial memory. In Hulse, S. H., Fowler, H., and Honig, W. K. (eds.), *Cognitive Processes in Animal Behavior,* Erlbaum, Hillsdale, New Jersey, pp. 341–373.
Pepperberg, I. (1981). Functional vocalizations of an African grey parrot (*Psittacus erithacus*). *Z. Tierpsychol.* **55**:139–151.
Pepperberg, I. (1987). Evidence for conceptual quantitative abilities in the African grey parrot: labeling of cardinal sets. *Ethology,* **75**:37–61.
Peterson, C., and Seligman, M. E. P. (1984). Causal explanations as a risk factor for depression: theory and evidence. *Psychol. Rev.* **91**:347–374.
Premack, D. (1976). *Intelligence in Ape and Man,* Erlbaum, Hillsdale, New Jersey.
Riley, D. A., and Roitblat, H. L. (1978). Selective attention and related processes in pigeons. In Hulse, S. H., Fowler, H., and Honig, W. K. (eds.), *Cognitive Processes in Animal Behavior,* Erlbaum, Hillsdale, New Jersey, pp. 249–276.

Roitblat, H., Bever, T., and Terrace, H. (eds.) (1984). *Animal Cognition*, Hillsdale, Erlbaum, Hillsdale, New Jersey.

Rumbaugh, D. M. (ed.) (1977). *Language Learning by a Chimpanzee: The Lana Project*, Academic Press, New York.

Schwartz, E. L. (1980). Computational anatomy and functional architecture of striate cortex: a spatial mapping approach to perceptual coding. *Vis. Res.* **20**:645–669.

Seligman, M. E. P. (1975). *Helplessness: On Depression, Development, and Death*, Freeman, San Francisco.

Shepard, R. N. (1987). Evolution of a mesh between principles of the mind and regularities of the world. In Dupré, J. (ed.), *The Latest on the Best: Essays on Evolution and Optimality*, Bradford Books, Cambridge, Massachusetts, pp. 251–275.

Skinner, B. F. (1966). The phylogeny and ontogeny of behavior. *Science* **153**:1205–1213.

Staddon, J. E. R. (1987). Optimality theory and behavior. In Dupré, J. (ed.), *The Latest on the Best: Essays on Evolution and Optimality*, Bradford Books, Cambridge, Massachusetts, pp. 179–198.

Terrace, H. S. (1979). *Nim*, Knopf, New York.

Visintainer, M., Volpicelli, J. R., and Seligman, M. E. P. (1982). Tumor rejection in rats after inescapable or escapable shock. *Science* **216**:437–439.

Chapter 6

RECOGNITION LEARNING IN BIRDS

Milton D. Suboski

Department of Psychology
Queen's University
Kingston, Ontario K7L 3N6, Canada

I. ABSTRACT

When a naive bird learns to recognize food, reciprocal interactions between learner and experienced conspecific (usually a parent) are often involved. While much food-recognition learning is consistent with the operation of traditional conditioning processes, evidence indicates the function of a more general mechanism, termed "released-image recognition learning." In the released-image model, phylogenetically "recognized" invariant stimuli (releasing stimuli) mediate recognition learning of the characteristics of variable stimuli. The occurrence of a particular primary releasing stimulus results in the selective narrowing of an initially large class of potential releasers. The potential releasers are originally either weak releasers or behaviorally inert stimuli. Released-image recognition learning occurs when a subset of the generic class acquires releasing power or "valence" as a result of a phylogenetically adaptive temporal, spatial, or dynamic relationship with the original releaser. Stimuli that acquire releasing valence thereby become capable of eliciting and directing the same behavior pattern as that evoked by the original releaser. It appears that the released-image recognition model may account for individuals' learning to recognize food, prey, noxious stimulus, predator, parent, egg, offspring, mate, and individual conspecific. It may also account for filial and sexual imprinting, social learning, Pavlovian conditioning, and simple forms of instrumental conditioning.

II. INTRODUCTION: PURPOSE AND SCOPE

The first of the two major parts of this review is concerned with the acquisition of food recognition by birds considered as a prototypical system whereby stimulus recognition can occur. Complex releasing-stimulus–released-response interactions between parent and offspring that bring the offspring into direct contact with food are described. Such interactions may permit the operation of classical (Pavlovian) and instrumental conditioning processes so that a permanent change is effected in food recognition by the young bird. Mechanisms for social transmission of food recognition that indirectly lead the naive bird into contact with food are also described. In spite of their diversity, these mechanisms do not completely account for the acquisition of food recognition. This is shown by a review of evidence that Pavlovian and instrumental processes are not always either necessary or sufficient to produce food recognition. The failure of the traditional conditioning paradigms to adequately account for the acquisition of food recognition, as documented in the first major section of this review, led to the development of the "released-image recognition" model that is proposed in the second section. This model incorporates classical ethological concepts to yield a recognition learning model that includes the Pavlovian conditioning, imprinting, and social learning models as special cases. Evidence then considered indicates that the released-image recognition model provides a mechanism that organizes and integrates a substantial number of learning and recognition phenomena.

In general, recognition of environmental objects can occur in two quite different ways. The stimulus characteristics of objects may be phylogenetically preprogrammed as releasing stimuli for an animal, with phylogenetically adapted motor acts as the released behavior. Alternatively, stimulus recognition may be acquired through experience. These recognition modes are not mutually exclusive, since they can operate separately or simultaneously. Although phylogenetically preprogrammed recognition of stimulus objects probably occurs to some extent in all birds and may be the most important mechanism for some birds, it is likely to be the sole mechanism for very few birds. This review is primarily concerned with mechanisms alternative to unlearned recognition: those mechanisms that underlie the experiential development of stimulus recognition.

III. THE ACQUISITION OF FOOD RECOGNITION

Food recognition is considered initially and at length for several reasons in addition to its role in stimulating development of the released-image recognition model. First, food is often not initially recognized by birds (however, cf. Mueller, 1974) and starvation in the presence of ample but unrecognized food is not uncommon (Hogan, 1971; Rabinowitch, 1968; Shreck et al., 1963). Another reason is that food recognition is immediately essential to survival and thus may result from prototypic recognition processes. Third, food recognition should be relatively simple to explain, since it is usually assumed to be easily accommodated within the framework of Pavlovian conditioning and instrumental learning. Thus, at least superficially, acquisition of food recognition does not appear to be a serious problem. Only rarely (e.g., Rozin, 1976) has it been considered a matter of importance. Presumably, all that is required is for a bird to be brought into sufficiently close contact with food so that eating occurs. Then incentive and reinforcement mechanisms can be expected to strengthen the response-eliciting ability of food-associated stimuli and food recognition will be learned as indicated by conditional responses. Furthermore, ultimate food preferences appear to be heavily dependent upon shaping by instrumental conditioning processes. Nevertheless, despite the apparent simplicity, complex releasing-stimulus–released-response interactions between parents and young are required in order that food recognition may occur by means of simple conditioning mechanisms. Moreover, as discussed in Section III.C, unlearned factors and conditioning processes do not provide a complete account of the acquisition of food recognition by birds.

A. Food Recognition from Parent–Offspring Interactions

Parent–offspring vocal interactions begin before hatching in precocial species (e.g., Shapiro, 1981; Vince, 1969). Prehatch vocal interactions have been found to relate to hatching synchrony (Shapiro, 1981; Vince, 1969), subsequent reciprocal species recognition (Bailey, 1983; Beer, 1966; White and Guadalupe del Rio Pasado, 1983), postnatal feeding behavior (Impekoven, 1971), and perhaps other factors. As an example, Nelson (1965) noted that prenatal auditory stimuli may stimulate the brooding gannet (*Sula bassana*) to move its egg from below to above its webbed feet just prior to hatch. In addition, Templeton (1983) suggested that vocalizations of herring gull (*Larus argentatus*) chicks just before hatching serve as a releasing stimulus to the parents so that they shift from incubation to brooding and feeding (see also Impekoven, 1973). Failure of adult herring gulls to make an adequate transition from incubation of eggs to care of young is cited as the major cause of chick mortality (Kadlec et al., 1969).

Vocalization and perhaps visual features of the hatchlings appear to release food gathering by adults. For example, Harris (1983) reported that increasing the food-begging calls of puffin (*Fratercula arctica*) chicks increased parental trips to the nest burrow. Miller and Conover (1979, 1983) also found chick vocalizations critical to the instigation of parental feeding of ring-billed gull (*Larus delawarensis*) chicks. Chicks muted early were not fed well, with some 50% dying because parents either ignored or actively rejected them (Miller and Conover, 1979). Interestingly, blindfolding chicks had no comparable effect (Miller and Conover, 1979), suggesting that pecking at the parent's bill by the chick was unnecessary to the occurrence of parental feeding. This result could be just another example of biological redundancy since bill pecking by gull chicks appears to be an effective releaser of parental feeding (e.g., N. Tinbergen, 1960). Another possibility will be discussed in Section VE3: that chick vocalizations are essential to acquisition of chick recognition by parents.

Parental feeding of young certainly requires a close harmony of releasing-stimulus–released-response interactions between parent and young. Three well-studied examples may be found: gulls (e.g., Hailman, 1967; Miller and Conover, 1979, 1983; N. Tinbergen, 1960), oystercatchers (*Haematopus ostralegus*) (e.g., Hørlyk and Lind, 1978; Lind, 1965; Norton-Griffiths, 1969), and the chicken (*Gallus gallus spadiceus*) (e.g., Sherry, 1977; Stokes, 1971). Although many details differ among the species, a common pattern may be described. In gulls, the presence of the parent (touch, a parent-sized object) initiates food-seeking behavior by the young (Miller and Conover, 1983). In the first instance these are food-begging vocalizations followed, in gulls and oystercatchers, by pecks directed toward the parent's bill. The visual characteristics of stimuli that release bill pecking in gulls have been extensively investigated (e.g., Hailman, 1967; N. Tinbergen, 1960). Oystercatchers also peck at the lower end of objects that are roughly bill-shaped, particularly if a contrasting spot is located there (Hørlyk and Lind, 1978). Pecking at the parent's bill appears common among bird species that engage in parental feeding. These include arctic terns (*Sterna macrura*) (Quine and Cullen, 1964) and ring doves (*Streptopelia risoria*) (Wortis, 1969). Turner (1964) found that newly hatched domestic chickens (*Gallus gallus*) also selectively pecked at angular objects. Again, such pecking was enhanced if a contrasting spot was located at the apex of the angle and/or if the angle was moved up and down in pecking movements.

In the gull, pecking at the parent's bill apparently stimulates regurgitation of food for the young. In the oystercatcher, parents hold food at the tip of the beak so that pecking by the chick will usually yield contact with the food. If, as frequently happens initially, the chick fails to peck or fails to obtain the food, the food is dropped and the parent "points" to it by holding the bill in a vertical position over the food. If this also fails to stimulate eating by the chick, the parent may pick up the food and repeat the sequence (Lind, 1965; Norton-Griffiths, 1969). In the Burmese red junglefowl (Stokes, 1971) and domestic chicken (personal observation), chicks often peck at objects held in the hen's beak. Stokes (1971) also reported that junglefowl hens would pick up, drop, and hold out food for chicks to peck at. According to Hogan (1966), food otherwise avoided by junglefowl chicks is readily accepted if presented by the hen. The parental food call is an important feature of feeding behavior, particularly in species (termed "nidifugous") that leave the nest early. Oystercatchers (Norton-Griffiths, 1969) and laughing gulls (*Larus atricilla*) (Beer, 1970b) have two different feeding calls depending on how far away the chick is. Just as for the junglefowl, the food call attracts the chicks to the vicinity of parentally identified food.

It is important to note that releasing-stimulus–released-response interactions involve both activation and inhibition of behavior. Released inhibitory processes are not uncommon in food recognition, as well in as other forms of recognition. Thus, the presence of chicks can inhibit immediate eating of food by a parent (Norton-Griffiths, 1969; Stokes, 1971). Avoidance of intraspecies predation by raptors and other predators often seems to occur because of inhibition released by species-specific features (e.g., Cushing, 1944). Functioning of released inhibitory processes is shown by observations of Yom-Tov (1976) on carrion crows (*Corvus corone*). Eggs and nestlings of various species placed in crows' nests were neither eaten nor ejected by the crows. The same objects—including crows' eggs, but not a crow nestling—placed nearby outside of the nest were promptly seized. Obviously, the stimulus complex provided by the nest strongly releases inhibition against eating any objects within it.

The interactions discussed so far are undoubtedly far more complex in reality than is indicated by the present summary. Furthermore, other mechanisms may be found whereby parental behavior leads to pairing of sensory (primarily visual) and reinforcing properties of food. Noteworthy is Wortis' (1969) observation that ring dove parents change their response to food begging by squabs as the squabs approach fledging. The parents switch from regurgitation feeding to pecking at grain, with the result that food-begging pecks directed at the parents' beaks bring the squab into oral contact with the grain. Parents often alter the diet fed to young during the period from hatch to fledging so that by fledging the young are receiving the adult diet. The crop milk of ring doves is reputed to contain increasing amounts of undigested grain as the squabs approach fledging (Wortis, 1969). Bond (1936) reported that falcon (genus *Falco*) parents feed their young less and less well-prepared game as they approach fledging. Initially the young receive small pieces of plucked or skinned game; by fledging they are presented with the entire dead animal. British finches (*Pyrrhula pyrrhula*) (Newton, 1967) and European jays (*Garrulus glanderius*) (Bossema, 1979) feed hatchlings primarily with invertebrates, whereas by fledging their diet is mostly seeds (for finches) or acorns (for jays), as is the adult diet.

Lorenz (e.g., 1965) emphasized the importance of released responses by parents that bring the young into direct contact with food, describing these responses as "innate teaching mechanisms" for the acquisition of food recognition. This description is somewhat incomplete since "innate learning mechanisms" (Ewer, 1969)—that is, reciprocal released responses by the young—are equally necessary and, in the final analysis, the acquisition of food recognition still depends on conditioning processes for explanation.

B. Social Transmission of Food Recognition

The releasing-stimulus–released-response interactions which induce direct contact with food are hypothesized (Thorpe, 1963) to be largely composed of three mechanisms for social transmission of food recognition. These are social facilitation, local enhancement, and imitation (Thorpe, 1963). Social facilitation, "contagious behavior," occurs when the behavior of one bird releases the same behavior in an observing companion. Local enhancement was defined by Thorpe (1963) as directing an animal's attention to a particular object or part of the environment. Thorpe reserved the term "imitation" to an active or intentional copying by an observer of demonstrated behavior.

1. Social Facilitation

There is no shortage of examples of social facilitation in birds. Feeding releases feeding (e.g., Evans and Patterson, 1971), flight releases flight (e.g., Crook, 1960; Evans and Patterson, 1971), aggression releases aggression (e.g., Altmann, 1956; Crook, 1960), preening prompts preening (e.g., Evans and Patterson, 1971; Moynihan and Hall, 1953), and copulation begets copulation (e.g., Gochfeld, 1980; Southern, 1974). At least some socially facilitated responses are considered to constitute a class of released behaviors that have the defining characteristic of topographical similarity between releasing stimulus and released response. However, it is not necessarily the case that behavioral similarity particularly distinguishes social facilitation. As is the case for many releasing stimuli, the critical feature of the releasing stimulus may be a limited aspect of the stimulus complex. A tapping finger (Collias, 1952) or pencil (Tolman, 1964), a crude "pecking" model of a hen (Turner, 1964), and a "pecking" arrow (Strobel and Macdonald, 1974; Suboski, 1984) have all served to release pecking by hatchling chickens. Thus, social facilitation appears to be an arbitrary category of behavior referring to cases where the natural releaser happens to be topographically similar to the released response rather than constituting a functionally different class.

Social facilitation can lead, indirectly, to the development of food recognition. If the releasing stimulus for feeding arises from the behavior of a parent or other experienced companion, then feeding will tend to occur in the presence of food (see, e.g., Hailman, 1967). The released pecking will be further directed (taxic) by the releasing properties of foodlike objects. Such objects may in fact be, as a result of the experience of the companion, food. Again, reinforcement processes which follow ingestion of that food are presumably required for the occurrence of permanent changes in food recognition. As Klopfer (1959, 1961) pointed out, a distinguishing characteristic of social facilitation is the requirement for the continued presence of an appropriately behaving companion. Released behavior usually occurs only in immediate response to releasing stimuli and ceases in the absence of the releasing stimulus unless some other process intervenes.

2. Local Enhancement

Modification of the definition of local enhancement by replacing "attention" with "behavior" does no harm and gains some objectivity. Domestic chicken (Stokes, 1971; personal observation) and willow grouse (*Lagopus l. lagopus*) (Allen et al., 1977) chicks tend to peck at the same objects as those pecked by the hen. Two or more chicks also tend to peck at the same thing (Brown, 1975; Frank and Meyer, 1970; Hailman, 1967; Tolman, 1964).

Local enhancement is a form of social facilitation wherein matching behavior is directed to the *same* stimulus by demonstrator and observer. It is important to note that if observer and demonstrator direct identical behavior to *matching* stimuli, then the phenomenon exemplified is considerably more complex than simple local enhancement—whether or not observer and demonstrator are simultaneously present. Local enhancement can be quite powerful. Brown (1975), Frank and Meyer (1970), and Meyer and Frank (1970) trained groups of chicks to peck one of two distinctive stimuli using food reinforcement. Hundreds of pecks to that stimulus were recorded in subsequent unrewarded tests with groups of chicks, whereas chicks tested individually quickly ceased to respond. However, as with social facilitation, local enhancement does not contain a mechanism for the acquisition of a permanent change in food recognition. Only if ingestion of food results from responding to the same environmental feature as responded to by an experienced companion will food recognition occur, again presumably by means of reinforcement processes. Finally, the distinction between social facilitation and local enhancement is probably worth maintaining. If so, the use of the term "local enhancement" to refer to indiscriminate aggregative behavior (e.g., Waite, 1981) at best obscures the distinction.

3. Imitation

Thorpe's (1963) stringent definition restricts imitation to active copying of novel or improbable acts and eliminates any behavior for which an instinctive tendency exists, thus virtually assuring that imitation so defined will not easily be found in infrahumans. The definition almost certainly guarantees that all behaviors that can possibly be classified as social facilitation or local enhancement will be so classified. This may be useful in a search for an unequivocal example of imitation, but is of little help in distinguishing between socially released responses and imitation when either might be occurring. Furthermore, the definition of imitation does not contain an explicit construct for producing permanent behavior change. Klopfer (1961) implied such a construct by asserting that imitation is differentiated from social facilitation by persistence of the effects of observation, but failed to specify the mechanism by which such changes might be rendered permanent. The prospects for finding bird behavior that meets both Thorpe's and Klopfer's specifications seem somewhat remote. The issue is certainly not closed. Evidence that birds may actively copy behavior of conspecifics solely as a result of observation of that behavior can be found (e.g., Palameta and Lefebvre, 1985). Nevertheless, imitation, as defined by Thorpe and Klopfer, does not yet provide a viable mechanism for the acquisition of food recognition.

C. Food Recognition without Reinforcement

The foregoing overview found no lack of mechanisms whereby naive birds may acquire the ability to recognize food, including mechanisms by which food recognition may be transmitted from an experienced to an inexperienced conspecific. All of these mechanisms have one common feature. Each brings the naive bird into contact with food and then depends upon conditioning and/or reinforcement processes to transform that experience into permanent food recognition. In spite of the diversity of such mechanisms among and within species, there is evidence that conditioning and reinforcement processes are neither necessary to, nor directly involved in, at least some forms of food recognition learning.

Hess (1964) first noted that reinforcement with food had little effect on the direction of pecking by domestic chicks within the first 2 days posthatch. In an extensive series of studies, Hogan (summarized in Hogan, 1977) showed that 1- and 2-day-old chicks pecked indiscriminately at food and sand and that such experience did not modify pecking preferences (see also Hogan-Warburg and Hogan, 1981; Hale and Green, 1979). Only by the third day of life did experience with pecking at food lead to selectively increased pecking of food. Paradoxically, then, reinforcement by ingestion of food, the archetypical reinforcer, is not effective in modifying initial pecking preferences in chicks.

Recent evidence (Bartashunas and Suboski, 1984; Suboski, 1984, 1987; Suboski and Bartashunas, 1984) demonstrated that initial acquisition of food recognition in neonatal chicks does not involve reinforcement from ingestion of food or any other environmental change. Our work was based on Turner's (1964) report of a simple, yet elegant, method of studying the development of food recognition in young chicks. Turner hand-operated a crude outline of a hen on a pivot to make pecking movements. A red or green colored wheat grain was glued to the floor at the tip of the "hen's" beak. Separated from the "hen" by wire netting was a series of three red grains alternated in a row with three green grains, which were available to a hatchling chick. These chicks made significantly more pecks to the grain of the color that matched the grain being pecked by the "hen" than to the other colored grain.

We (Suboski and Bartashunas, 1984) replicated Turner's basic results using a simple black arrow, motor-driven to make "pecking" movements. The results were quite clear: Chicks given brief (20 presentations of 10 sec each) exposure to the "pecking" arrow rapidly developed a strong preference for pecking at stimuli that matched in color those "pecked" by the arrow. The stimuli "pecked" by the arrow were colored plastic pinheads attached to the floor at the tip of the arrow or attached to the tip of the arrow. Minature lamps which emitted either red or green light were also effective in place of the pinheads, despite random switching of the location of the lamp color which matched the arrow-"pecked" color. The phenomenon occurred in unfed chicks as young as 9 hr posthatch, with stimulus presentations started 2 min after initial removal from the unilluminated, noise-masked incubator. Pecking by chicks that was selectively directed by the arrow was found to develop over trials, persisted after cessation of arrow movement (even when the colored pinheads were removed from the arrow), and required for acquisition only visual exposure to the display of the arrow "pecking" at the pinheads. The altered preference appeared permanent. The color preference was little diminished 3 days later when the chicks were tested with the arrow "pecking" but without any associated pinheads.

The phenomenon underlying arrow-assigned color preferences by chicks undoubtedly functions as a mechanism for social transmission of food recognition from hen to chicks (cf. Hogan, 1973). It provides a means by which a hen can rapidly transmit information about the visual characteristics of available food to her chicks. The resultant acquisition of food recognition by the chicks does not depend upon the instrumental consequences of pecking by the chicks. In our experiments the stimuli available to the chicks were firmly attached to the floor. Thus, pecking by the chicks produced absolutely no environmental change, positive or negative, and was in fact irrelevant to the development of arrow-directed pecking. Consequently, instrumental reinforcement processes were unlikely to have been involved. In fact, none of the known characteristics of reinforcement, in either the Pavlovian or instrumental conditioning senses, provide any reason why chicks should develop a permanent pecking preference for a class of stimuli as a result of watching an arrow "peck" at one such stimulus.

Furthermore, neither social facilitation nor local enhancement is an adequate explanation for arrow-assigned color preferences in pecking by chicks. Social facilitation fails on a number of accounts. First, a "pecking" arrow is not a conspecific companion (but may, nevertheless, provide the essential releasing-stimulus features of a conspecific). Second, the enhancement of stimulus direction of pecking by chicks induced by arrow movement persists after the removal of demonstration conditions. Pecking by the chicks continues to be directed to a specific target stimulus even when all stimuli have been removed from the arrow. "Social" facilitation can only predict taxic pecking by chicks that is directed to both colored stimuli, unless reinforcing consequences have followed such pecking—clearly this was not the case in the present example.

Neither can the phenomenon be satisfactorily considered as local enhancement. Besides the remarkable persistence of the change in pecking preferences, the chicks do not peck at the *same* thing at which the arrow "pecks" (although they often try). Rather, the chicks peck at stimuli which *match* the one "pecked" by the arrow. To view pecking at matching objects by observer and "demonstrator" as stimulus generalization may have some appeal but simply begs the question. An explanation in terms of stimulus generalization presupposes that there is some learning about the object being "pecked" by the arrow to be generalized to other stimuli, but such learning, after all, is the very phenomenon whose explanation is being sought. Pecking by chicks at the same time and at the same stimulus that the arrow is "pecking" is subtly but profoundly different from selectively pecking at a visually matching stimulus 3 days later when the arrow no longer has any associated stimuli (Bartashunas and Suboski, 1984).

A case could be made for viewing arrow-assigned pecking as a form of imitation. Certainly the effect of the mechanism is that chicks duplicate the pecking preferences of the hen. Furthermore, some kind of active observing and perceptual matching process on the part of the chicks is required in order that arrow-assigned response tendencies be generated. Nevertheless, it appears to be extremely unlikely that chicks actively copy the "pecking" movements of the arrow, and the "cognitive" process involved is probably rather primitive. Unsurprisingly, this phenomenon does not meet all of Thorpe's (1963) criteria for imitation since it involves an instinctive response lacking novelty and intentionality, although it does meet Klopfer's (1961) criterion of permanent behavior change.

IV. RELEASED-IMAGE RECOGNITION

A new approach to the understanding of basic recognition processes emerges from a functional analysis of arrow-assigned pecking. The "pecking" arrow, like other vertically oscillating pointed stimuli (Collias, 1952; Strobel and MacDonald, 1974; Tolman, 1964; Turner, 1964), is a releaser of pecking by chicks. The released pecking is not totally indiscriminate but rather is further directed (taxic) toward small, high-contrast objects with a relatively low environmental frequency. In addition, stimuli matching those placed on or under the tip of the arrow receive enhanced releasing power or increased positive valence from movement of the arrow—such stimuli are pecked by chicks in preference to otherwise equally preferred stimuli. Releasing valence from the "pecking" arrow may thus be assigned to stimuli that are in a particular spatial relationship to a particular dynamic stimulus. Pecking responses of chicks that were previously released by arrow movement came to be released as well by the stimulus that had been "pecked" by the arrow. The pecking arrow not only released pecking behavior by the chicks but also activated a phylogenetically preprogrammed mechanism for stimulus-recognition learning.

A. The Released-Image Recognition Model

Figure 1 displays a model which summarizes the released-image mechanism of food recognition by neonatal chicks. In the released-image model, valence to release particular behaviors is originally possessed only by primary releasing stimuli. The releasing valence from some stimuli, however, can be assigned to and shared with other environmental stimuli that may initially lack any appreciable control over behavior. The upper part of Fig. 1 illustrates the main effect of a

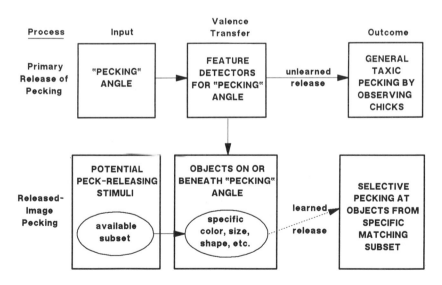

Fig. 1. Model of the development of released-image pecking by neonatal chicks. The upper part of the figure summarizes the primary releasing mechanism for taxic pecking. A second effect of the releaser in assigning releasing valence to stimuli that have particular spatial relationships to the moving releaser (angular stimulus) is diagrammed at the bottom of the figure, where the outcome is shown: learned release (dashed line) of pecking at specific stimuli. (See text for further details.)

vertically "pecking" angular stimulus, a primary releaser of pecking behavior by neonatal chicks. Most releasing stimuli elicit either autotaxic responses (directed at the releaser) or antitaxic responses (directed away from the releaser), whereas a "pecking" angle releases allotoxic responses (directed toward some other stimulus). In the present instance, released pecks by chicks are sometimes directed toward the tip of the "pecking" angle, but are usually directed toward small foodlike objects that are themselves releasers of autotaxic pecking.

The valence-transferring function of a "pecking" stimulus for neonatal chicks is diagrammed in the lower section of Fig. 1. In addition to the release of allotaxic pecking by chicks, a "pecking" angle activates (or primes) a second set of feature detectors generic for an image (or template) conditional on a second stimulus input. The conditional image is initially relatively nonspecific and may be satisfied by various environmental stimuli. In the case of pecking by chicks, the conditional image produced by a "pecking" angle is for foodlike objects located on or near the tip of the moving angle. Any available environmental stimulus that activates some subset of the primed feature detectors in the conditional image set may "satisfy" the conditional image and thereby generate a "specific assigned image" for those visual features. Subsequent matching stimuli encountered by the chick function as specific assigned-image releasers. The bottom part of Fig. 1 diagrams the final outcome: Initially unpreferred objects acquire an enhanced capacity to release pecking, the same response as was originally elicited by the "pecking" angle. Both primary and acquired-valence releasers come to share the power to release the same phylogenetically preorganized responses.

The released-image recognition model shown in Fig. 1 is essentially an effort to provide an abstract description of the outcome of a process whereby newly hatched chicks learn to recognize food. However, acquisition of food recognition may be but one instance of a general recognition-learning mechanism that transforms generic classes of potential releasers into specific instances of learned releasers. Schöne (1964) first viewed learning as a transfer of releasing ability from one stimulus to another. The possibility that a phylogenetically preorganized learning process may be initiated by releasing stimuli was later discussed by Gould and Gould (1981). A concept of Pavlovian conditioning as a process of learned release was suggested by Segal (1972) and expanded by Woodruff and Williams (1976), Woodruff and Starr (1978), Jenkins et al. (1978), and Timberlake (1983). Figure 2 presents a generalized released-image recognition model. Apart from the substitution of general terms for specific labels, the model contains several additional features.

First, unlearned and learned released responses are shown in Fig. 2 to be identical. The primary releaser and specific assigned image come to share valence, i.e., to elicit the same released responses. Nevertheless, responses to original and valence-acquired releasing stimuli will not necessarily be topograph-

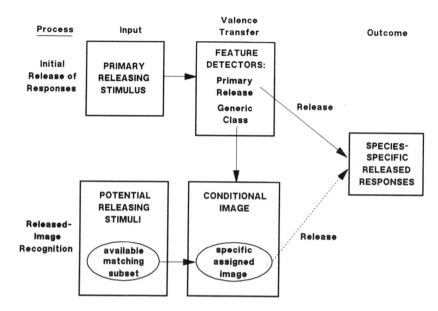

Fig. 2. General model of released-image recognition. Interpretation of the model parallels that for released-image pecking.

ically identical. Original releasers may have peripheral effects that are not mediated centrally and therefore are not part of the released responses for which releasing valence may be transferred. Another consideration is that releasing valence, and not specific components of the released-response sequence, is transferred. If the initial releaser is an isolated terminal element (e.g., food in the mouth) in a phylogenetically organized releasing-stimulus–released-response sequence, then the specific assigned image generated by that releaser may acquire valence to release the entire behavioral sequence of approach and consummatory contact. In addition, taxic direction of released responses is an integral part of transferred releasing valence. Thus, taxic releasing valence may be transferred to a diffuse stimulus that nonetheless cannot elicit the behavior, for want of a focal stimulus to direct the behavior (see, e.g., Rescorla, 1980).

Second, it is unlikely that conditional images are satisfied only by stimuli with preexisting releasing valence. Although this was the case for the present instance of pecking by chicks, where releasing valence was enhanced rather than created, it is not likely to be a general requirement. A third point may be noted here. For the sake of simplicity, the released-image model has up until now been discussed as functioning to transfer releasing valence only from original releasers to other stimuli. Again, there is no general requirement for an unlearned releaser. The second-order conditioning phenomenon (e.g., Rescorla, 1980), for example, clearly shows that learned releasers can also serve as valence-assigning releasers.

The term "image" is used for convenience in the released-image model to refer to a neural representation of a stimulus object. The representation then serves as a template for comparison with later stimulus input. Matching stimuli are recognized in the sense that they release the appropriate behavior. Such images are not necessarily pictorial nor are they restricted to the visual modality. Green (1982) pointed out the difficulties attendant to internal representation models that require a pictorial template. However, conceptualizing such images as feature detectors for releasing stimuli simplifies the problem considerably. As is well known and previously discussed, the effective releasing stimulus is rarely, if ever, the entire stimulus complex of the natural releaser, but rather is some limited part of it. Whether or not chicks form an object-constant image of the mother hen, for example, they follow her leading call and peck at objects designated by moving angular stimuli. By doing so, they display a substantial part of chick filial behavior. Moreover, just as releasing stimuli can occur in separate or combinations of sense modalities, conditional and assigned images are not restricted to the visual modality alone. Gustatory images are necessary to taste-aversion learning and olfactory images are important in mammals and lower vertebrates. Auditory images are common in birds and will be identified and discussed in Section VE.1.

Conceptually, released-image recognition is a mechanism whereby phylogenetically preprogrammed recognition of stimuli that are invariant over time and space mediates learned recognition of biologically important but environmentally variable stimuli. Every specific manifestation of the two-stage (conditional-image activation and specific-image assignment) released-image mechanism is shaped by convergent and/or divergent evolutionary processes to serve some phylogenetically adaptive function (cf. Hollis, 1982). Biological constraints on learning can occur at either stage. Not all releasing stimuli generate a conditional image and therefore enter into a valence-transferring relationship. When a conditional image is activated, only a particular subset of the total environment stimuli available can satisfy that conditional image. If no such subset is available, no transfer of releasing valence can occur.

B. Comparison with Pavlovian Conditioning

The major purpose of this review is to present and explicate the released-image recognition learning model within the context in which it was developed. A brief digression to compare the released-image model with Pavlovian conditioning may further that purpose, although such a comparison is worthy of separate review. The released-image recognition model originated from the realization that arrow-directed pecking by hatchling chicks reflected releasing valence that had been transferred from the "pecking" arrow to the arrow-"pecked" stimulus. The resulting abstract description, as shown in Fig. 2, suggests that three elements are involved in releasing valence transfer: a releasing-valence-assigning stimulus, a releasing-valence-receiving stimulus, and a releasing-valence-transferring relationship. Parallel elements are to be found in the Pavlovian conditioning model: the unconditional stimulus (US), the conditional stimulus (CS), and conditional response (CR) acquisition from temporal prediction of CS by US.

When two models purport to account for the same phenomena, the finding of many common features should not be surprising, particularly if one model is as mature as is Pavlovian conditioning. Nevertheless, the two models do differ in a number of ways that will be discussed briefly. The most basic difference is that released-image recognition learning involves releasing-valence transfer as a consequence of spatial, configural, and dynamic relationships among stimuli in addition to the temporal relationships that distinguish Pavlovian conditioning. Thus, Pavlovian conditioning can be viewed as that subset of released-image recognition processes in which releasing-valence transfer occurs by temporal prediction of the primary releaser by the specific assigned image.

Although extension of the Pavlovian model to encompass learning from nontemporal relationships may superficially appear to be simple, Pavlovian CRs are in fact defined by temporal relationships between CS and US. Redefinition of CRs in terms of spatial relationships, for instance, means developing appropriate control conditions for spatial relatedness. One implication of such redefinition is the need for careful analysis of the critical valence-assigning features of releasing stimuli—an enterprise very much in the tradition of classical ethology.

The released-image view of responses as phylogenetically preorganized and released rather than learned has several interesting consequences. Generation of relevant behavior is intrinsic to the released-image recognition model. Many interpretations of Pavlovian conditioning differ by involving concepts such as "conditioned central motivational state" or "expectancy" that require extrinsic processes such as reinforcement and/or instrumental conditioning to account for the production of CRs.

In contrast, the released-image recognition model sees reinforcement as redundant and therefore unnecessary. Both apparent reinforcement and generation of behavior originate from taxic direction of released responses. "Positive reinforcement" results when approach, contact, and consummatory responses are elicited by autotaxic releasers; a "negative reinforcer" is found when rejection, escape, and avoidance responses are the product of an antitaxic releaser. Instrumental behavior is generated by the taxic effect of releasers that attract or repel an animal in preorganized sequences of released responses. Thus, at least some forms of instrumentally conditioned responses are actually intrinsic to released-image functioning. Occurrence of these responses is independent of reinforcement and under the control of an assigned-image releaser. In the released-image model, instrumental conditioning is seen as a potentially explainable phenomenon rather than as an explanatory model.

Autotaxic and antitaxic categories of releasing stimuli leave a third class: "allotaxic" releasers that elicit behavior directed at some stimulus other than the releaser. Such releasers should equally enter into the released-image recognition mechanism, whereas no comparable category of reinforcers, and presumably no such CRs, occur in the Pavlovian model. Thus, selective pecking by neonatal chicks learned from arrow movement directed toward a distinctive small object confounds the Pavlovian but not the released-image model.

Finally, responses learned within the released-image model are clearly identical to the original released responses (Fig. 2). This apparently foolhardy specificity in identifying learned behavior contrasts with Pavlovian conditioning, where the question of what is learned remains an intractable problem with a multiplicity of unsatisfactory answers. Released-image recognition learning also contrasts with the classical imprinting model, which has been largely unconcerned with mechanisms whereby imprinting techniques yield appropriate behavior.

V. RECOGNITION LEARNING PHENOMENA

The balance of this review is devoted to development of the released-image recognition model in the context of a survey of a number of stimulus-recognition phenomena in birds. Relevant research is examined in order to evaluate the possibility that the released-image recognition model can serve as a broad unifying principle in learning.

A. Imitation or Observational Learning of Food Recognition

Initial food recognition by neonatal domestic chicks appears to occur by released-image learning. A similar mechanism seems to produce recognition of new foods by adult birds of other species. Released-image recognition is not asserted to account for instances of imitation or observational learning that involve active copying of conspecific behavior (see, e.g., Palameta and Lefebvre, 1985), if any exist. However, many reported instances of observational learning of food recognition by birds are easily accommodated within the released-image framework. That a feeding bird is a releaser for feeding responses of observing conspecifics is documented by numerous reports of social facilitation. Subsequent selective feeding on an otherwise nonpreferred or unrecognized novel food, based only on visual observation of feeding conspecifics (and therefore in the absence of primary reinforcement), suggests that released-image recognition has occurred. All that is required is the assumption that feeding behavior of a conspecific activates, for an observer bird, a conditional image that is satisfied by the visual characteristics of an object eaten by a demonstrator bird. Satisfaction of the conditional image assigns a specific image for the observer so that similar objects subsequently release feeding responses.

With no attempt to be exhaustive, creditable learning of food or prey recognition by observation of feeding conspecifics has been reported for forktailed flycatchers (*Muscivora tyrannus*) (Alcock, 1969), red-winged blackbirds (*Agelaius phoeniceus*) (Mason and Reidinger, 1981), chaffinches (*Fringilla coelebs*) (Turner, 1964), and house sparrows (*Passer domesticus*) (Turner, 1964). Reports of rapid propagation of a new food in the diet of groups of birds (e.g., Buxton, 1948; Newton, 1967; Petterson, 1956) are suggestive of social transmission by released-image recognition, but are nevertheless indeterminate as to the method of propagation.

B. Search-Image Formation

Another phenomenon that appears readily explicable in terms of the released-image recognition model is the rapid exploitation of cryptic prey following initial discovery. In one sense, this can be viewed as just another example of learned food recognition, easily accounted for by Pavlovian conditioning, reinforcement, and stimulus generalization. As Shettleworth (1972) noted, selective predation implies nothing more than preference. The status of a separate concept for the phenomenon, however, was suggested because of the rapidity with which a "search image" is formed and generalized. According to de Ruiter (1952), after European jays had accidentally found the first of a number of protectively colored stick caterpillars, "they immediately began to peck up sticks and caterpillars indiscriminately" (p. 228), quickly finding all of the caterpillars. L. Tinbergen (1960) found the phenomenon suggestive of the formation of a template for the prey item that permitted rapid identification of individual prey. Other instances of search-image formation by birds have been reported by Dawkins (1971), Kettlewell (1955), and others.

In order to account for search-image formation within the released-image model, first note that exteroceptive features of the prey form the initial releasing stimulus that elicits taxic pecking and seizing of the prey. However, oral stimuli from the prey object serve as further releasing stimuli for either approach and consumption or rejection and withdrawal, depending on the nature of those oral stimuli. The oral stimuli also generate a conditional image that is satisfied by the exteroceptive sensory features of the prey. Such features alone subsequently suffice to elicit consummatory or avoidance behavior. The concept of reinforcement is thus not required in the released-image account of food recognition learning.

Primacy effects in the development of food preferences by some birds and so-called "food imprinting" (Hess, 1964) appear to be related to search images and therefore to released-image recognition. Primacy effects in food preferences can be quite powerful. Hogan (1977) described the behavior of many domestic chicks given early exposure to mealworms as food. Such birds appear to avidly seek mealworms and die of starvation if mealworms are unavailable. Strong primacy effects have also been shown for herring and ring-billed gull chicks (Rabinowitch, 1968), and laughing gull chicks (Hailman, 1967). The main difference between food imprinting and other preference primacy effects is that food imprinting in domestic chicks supposedly occurs only during a "sensitive period" from 3 to 5 days of age (Hess, 1964).

Primacy effects that limit subsequent learning are quite common and include "blocking" in Pavlovian conditioning and effects of initial filial imprinting to inappropriate stimuli (e.g., Graves, 1973). Permanent primacy effects are accounted for quite nicely by Bateson's (1981) competitive exclusion model. Complementary to released-image recognition learning, the competitive exclusion model proposes that primacy effects result from limited access to the executive system controlling the output of relevant behavior.

C. Avoidance-Image Formation

If the oral stimuli resulting from a feeding response release rejection behavior, the resulting transfer of releasing valence to the sensory features of the consumed object may be termed, for the sake of symmetry, "avoidance-image formation," that is, an image for a stimulus to be avoided is formed. Chicks learn very rapidly, often in one trial, to avoid foul-tasting stimuli (e.g., Lee-Teng and Sherman, 1966) or food that is followed by aversive somatic effects (e.g., Hale and Green, 1979). Again, conditioning processes can more or less satisfactorily account for such avoidance learning. The purpose of this section is to point out that a released-image recognition model does at least as well as a conditioning model, and certainly better if observational avoidance is considered. Social transmission of food rejection would be consistent with, indeed expected from, consideration of the released-image recognition model. Warning or alarm calls and displays are releasers of avoidance behavior that form a conditional image for the stimulus object at the focus of the alarm behavior. As a consequence, that stimulus becomes a specific assigned avoidance image, capable, by itself, of eliciting rejection by the observer.

Rothschild and Lane (1960) reviewed anecdotal reports and observations on the behavior of birds that seize a distasteful insect. Such birds subsequently not only fail to attack the insect and its mimics, but also alarm call and/or alarm display and attempt to escape the vicinity of the offending object. Rothschild and Lane concluded that it is "highly probable that young birds can learn to avoid certain aposematic insects directly from the alarm cries or warning signals of their parents or other adult birds without the need of actual experience" (Rothschild and Lane, 1960, p. 329), but very little evidence is available. McBride *et al.* (1969) observed alarm calling by a semiferal domestic brood hen in response to a large spider, but it is possible that the hen responded to the spider (larger than her chicks) as a potential predator rather than as an unpalatable prey.

Rothschild and Ford (1968) reported a single anecdotal instance of interspecific communication of prey avoidance. A mistle thrush (*Turdus viscivorus*) was given, and promptly attacked, a poisonous (presumed unpalatable) grasshopper. The bird quickly ceased the attack and wiped its bill (presumed to be a warning display). The grasshopper was then offered to an observing starling (*Sturnus vulgaris*) that immediately bill-wiped and moved away without contacting the grasshopper. Mason and Reidinger (1982) allowed demonstrator redwinged blackbirds access to food paired with a distinctive visual stimulus and then intubated them with an illness-producing agent. Observer red-wings that watched the demonstrators during access to the distinctive food and during illness subsequently avoided consumption of the food when it was presented along with the distinctive stimulus.

Klopfer (1957) provided another example. In his experiments, mallard ducks (*Anas platyrhynchos*) observed conspecific demonstrators as they were given a painful electric shock at a feeding dish. In a subsequent test with demonstrator removed, observer birds avoided the previously electrified dish. Naive ducks placed with the observers also avoided the dish after observational experience, even when the observers were subsequently removed. Later experiments failed to find observation-based avoidance of mildly unpalatable food (Klopfer, 1958, 1959, 1961; Sexton and Fitch, 1967).

D. Predator Recognition

Good examples of learning by released-image recognition can also be found in the development of predator recognition, an important form of avoidance-image formation. Lorenz (1952) vividly described the acquisition of predator recognition by juvenile jackdaws (*Corvus monedula*), an account that appears to be a straight forward description of released-image assignment. Adult jackdaws emit a "rattle" warning call on the appearance of a predator. The young thereafter avoid the predator and rattle call upon its appearance. Furthermore, according to Lorenz (1952), persons or animals carrying a dead jackdaw or other limp black object are releasers of the rattle call and can become permanently labeled as a predator throughout the area. Jackdaws that recognize the new assigned-image predator emit the warning call, which is then taken up and disseminated by other jackdaws.

Lorenz's anecdotal account has been substantiated by field study and experimentation and extended to other species. Kruuk (1976) found similar effects in a brood colony of gulls (*L. argentatus, L. fuscus*). Placing various stuffed predators in the colony produced predator-appropriate behavior (flocking in the air above, avoidance on the ground) that rapidly habituated. However, placing a dead gull near the predator's head caused a greater effect on the gulls than did predator or gull alone—indeed, greater than the sum of the separate effects. Most important, this inflated effect was maintained in a later test with the predator again alone. Just as for the jackdaws, a dead conspecific held by a predator released antipredator responses by observing gulls as well as a conditional image for the features of the predator.

Major support for Lorenz's (1952) description came from the report of Curio *et al.* (1978; see also Vieth *et al.*, 1980). They arranged circumstances so that a demonstrator blackbird (*Turdus merula*) appeared to mob objects that do not normally arouse mobbing in blackbirds—objects such as an innocuous bird (Australian honeyeater, *Philemon corniculatus*) or a plastic bottle. An observer blackbird that had been exposed to the deceptive demonstration subsequently mobbed the newly designated predator. Furthermore, using the previous observer as a demonstrator for a new observer permitted the transmission of mobbing that was little diminished over a series of six birds. Curio *et al.* (1978) also provided anecdotal evidence similar to that of Lorenz (1952). One author approached a nest of fledgling blackbirds that showed no evidence of alarm. That author was then mobbed by the parents, departed, and returned to the nest a few minutes later, whereupon the fledglings fled while giving mobbing calls. Social transmission of predator recognition is easily accommodated by the released-image model. For the observing birds, the releasing valence of the mobbing display is directly transferred to the object at the focus of the mobbing behavior.

E. Imprinting

Imprinting initially referred only to filial behavior in galliform and other nidifugous birds and was interpreted as a largely innate process that required experience but not learning (Lorenz, 1937). This view has yielded to the "imprinting model" where stimulus recognition results from a "perceptual learning" process assumed separate in procedure and outcome from other forms of learning, Pavlovian conditioning in particular. Outcome differences often cited include contrary effects of punishment and extinction, primacy rather then recency effects, limited plasticity (sensitive periods), and imperviousness of imprinted responses to later modification (irreversibility). Procedurally, the imprinting model has been invoked in explanation of a diversity of recognition phenomena where altered response occurs to a unitary stimulus apparently based only on simple perceptual experience with that stimulus. As Bateson (1979, p. 472; see also Rajecki, 1973, p. 49; Shettleworth, 1983, p. 6) noted, "even if the imprinting process turns out to be associative, the imprinting procedure is not. It simply involves exposing a bird to a single stimulus without explicit pairing of neutral and significant events." In contrast, successful application of either released-image or Pavlovian conditioning models to imprinting phenomena requires explicit identification of two separate stimuli.

1. Filial Imprinting

Filial imprinting occurs when a stimulus selectively releases following and other filial behavior in neonatal birds as a result of prior experience with that stimulus. Hoffman (e.g., 1978; Hoffman and Ratner, 1973; Hoffman and Segal, 1983) has argued that when the imprinting stimulus is a moving object, what appears to be a unitary stimulus is really two stimuli: motion and static components of the imprinting object (in the Pavlovian model, US and CS, respectively). The most direct evidence supporting this assertion is that display of duckling (*A. platyrhynchos*) filial behavior toward immobile objects depends upon previous experience with that object in motion (Hoffman, 1978; Hoffman and Ratner, 1973; Hoffman and Segal, 1983; however, see Eiserer, 1980). Other research also indicates that the visual characteristics of an imprinting stimulus have effects on observing hatchlings that are disassociable from the effects of movement of the same object (e.g., Gervai and Csányi, 1973; Klopfer, 1965, 1967; Klopfer and Hailman, 1964a,b).

Parallel to the Pavlovian interpretation of imprinting, the released-image model identifies movement of the object as a releasing stimulus that assigns filial-behavior-releasing valence to the visual characteristics of that same object. However, the released-image and Pavlovian models diverge on the nature of the valence-transferring relation. Pavlovian conditioning is constrained to emphasize the temporal association, whereas the expectation from the released-image model of a functionally adaptive relationship suggests that spatial congruity is the operative factor.

If motion and visual aspects are separate components of an imprinting stimulus, then what happens if they are separated? This issue was addressed in research reported by James (1959, 1960a,b; Abercrombie and James, 1961) who used a pattern of flashing lights as the imprinting stimulus. For groups of domestic chicks, a turquoise ball was mounted on a wall amid flashing lights. The same pattern of lights, constantly illuminated (i.e., not flashing), was located at the other end of the alley apparatus. For other groups of chicks, the only difference was that the ball was located amid the constant lights. When tested with the ball alone, only the first groups showed strong filial attachment (approach, following, pecking, contentment calls) to the ball. The flashing light presumably simulated movement which, in released-image terms, assigned filial-attachment-releasing valence to the ball. It may be noted that only a difference in spatial location of the imprinting object relative to the imprinting stimulus occasioned the difference in attachment behavior, since flashing lights, steady lights, and ball were simultaneously present for both groups (granted, however, that the greater attractiveness of the flashing lights might have resulted in greater exposure to the ball in that group).

Transfer of releasing valence for filial attachment from movement to visual features of the moving object is clearly only one aspect of the process whereby recognition of parent is acquired by progeny. In fact, visual features of the parent may be of secondary importance in filial recognition (e.g., Johnston and Gottlieb, 1981). Certain species-specific parental calls (exodus, leading, rallying) are powerful releasers of following and other filial behavior in hatchlings (e.g., Gottlieb, 1965, 1968). The role in imprinting of such calls is far from clear at the present time (e.g., Evans, 1982). Responsiveness to calls sometimes dissipates in the absence of experience with such calls (e.g., Case and Graves, 1978; Graves, 1973), whereas responsiveness is often enhanced by pre- or postnatal experience with the calls (e.g., Allen, 1980; Bailey, 1983; Cowan, 1973; Evans, 1975; Impekoven, 1976). In addition, responsiveness to a moving model may be enhanced by having the calls emanate from the model (e.g., Fischer, 1966; Gottlieb, 1965, 1968).

Availability of species-specific calls, powerful releasers of filial behavior, provides a basis for released-image transfer of recognition of parent to young. Nevertheless, nonanecdotal evidence that parental calls serve to assign filial-behavior-releasing valence to the visual features of the parent is difficult to find. Part of the reason may be a paucity of relevant studies, but some direct evidence comes from experiments with domestic chickens in which released-image recognition has not been found to occur. Fischer (1966), Graves (1973), and Suboski (unpublished results, 1985) all found that exposure to broody hen calls emanating from a moving model failed to enhance filial responding to the silent model at a later test.

In contrast to the apparent failure of transfer of imprinting-releasing valence from species-specific calls to visual features of the caller, movement of visual stimuli has been shown to mediate recognition by young of the calls of a specific parent. Evans and Mattson (1972) had the audiotaped leading call of a particular domestic brood hen consistently emanate from the location of a moving pendulum. The call of a second hen was spatially located at a still pendulum. Upon testing with calls alone, the chicks preferentially went to the call which had been presented with the moving pendulum. Similar results were obtained by Cowan for chicks (1973a, 1974), by Cowan for giant Canada geese (*Branta canadensis maxima*) (1973b), and by Evans for ring-billed gulls (1977; however, cf. Evans, 1975; Cowan and Evans, 1974). Thus, for a number of bird species, moving visual stimuli seem to assign filial-attachment-releasing valence to the call of a particular parent.

2. Egg Recognition

Evidence that some birds must learn the visual characteristics of their own eggs is interesting albeit dependent on a fragile data base. According to Rothstein (e.g., 1974), species of birds subject to brood parasitism are either complete acceptors of parasite eggs (most species) or complete rejectors. Grey catbirds (*Dumetella carolinensis*) are rejectors of the eggs of a brood parasite, the brown-headed cowbird (*Molothrus ater*). Rothstein (1974) replaced a single catbird egg with a real or artificial cowbird egg either within 4.5 hr after the first egg was laid, or after two or more eggs were already present in the nest. The cowbird egg was rejected at a much lower rate when the first catbird egg was the one replaced.

Similar results were obtained for another rejector species, the northern oriole (*Icterus galbula*) (Rothstein, 1978). In this species, yearling females are discriminable from older females, thus permitting distinction between nests of first-time breeders and experienced mothers. Naive birds usually accepted foreign eggs, whereas older birds did not. Foreign eggs were accepted by naive females provided that the egg was placed in the nest within a period that began several days before, and ended shortly after, the onset of laying. Thus, for some bird species, the early eggs observed in the context of releasing stimuli from the nest and egg-laying/brooding process apparently serve to specify an egg image. Subsequent stimuli from matching eggs inhibit egg rejection and release maternal care.

3. Chick Recognition

The evidence is somewhat stronger that recognition of young by parents sometimes proceeds by means of released-image recognition. Schleidt *et al.* (1960, as elaborated by Lorenz, 1963) deafened hen turkey poults (*Meleagris gallopavo*), which nevertheless appeared to develop normally. Such hens successfully incubated egg clutches but pecked the chicks to death immediately after hatch without showing any signs of visual recognition of the chicks. According to Lorenz (1963), normal primiparous turkey hens similarly pecked at a stuffed chick moved toward the nest unless the model was emitting turkey chick distress calls. A stuffed predator (polecat), on the other hand, elicited maternal behavior if it was accompanied by turkey chick calls. Normal experienced turkey brood hens accept both calling and silent chicks. Thus, the distinctive species-specific calls of turkey chicks appear to release maternal responses and assign a maternal-response-releasing image to the visual features of the calling chicks.

At this point, the findings of Miller and Conover (1979) may be reconsidered. They found that muted ring-billed gull chicks were poorly fed and often rejected by the parents. Their conclusion was that chick vocalizations were necessary to release parental feeding behavior. From the present perspective, however, a failure by the chicks to release visual recognition from the parents is more likely to be involved (cf. Beer, 1966). In the absence of chick vocalization to assign releasing valence to chick visual features, the parents might not learn to recognize the chicks. This interpretation is supported by an additional finding of Miller and Conover (1979)—that muting gull chicks when they were 3 days old had no effect on parent behavior. By the third day posthatch, releasing valence for parental responses will have already been transferred to the visual features of the young. This explanation accounts for the fact that chick vocalizations, presumed to be critical releasing stimuli for parental feeding on day one posthatch, are completely unnecessary by day three. Finally, as a specific implication of the released-image model, mostly first-time parents would be expected to fail to feed mute chicks.

Dilger (1960) reported a particularly striking example of released-image offspring recognition in African parrots (genus *Agapornis*). Different species have young with either red or white down. Numerous attempts to cross-foster among experienced parents revealed that "none of the species normally having red-downed young would feed newly hatched white-downed young and *vice-versa*" (Dilger, 1960, p. 681). However, an attempt to get a pair of parentally naive *A. roseicollis* (red-downed young) to raise white-downed young was completely successful. Furthermore, the same pair refused to feed red-downed young of the same age as the white-downed young they were rearing. Mayr (1974, p. 654), without reference or further elaboration, stated: "In several species of wild pheasants. . . the downy young have uniquely characteristic color patterns on the crown and the back. If one places the eggs of the wrong species into the nest of a hen pheasant, the mother treats the chicks like alien intruders as soon as they hatch and kills them." One can but wonder whether this is true of inexperienced as well as experienced mothers.

4. Sexual Imprinting

Sexual imprinting will be discussed only briefly here. Many of the empirical issues involved are presently in the process of clarification. Most studies of sexual imprinting involve cross-fostering or cross-rearing of congeners or different color morphs of the same species, followed by sexual preference tests at maturity. The matters at issue include: (1) confounding effect on sexual preference tests of morphic or species differences in the behavior of the test bird (e.g., Kruijt *et al.*, 1982; ten Cate, 1982); (2) possible differences in sexual imprintability between females of sexually monomorphic and dimorphic species (e.g., Sonneman and Sjölander, 1977); and (3) the relationship between filial and sexual imprinting (e.g., Vidal, 1980).

It seems clear that experientially dependent sexual imprinting occurs in some bird species. Such effects have been obtained for zebra finch (*Poephila guttata*) and Bengalese finch (*Lonchura striata*) cross-fostering (e.g., Immelman, 1972; Sonneman and Sjölander, 1977) and between color morphs of the Japanese quail (*Coturnix coturnix japonica*) (e.g., Gallagher and Ash, 1978; Truax and Siegal, 1982). Vidal (1980) convincingly demonstrated that, for domestic chickens, the development of sexual preference is not a direct result of filial imprinting. Both filial and sexual imprinting were produced by early exposure to an inanimate object, whereas later exposure produced weak filial but strong sexual imprinting. No evidence directly implicates released-image recognition in sexual imprinting. In principle, however, sexual imprinting is as easily subsumed by the released-image model as is any other form of conspecific identification.

5. Individual Conspecific Recognition

Birds not only recognize conspecifics by categories, they come to recognize individuals: territorial neighbors (e.g., Weeden and Falls, 1959), mates (e.g., Beer, 1970a; Mundinger, 1970; White, 1971; Wooller, 1978), siblings (Evans, 1980), as well as parents and offspring. The mutual recognition of individual chicks and parents, largely by vocalization, has been most widely investigated.

Individual recognition is usually inferred from differential responses to familiar and unfamiliar conspecifics. However, the logic of inferring recognition occurrence or failure contains sometimes unappreciated hazards and assumptions (cf. Alley and Boyd, 1950; Evans, 1980). A conspecific may be rejected, ignored, or accepted but isomorphic categories of recognition do not automatically follow. A major problem is that acceptance may signify indifference rather than familial recognition. For example, Newton (1967) interchanged bullfinch (*P. pyrrhula*) broods of 3 and 10 days. The adults of both broods not only accepted the fosterlings (implying recognition failure), they appropriately altered the diets, feeding more invertebrates to younger nestlings, less to older (implying recognition).

A second problem lies in distinguishing recognizor from recognizee. Beer

(1979) observed that laughing gull chicks seemed better at recognizing parents than were the parents at recognizing chicks, and suggested the possibility that only one-way recognition occurred. Perhaps chicks identified themselves to parents by responses to adults that differed to parents and nonparents. Obviously, the reciprocal relation could as easily hold: Adult birds certainly respond very differently to their own and other offspring.

A very substantial body of evidence is in essential agreement on several points. First, mutual parent–offspring recognition occurs when required, that is, shortly before the young are capable of leaving the nest area and mixing with young of other broods. Evans (1980) found close correspondence between age of nest-leaving and development of recognition in sea birds. The correlation seems to exist for many other species as well. In the cockatoo (*Cacatua roseicapilla*), parents apparently do not initially recognize their own young, since fosterlings are readily accepted until about day 40, just before fledging (about day 46), but are thereafter rejected (Rowley, 1980). Bank swallow (*Riparia riparia*) nestlings are also first recognized by parents just before fledging (Beecher, 1981; Beecher *et al.*, 1981a, b; Hoogland and Sherman, 1976), as are tree swallow (*Iridoprocne bicolor*) and barn swallow (*Hirundo rustica*) (Burtt, 1977), and piñon jay (*Gymnorhinus cyanocephalus*) (McArthur, 1982) young.

Much evidence also agrees that at least initial parent–chick recognition is of individual calls, with recognition of visual features occurring later if at all (e.g., Miller and Emlen, 1975). Buckley and Buckley (1972) switched 1- or 2-day-old chicks of the royal tern (*Sterna maxima maxima*) between adjacent nest sites. Of three adults observed, each was unable to visually recognize its own chick from a distance of only 40 cm, but did so after the chick vocalized. The process of parental recognition of offspring often appears to be relatively abrupt, occurring within a day or two of nest-leaving but well established by nest-leaving, and dependent on the production of individualistic "signature" (Beecher *et al.*, 1981a; McArthur, 1982) or "location" (Nice, 1950; Peek *et al.*, 1972) calls. For a diversity of species the signature call is then used to locate offspring for feeding in crèches that sometimes contain hundreds of fledglings. Such species include many seabirds (see Evans, 1980), the cockatoo (Rowley, 1980), piñon jay (McArthur, 1982), and bank swallow (Beecher, 1981b). Offspring are fed only by their own parents, after emitting the individually recognized signature call.

According to Beecher (1981, p. 49), the signature call of the chick is learned by the parent in an "irreversible, imprinting-like process." However, except in the unlikely event that signature calls are innately recognized, then something very like released-image recognition must necessarily occur. The nest and nestlings are presumably original releasers of parental-care behavior. Upon production of the signature call at fledging, parental-care responses are abruptly restricted to the fledgling that utters a particular signature call, presumably initially at the nest site. The released-image description of the acquisition of signature call recognition has a number of testable implications. For example, identification of nest-site stimuli as the valence-assigning releasing stimuli suggests that parental experience with a chick's signature call outside the nest site will not result in acquisition of recognition of that call. Similarly, presenting the signature call of another chick at the nest site of parents of a muted chick can be expected to result in those parents feeding the other chick in the crèche rather than their own.

VI. ASSESSMENT AND CONCLUSIONS

The released-image recognition model combines concepts from the psychology of learning and classical ethology to provide an abstract description of a mechanism hypothesized to underlie most or all of the various forms of recognition learning, including Pavlovian conditioning, imprinting, and social transmission of recognition of many different stimuli. A number of features of the released-image recognition model suggest that the model can provide a conceptual perspective of value in understanding recognition learning processes. Released-image recognition is an ethologically and ecologically oriented learning model, not a plea that one be devised (Johnston, 1981). The released-image mechanism is proposed to perform biologically adaptive functions wherein unlearned "recognition" of invariant stimuli mediates learned recognition of variable stimuli (cf. Lorenz, 1969; Schöne, 1964). Species- and context-specific calls and displays are among the important stimuli that are recognized without significant prior experience by birds and other animals. The fact that releasing valence from such calls may be transferred to related environmentally variable stimuli (e.g., predator or food) suggests that the released-image recognition mechanism may have provided the phylogenetic basis for development of semantic communication (cf. Seyfarth *et al.*, 1980).

Nevertheless, it remains to be addressed whether there is a real phenomenon to be accounted for by the released-image model. One possibility to be considered is that no common mechanism underlies the various phenomena purported here to be instances of released-image recognition. Perhaps the present scheme capitalizes on spurious similarities to produce a functionally amorphous organization. Lorenz (1950) provided some guidance on issues of this kind:

> all . . . professional students of animal behaviour have been guilty of the one unpardonable offence against the most fundamental law of inductive natural science: they have one and all formed a hypothesis *first* and proceeded to look for examples to confirm it *afterwards*. The

protean multiformity of organic nature and quite particularly of the behaviour of higher animals is such that a circumspect search for examples can never fail to detect a wealth of evidence for literally *any* theory, however arbitrarily you chose to invent one. The *facts* in themselves may be quite correct but *choice* of facts in itself is ever a falsification. (Lorenz, 1950, p. 233)

Later on we find:

In biological research, an all-too-cautious abstaining from forming a hypothesis would get us nowhere, and we must have the courage to formulate preliminary hypotheses, though we are well aware that these preliminary formulations are much too simplistic and correspond, at the best, to a particularly simple special case. In fact, the discovery of a natural law has been, in many instances identical with the discovery of a special case, in which it was actualized in a particularly simple manner. (Lorenz, 1950, p. 262)

Lorenz sets for us a narrow line! The released-image model was derived from research that explicated a special case leading to the formation of a hypothesis, followed by a (successful) search for examples to confirm it. It appears necessary to look elsewhere to establish the value of the model.

A number of the many differences between the released-image recognition and other major learning models have been pointed out. Divergent predictions can be derived from such differences and subjected to empirical test. Such tests are unlikely to confirm or disconfirm the model. However, if the model both generates sufficient successful empirical predictions and provides a paradigmatic framework that parsimoniously organizes and integrates a large body of recognition learning phenomena, it will satisfy the fundamental and ultimate criterion for a theory: usefulness.

VII. ACKNOWLEDGMENTS

This work was supported by Grant A-7059 from the Natural Sciences and Engineering Research Council of Canada and grants from the Advisory Research Committee of Queen's University. Development of the ideas contained herein benefited greatly from the interest and suggestions of Darwin Muir. Colleagues Rick Beninger and Ronald Weisman gave careful reading to drafts and provided many valued and helpful comments and criticisms. I am grateful to Evalyn F. Segal for an extensive and thoughtful review of an earlier version of this manuscript that provided a number of valuable suggestions that were incorporated into the present manuscript.

VIII. REFERENCES

Abercrombie, B., and James, H. (1961). The stability of the domestic chick's response to visual flicker. *Anim. Behav.* **9**:205–212.
Alcock, J. (1969). Observational learning by fork-tailed flycatchers (*Muscivora tyrannus*). *Anim. Behav.* **17**:652–657.
Allen, H. M. (1980). The response of willow grouse chicks to auditory stimuli. 3. Recognition of the incubating hen's voice. *Behav. Proc.* **5**:39–43.
Allen, H. M., Boggs, C., Norris, E., and Doering, M. (1977). Parental behaviour of captive willow grouse *Lagopus l. lagopus*. *Ornis Scandinavica* **8**:175–183.
Alley, R., and Boyd, H. (1950). Parent–young recognition in the coot *Fulica atra*. *Ibis* **92**:46–51.
Altmann, S. A. (1956). Avian mobbing behavior and predator recognition. *Condor* **58**:241–253.
Bailey, E. D. (1983). Influence of incubation calls on post-hatching responses of pheasant chicks. *Condor* **85**:43–49.
Bartashunas, C., and Suboski, M. D. (1984). Effects of age of chick on social transmission of pecking preferences from hen to chicks. *Dev. Psychobiol.* **17**:121–127.
Bateson, P. (1979). How do sensitive periods arise and what are they for? *Anim. Behav.* **27**:470–486.
Bateson, P. (1981). Control of sensitivity to the environment during development. In Immelmann, K., Barlow, G. W., Petrinovich, L., and Main, M. (eds.), *Behavioral Development*, Cambridge University Press, Cambridge, pp. 432–453.
Beecher, M. D. (1981). Development of parent–offspring recognition in birds. In Aslin, R. N., Alberts, J. R., and Petersen, M. R. (eds.), *Development of Perception*, Vol. 1, Academic Press, New York, pp. 45–65.
Beecher, M. D., Beecher, I. M., and Hahn, S. (1981a). Parent–offspring recognition in bank swallows (*Riparia riparia*): II. Development and acoustic basis. *Anim. Behav.* **29**:95–101.
Beecher, M. D., Beecher, I. M., and Lumpkin, S. (1981b). Parent–offspring recognition in bank swallows (*Riparia riparia*): I. Natural history. *Anim. Behav.* **29**:86–94.
Beer, C. G. (1966). Incubation and nest-building behaviour of black-headed gulls. V: The post-hatching period. *Behaviour* **26**:190–214.
Beer, C. G. (1970a). Individual recognition of voice in the social behavior of birds. *Adv. Stud. Behav.* **3**:27–74.
Beer, C. G. (1970b). On the responses of laughing gull chicks (*Larus atricilla*) to the calls of adults. II. Age changes and responses to different types of calls. *Anim. Behav.* **18**:661–677.
Beer, C. G. (1979). Vocal communication between laughing gull parents and chicks. *Behaviour* **70**:118–146.
Bond, R. M. (1936). Eating habits of falcons with special reference to pellet analysis. *Condor* **38**:72–76.
Bossema, I. (1979). Jays and oaks: An eco-ethological study of a symbiosis. *Behaviour* **70**:1–117.
Brown, R. T. (1975). Modification of chicks' pecking preferences: Food imprinting or instrumental conditioning? *Anim. Learn. Behav.* **3**:217–220.
Buckley, P. A., and Buckley, F. G. (1972). Individual egg and chick recognition by adult royal terns (*Sterna maxima maxima*). *Anim. Behav.* **20**:457–462.
Burtt, E. H., Jr. (1977). Some factors in the timing of parent–chick recognition in swallows. *Anim. Behav.* **25**:231–239.
Buxton, E. J. M. (1948). Tits and peanuts. *Brit. Birds* **41**:229–232.
Case, V. J., and Graves, H. B. (1978). Functional versus other types of imprinting and sensitive periods in *Gallus* chicks. *Behav. Biol.* **23**:433–445.

Collias, N. E. (1952). The development of social behavior in birds. *Auk* **69:**127–159.
Cowan, P. J. (1973a). Parental calls and the approach behavior and distress vocalization of young domestic chicken. *Can. J. Zool.* **51:**961–967.
Cowan, P. J. (1973b). Parental calls and the approach behavior of young Canada geese: A laboratory study. *Can. J. Zool.* **51:**647–650.
Cowan, P. J. (1974). Selective responses to the parental calls of different individual hens by young *Gallus gallus*: Auditory discrimination learning versus auditory imprinting. *Behav. Biol.* **10:**541–545.
Cowan, P. J., and Evans, R. M. (1974). Calls of different individual hens and the parental control of feeding behavior in young *Gallus gallus*. *J. Exper. Zool.* **188:**353–360.
Crook, J. H. (1960). Studies on the social behaviour of *Quelea q. quelea* (Linn.) in French West Africa. *Behaviour* **16:**1–55.
Curio, E., Ernst, U., and Vieth, W. (1978). The adaptive significance of avian mobbing. II. Cultural transmission of enemy recognition in blackbirds: Effectiveness and some constraints. *Z. Tierpsychol.* **48:**184–202.
Cushing, J. E., Jr. (1944). The relation of non-heritable food habits to evolution. *Condor* **46:**265–271.
Dawkins, M. (1971). Perceptual changes in chicks: Another look at the "search image" concept. *Anim. Behav.* **19:**566–574.
de Ruiter, L. (1952). Some experiments on the camouflage of stick caterpillars. *Behaviour* **4:**222–232.
Dilger, W. C. (1960). The comparative ethology of the African parrot genus *Agapornis*. *Z. Tierpsychol.* **17:**649–685.
Eiserer, L. A. (1980). Development of filial attachment of static visual features of an imprinting object. *Anim. Learn. Behav.* **8:**159–166.
Evans, R. M. (1975). Responsiveness to adult mew calls in young ring-billed gulls (*Larus delawarensis*): Effects of exposure to calls alone and in the presence of visual imprinting stimuli. *Can. J. Zool.* **53:**953–959.
Evans, R. M. (1977). Auditory discrimination-learning in young ring-billed gulls (*Larus delawarensis*). *Anim. Behav.* **25:**140–146.
Evans, R. M. (1980). Development of behavior in seabirds: An ecological perspective. In Burger, J., Olla, B. L., and Winn, H. E. (eds.), *Behavior of Marine Animals*, Vol. 4, Plenum Press, New York, pp. 271–322.
Evans, R. M. (1982). The development of learned auditory discriminations in the context of postnatal filial imprinting in young precocial birds. *Bird Behav.* **4:**1–6.
Evans, R. M., and Mattson, M. E. (1972). Development of selective responses to individual maternal vocalizations in young *Gallus gallus*. *Can. J. Zool.* **50:**777–780.
Evans, S. M., and Patterson, G. R. (1971). The synchronization of behaviour in flocks of estrildine finches. *Anim. Behav.* **19:**429–438.
Ewer, R. F. (1969). The "instinct to teach." *Nature* **222:**698.
Fischer, G. J. (1966). Auditory stimuli in imprinting. *J. Comp. Physiol. Psychol.* **61:**271–273.
Frank, L. H., and Meyer, M. E. (1970). Food imprinting in domestic chicks as a function of social contact and number of companions. *Psychon. Sci.* **19:**293–295.
Gallagher, J. E., and Ash, M. (1978). Sexual imprinting: The stability of mate preference in Japanese quail (*Coturnix coturnix japonica*). *Anim. Learn. Behav.* **6:**363–365.
Gervai, J., and Csányi, V. (1973). The effects of protein synthesis inhibitors on imprinting. *Brain Res.* **53:**151–160.
Gochfeld, M. (1980). Mechanisms and adaptive value of reproductive synchrony in colonial seabirds. In Burger, J., Olla, B. J., and Winn, H. E. (eds.), *Behavior of Marine Animals*, Vol. 4, Plenum Press, New York, pp. 207–270.

Gottlieb, G. (1965). Imprinting in relation to parental and species identification by avian neonates. *J. Comp. Physiol. Psychol.* **59**:345–356.
Gottlieb, G. (1968). Species recognition in ground-nesting and hole-nesting ducklings. *Ecology* **49**:87–95.
Gould, J. L., and Gould, C. G. (1981). The instinct to learn. *Science 81* **2**:44–50.
Graves, H. B. (1973). Early social responses in *Gallus*: A functional analysis. *Science* **182**:937–939.
Green, P. R. (1982). Problems in animal perception and learning and their implications for models of imprinting. In Bateson, P. P. G., and Klopfer, P. H. (eds.), *Perspectives in Ethology*, Vol. 5, Plenum Press, New York, pp. 243–273.
Hailman, J. P. (1967). The ontogeny of an instinct. The pecking response in chicks of the laughing gull (*Larus atricilla* L.) and related species. *Behav. Suppl.* **15**:1–159.
Hale, C., and Green, L. (1979). Effect of initial-pecking consequences on subsequent pecking in young chicks. *J. Comp. Physiol. Psychol.* **93**:730–735.
Harris, M. P. (1983). Parent–young communication in the puffin *Fratercula arctica*. *Ibis* **125**:109–114.
Hess, E. H. (1964). Imprinting in birds. *Science* **146**:1128–1139.
Hoffman, H. S. (1978). Experimental analysis of imprinting and its behavioral effects. *Psychol. Learn. Motiv.* **12**:1–37.
Hoffman, H. S., and Ratner, A. M. (1973). A reinforcement model of imprinting: Implications for socialization in monkeys and man. *Psychol. Rev.* **80**:527–544.
Hoffman, H. S., and Segal, M. (1983). Biological factors in social attachments: A new view of a basic phenomenon. In Zeiler, M. D., and Harzem, P. (eds.), *Advances in Analysis of Behaviour: Biological Factors in Learning*, Vol. 3, Wiley, New York, pp. 41–61.
Hogan, J. A. (1966). An experimental study of conflict and fear. An analysis of behavior of young chicks toward a mealworm. II. The behavior of chicks which eat the mealworm. *Behaviour* **27**:273–289.
Hogan, J. A. (1971). The development of a hunger system in young chicks. *Behaviour* **39**:128–201.
Hogan, J. A. (1973). How young chicks learn to recognize food. In Hinde, R. A., and Stevenson-Hinde, J. (eds.), *Constraints on Learning*, Academic Press, New York, pp. 119–139.
Hogan, J. A. (1977). The ontogeny of food preferences in chicks and other animals. In Barker, L. M., Best, M., and Domjan, M. (eds.), *Mechanisms in Food Selection*, Baylor University Press, Waco, Texas, pp. 71–97.
Hogan-Warburg, A. J., and Hogan, J. A. (1981). Feeding strategies in the development of food recognition in young chicks. *Anim. Behav.* **29**:143–154.
Hollis, K. L. (1982). Pavlovian conditioning of signal-centered action patterns and autonomic behavior: A biological analysis of function. *Adv. Stud. Behav.* **12**:1–64.
Hoogland, J. L., and Sherman, P. W. (1976). Advantages and disadvantages of bank swallow (*Riparia riparia*) coloniality. *Ecol. Monogr.* **46**:33–58.
Hørlyk, N.-O., and Lind, H. (1978). Pecking response of artificially hatched oystercatcher *Haematopus ostralegus* young. *Ornis Scandinavica* **9**:138–145.
Immelman, K. (1972). Sexual and other long-term aspects of imprinting in birds and other species. *Adv. Stud. Behav.* **4**:147–174.
Impekoven, M. (1971). Prenatal experiences of parental calls and pecking in the laughing gull chick (*Larus atricilla* L.). *Anim. Behav.* **19**:475–480.
Impekoven, M. (1973). The response of incubating laughing gulls (*Larus atricilla* L.) to calls of hatching chicks. *Behaviour* **46**:94–113.
Impekoven, M. (1976). Responses of laughing gull chicks (*Larus atricilla*) to parental attraction- and alarm-calls, and effects of prenatal auditory experience on the responsiveness to such calls. *Behaviour* **56**:250–278.
James, H. (1959). Flicker: An unconditioned stimulus for imprinting. *Can. J. Psychol.* **13**:59–67.

James, H. (1960a). Imprinting with visual flicker: Evidence for a critical period. *Can. J. Psychol.* **14:**13–20.

James, H. (1960b). Social inhibition of the domestic chick's response to visual flicker. *Anim. Behav.* **8:**223–224.

Jenkins, H. M., Barrera, F. J., Ireland, C., and Woodside, B. (1978). Signal-centered action patterns of dogs in appetitive classical conditioning. *Learn. Motiv.* **9:**272–296.

Johnston, T. D. (1981). Contrasting approaches to a theory of learning. *Behav. Brain Sci.* **4:**125–173.

Johnston, T. D., and Gottlieb, G. (1981). Development of visual species identification in ducklings: What is the role of imprinting? *Anim. Behav.* **29:**1082–1099.

Kadlec, J. A., Drury, W. H., Jr., and Onion, D. K. (1969). Growth and mortality of herring gull chicks. *Bird Banding* **40:**222–233.

Kettlewell, H. B. D. (1955). Selection experiments on industrial melanism in the *Lepidoptera*. *Heredity* **9:**323–342.

Klopfer, P. H. (1957). An experiment on empathic learning in ducks. *Am. Nat.* **91:**61–63.

Klopfer, P. H. (1958). Influence of social interactions on learning rates in birds. *Science* **128:**903.

Klopfer, P. H. (1959). Social interactions in discrimination learning with special reference to feeding behavior in birds. *Behaviour* **14:**282–299.

Klopfer, P. H. (1961). Observational learning in birds: The establishment of behavioral modes. *Behaviour* **17:**71–79.

Klopfer, P. H. (1965). Imprinting—A reassessment. *Science* **147:**302–305.

Klopfer, P. H. (1967). Is imprinting a Cheshire cat? *Behav. Sci.* **12:**122–129.

Klopfer, P. H., and Hailman, J. P. (1964a). Basic parameters of following and imprinting in precocial birds. *Z. Tierpsychol.* **21:**755–762.

Klopfer, P. H., and Hailman, J. P. (1964b). Perceptual preferences and imprinting in chicks. *Science* **145:**1333–1334.

Kruijt, J. P., Bossema, I., and Lammers, G. J. (1982). Effects of early experience and male activity on mate choice in mallard females (*Anas platyrhynchos*). *Behaviour* **80:**32–43.

Kruuk, H. (1976). The biological function of gulls' attraction towards predators. *Anim. Behav.* **24:**146–153.

Lee-Teng, E., and Sherman, S. M. (1966). Memory consolidation of one-trial learning in chicks. *Proc. Nat. Acad. Sci.* **56:**926–931.

Lind, H. (1965). Parental feeding in the oystercatcher (*Haematopus o. ostralegus* (L.)). *Dansk Ornithologisk Forenings Tidsskrift*, **59:**1–31.

Lorenz, K. (1937). The companion in the bird's world. *Auk* **54:**245–273.

Lorenz, K. Z. (1950). The comparative method in studying innate behaviour patterns. *Symp. Soc. Exper. Biol.* **4:**221–268.

Lorenz, K. (1952). *King Solomon's Ring*, Crowell, New York.

Lorenz, K. (1963). *On Aggression*, Harcourt, Brace & World, New York.

Lorenz, K. (1965). *Evolution and Modification of Behavior*, University of Chicago Press, Chicago.

Lorenz, K. Z. (1969). Innate bases of learning. In Pribram, K. H. (ed.), *On the Biology of Learning*, Harcourt, Brace & World, New York, pp. 13–93.

Mason, J. R., and Reidinger, R. F., Jr. (1981). Effects of social facilitation and observational learning on feeding behavior of the red-winged blackbird (*Agelaius phoeniceus*). *Auk* **98:**778–784.

Mason, J. R., and Reidinger, R. F. (1982). Observational learning of food aversions in red-winged blackbirds (*Agelaius phoeniceus*). *Auk* **99:**548–554.

Mayr, E. (1974). Behavior programs and evolutionary strategies. *Am. Sci.* **62:**650–659.

McArthur, P. D. (1982). Mechanisms and development of parent–young vocal recognition in the piñon jay (*Gymnorhinus cyanocephalus*). *Anim. Behav.* **30:**62–74.

McBride, G., Parer, I. P., and Foenander, F. (1969). The social organization and behaviour of the feral domestic fowl. *Anim. Behav. Monogr.* **2**:127–181.

Meyer, M. E., and Frank, L. H. (1970). Food imprinting in the domestic chick: A reconsideration. *Psychon. Sci.* **19**:43–45.

Miller, D. E., and Conover, M. R. (1979). Differential effects of chicks' vocalizations and billpecking on parental behavior in the ring-billed gull. *Auk* **96**:284–295.

Miller, D. E., and Conover, M. R. (1983). Chick vocal patterns and non-vocal stimulation as factors instigating parental feeding behaviour in the ring-billed gull. *Anim. Behav.* **31**:145–151.

Miller, D. E., and Emlen, J. T., Jr. (1975). Individual chick recognition and family integrity in the ring-billed gull. *Behaviour* **52**:124–144.

Moynihan, M., and Hall, M. F. (1953). Hostile, sexual, and other social behaviour patterns of the spice finch (*Lonchura punctulata*) in captivity. *Behaviour* **7**:33–76.

Mueller, H. C. (1974). The development of prey recognition and predatory behaviour in the American kestrel *Falco sparverius*. *Behaviour* **49**:313–324.

Mundinger, P. C. (1970). Vocal imitation and individual recognition of finch calls. *Science* **168**:480–482.

Nelson, J. B. (1965). The behaviour of the gannet. *Brit. Birds* **58**:233–288, 313–336.

Newton, I. (1967). The adaptive radiation and feeding ecology of some British finches. *Ibis* **109**:33–98.

Nice, M. M. (1950). Development of a red-wing (*Agelaius phoeniceus*). *Wilson Bull.* **62**:87–93.

Norton-Griffiths, M. (1969). The organisation, control and development of parental feeding in the oystercatcher (*Haematopus ostralegus*). *Behaviour* **34**:55–114.

Palameta, B., and Lefebvre, L. (1985). The social transmission of a food-finding technique in pigeons: What is learned? *Anim Behav.* **33**:892–896.

Peek, F. W., Franks, E., and Case, D. (1972). Recognition of nest, eggs, nest site, and young in female red-winged blackbirds. *Wilson Bull.* **84**:243–249.

Petterson, M. (1956). Diffusion of a new habit among greenfinches. *Nature* **177**:709–710.

Quine, D. A., and Cullen, J. M. (1964). The pecking response of young arctic terns *Sterna macrura* and the adaptiveness of the "releasing mechanisms." *Ibis* **106**:145–173.

Rabinowitch, V. E. (1968). The role of experience in the development of food preferences in gull chicks. *Anim. Behav.* **16**:425–428.

Rajecki, D. W. (1973). Imprinting in precocial birds: Interpretation, evidence, and evaluation. *Psychol. Bull.* **79**:48–58.

Rescorla, R. A. (1980). *Pavlovian Second-Order Conditioning: Studies in Associative Learning*, Erlbaum Associates, Hillsdale, New Jersey.

Rothschild, M., and Ford, B. (1968). Warning signals from a starling, *Sturnus vulgaris*, observing a bird rejecting unpalatable prey. *Ibis* **110**:104–105.

Rothschild, M., and Lane, G. (1960). Warning and alarm signals by birds seizing aposematic insects. *Ibis* **102**:328–330.

Rothstein, S. I. (1974). Mechanisms of avian egg recognition: Possible learned and innate factors. *Auk* **91**:796–807.

Rothstein, S. I. (1978). Mechanisms of avian egg recognition: Additional evidence for learned components. *Anim. Behav.* **26**:671–677.

Rowley, I. (1980). Parent-offspring recognition in a cockatoo, the galah, *Cacatua roseicapilla*. *Austral. J. Zool.* **28**:445–456.

Rozin, P. (1976). The selection of food by rats, humans, and other animals. *Adv. Stud. Behav.* **6**:21–76.

Schleidt, W. M., Schleidt, M., and Magg, M. (1960). Störung der Mutter-Kind-Beziehung bei Truthühnern durch Gehörverlust. *Behaviour* **16**:254–260.

Schöne, H. (1964). Release and orientation of behaviour and the role of learning as demonstrated in crustacea. *Anim. Behav.* **Suppl. 1**:135–144.
Segal, E. G. (1972). Induction and the provenance of operants. In Gilbert, R. M., and Millenson, J. R. (eds.), *Reinforcement: Behavioral Analyses*, Academic Press, New York, pp. 1–34.
Sexton, O. J., and Fitch, J. (1967). A test of Klopfer's empathic learning hypothesis. *Psychon. Sci.* **7**:181–182.
Seyfarth, R. M., Cheney, D. L., and Marler, P. (1980). Monkey response to three different alarm calls: Evidence of predator classification and semantic communication. *Science* **201**:801–803.
Shapiro, L. J. (1981). Pre-hatching influences that can potentially mediate post-hatching attachments in birds. *Bird Behav.* **3**:1–18.
Sherry, D. F. (1977). Parental food-calling and the role of the young in the Burmese red jungle fowl (*Gallus gallus spadiceus*). *Anim. Behav.* **25**:594–601.
Shettleworth, S. J. (1972). Constraints on learning. *Adv. Stud. Behav.* **4**:1–68.
Shettleworth, S. J. (1983). Function and mechanism in learning. In Zeiler, M. D., and Harzem, P. (eds.), *Advances in Analysis of Behaviour: Biological Factors in Learning*, Vol. 3, Wiley, New York, pp. 10–39.
Shreck, P. K., Sterritt, G. M., Smith, M. P., and Stilson, D. W. (1963). Environmental factors in the development of eating in chicks. *Anim. Behav.* **11**:306–309.
Sonneman, P., and Sjölander, S. (1977). Effects of cross-fostering on the sexual imprinting of the female zebra finch *Taenopygia guttata*. *Z. Tierpsychol.* **45**:337–348.
Southern, W. E. (1974). Copulatory wing-flagging: A synchronizing stimulus for nesting ring-billed gulls. *Bird Banding* **45**:210–216.
Stokes, A. W. (1971). Parental and courtship feeding in red jungle fowl. *Auk* **88**:21–29.
Strobel, M. G., and MacDonald, G. E. (1974). Induction of eating in newly hatched chicks. *J. Comp. Physiol. Psychol.* **86**:493–502.
Suboski, M. D. (1984). Stimulus configuration and valence-enhanced pecking by neonatal chicks. *Learn. Motiv.* **15**:118–126.
Suboski, M. D. (1987). Environmental variables and releasing-valence transfer in stimulus-directed pecking of chicks. *Behav. Neur. Biol.* **47**:262–274.
Suboski, M. D., and Bartashunas, C. (1984). Mechanisms for social transmission of pecking preferences to neonatal chicks. *J. Exper. Psychol.: Anim. Behav. Proc.* **10**:182–194.
Templeton, R. K. (1983). Why do herring gull chicks vocalize in the shell? *Bird Study* **30**:73–74.
ten Cate, C. (1982). Behavioural differences between zebrafinch and Bengalese finch (foster) parents raising zebrafinch offspring. *Behaviour* **81**:152–172.
Thorpe, W. H. (1963). *Learning and Instinct in Animals*, 2nd ed., Harvard University Press, Cambridge, Massachusetts.
Timberlake, W. (1983). The functional organization of appetitive behavior: Behavior systems and learning. In Zeiler, M. D., and Harzem, P. (eds.), *Advances in Analysis of Behaviour: Biological Factors in Learning*, Vol. 3, Wiley, New York, pp. 177–221.
Tinbergen, L. (1960). The natural control of insects in pinewoods. I. Factors influencing the intensity of predation by song birds. *Arch. Nèerl. Zool.* **13**:265–343.
Tinbergen, N. (1960). *The Herring Gull's World*, revised ed., Basic Books, New York.
Tolman, C. W. (1964). Social facilitation of feeding behaviour in the domestic chick. *Anim. Behav.* **12**:245–251.
Truax, R. E., and Siegel, P. B. (1982). Plumage phenotypes and mating preferences in Japanese quail. *Behav. Proc.* **7**:211–222.
Turner, E. R. A. (1964). Social feeding in birds. *Behaviour* **24**:1–46.
Vidal, J. -M. (1980). The relations between filial and sexual imprinting in the domestic fowl: Effects of age and social experience. *Anim. Behav.* **28**:880–891.
Vieth, W., Curio, E., and Ernst, U. (1980). The adaptive significance of avian mobbing. III.

Cultural transmission of enemy recognition in blackbirds: Cross-species tutoring and properties of learning. *Anim. Behav.* **28:**1217–1229.

Vince, M. A. (1969). Embryonic communication, respiration, and the synchronization of hatching. In Hinde, R. A. (ed.), *Bird Vocalizations*, Cambridge University Press, Cambridge, pp. 211–260.

Waite, R. K. (1981). Local enhancement for food finding by rooks (*Corvus frugilegus*) foraging on grassland. *Z. Tierpsychol.* **57:**15–36.

Weeden, J. S., and Falls, J. B. (1959). Differential responses of male overbirds to recorded songs of neighboring and more distant individuals. *Auk* **76:**343–352.

White, N. R., and Guadalupe del Rio Pesado, M. (1983). Effects of visual and auditory stimuli on following in chicks. *Bird Behav.* **4:**82–85.

White, S. J. (1971). Selective responsiveness by the gannet (*Sula bassana*) to played-back calls. *Anim. Behav.* **19:**125–131.

Woodruff, G., and Starr, M. D. (1978). Autoshaping of initial feeding and drinking reactions in newly hatched chicks. *Anim. Learn. Behav.* **6:**265–272.

Woodruff, G., and Williams, D. R. (1976). The associative relation underlying autoshaping in the pigeon. *J. Exp. Anal. Behav.* **26:**1–13.

Wooller, R. D. (1978). Individual vocal recognition in the kittiwake gull, *Rissa tridactyla* (L.). *Z. Tierpsychol.* **48:**68–86.

Wortis, R. P. (1969). The transition from dependent to independent feeding in the young ring dove. *Anim. Behav. Monogr.* **2:**1–54.

Yom-Tov, Y. (1976). Recognition of eggs and young by the carrion crow (*Corvus corone*). *Behaviour* **59:**247–251.

Chapter 7

PSYCHOIMMUNOLOGY: RELATIONS BETWEEN BRAIN, BEHAVIOR, AND IMMUNE FUNCTION

Paul Martin

Sub-Department of Animal Behaviour
University of Cambridge
Madingley, Cambridge CB3 8AA, England

I. ABSTRACT

The purpose of this review is to draw the attention of ethologists to an exciting field of research that is of fundamental scientific interest and immense practical potential, yet of which most ethologists seem to be astonishingly unaware. Research into how the central nervous system and the immune system influence one another is casting new light on how psychological and emotional factors can affect an individual's susceptibility to various forms of illness, including infectious diseases, autoimmune disorders, and certain forms of cancer. This rapidly developing field is referred to as *psychoimmunology*. Some representative research findings are reviewed here. There is strong empirical support for the idea that behavioral factors, via the central nervous system, can influence the immune system. The neural and hormonal mechanisms whereby these influences operate are beginning to be understood. As a detailed understanding of the underlying mechanisms emerges, psychoimmunology is at last beginning to attain the academic respectability it has previously lacked. Psychoimmunology is essentially interdisciplinary, and requires the combined talents of immunologists, neuroendocrinologists, and behavioral scientists. There is great scope for an important contribution from ethologists, with their sophisticated methods for directly observing and analyzing complex behavior and their ability to deal with whole systems.

II. INTRODUCTION

The notion, long enshrined in folklore, that psychological or emotional factors can influence an individual's susceptibility to illness is far older than modern biology and medicine. The idea that emotional disorders and trauma can affect disease susceptibility was widely accepted in the 18th and 19th centuries, for example (LeShan, 1959). Many nonscientists readily believe that individuals who are depressed or under great stress are more likely to become ill. Yet these phenomena have, until recently, received surprisingly little attention from medical and biological science—despite a huge body of clinical and epidemiological evidence showing clear links between psychological states and disease susceptibility. Indeed, empirical data that demonstrate such links have often been treated with unwarranted skepticism. One reason for the unwillingness of scientists to look further at these phenomena is the traditional view of the immune system as an isolated and self-regulating entity. As we shall see, this view is now clearly untenable.

A second problem has been the tendency for psychoimmunology to be tarred with the same brush as is psychosomatic medicine—a field of research that has suffered from an uncertain academic status, especially in Britain. In fact, the thinking behind psychoimmunology differs fundamentally from the old notion that certain "psychosomatic diseases" are caused solely by psychological (as opposed to "organic") factors. Psychoimmunology is based on the more subtle idea that the central nervous system can influence immune function and, therefore, that psychological factors are included among the many causal factors that affect disease susceptibility.

Thus, the issue of psychological influences on disease has been ignored by mainstream biology and medicine for reasons that are no longer applicable. This situation is being rapidly transformed by emerging knowledge about how the immune system works and, in particular, by an increasing awareness of mechanisms whereby the immune system and central nervous system (CNS) may interact. Now that the mechanisms are being elucidated, the phenomena that they underlie are, at last, beginning to be taken seriously.

The developing field of research concerned with the relations between the brain, behavior, and immune function is referred to as *psychoneuroimmunology* (Ader, 1981a, 1983) or *psychoimmunology* (Maddox, 1984). The central idea is that reciprocal influences operate between the body's three principal control systems: the CNS, the endocrine system, and the immune system (Cunningham, 1981; see Fig. 1).

The primary role of the immune system is, of course, to recognize foreign materials (antigens), such as bacteria and viruses, and inactivate or destroy them (see Section XI for a brief glossary of immunological terms). In addition to

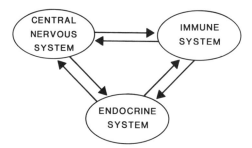

Fig. 1. Reciprocal influences among the central nervous system, endocrine system, and immune system.

dealing with infectious diseases, the immune system is also involved in the body's responses to certain forms of cancer, and in autoimmune disorders such as rheumatoid arthritis. Susceptibility to all these types of disease therefore depends on how well the immune system works. If changes in the brain can alter the way in which the immune system functions, either directly or via the endocrine system, then susceptibility to illness can be influenced by psychological and emotional factors.

In this review I shall describe only a small sample of the evidence showing that brain, behavior, and immune function are interlinked, and outline some of the mechanisms that have been proposed to account for these effects. By giving just a taste of the type of research that has been done, I hope to draw attention to an exciting and rapidly expanding field of research that most ethologists so far seem to have ignored. Because of its unashamedly proselytizing purpose this is not a *critical* review, and I have not always attempted to dissect every complexity and ambiguity to be found in the evidence. My intended reader is the receptive ethologist who may be excited by the ideas and findings, rather than the hardened immunologist who may be irritated by the simplifications and generalizations.

Before proceeding any further, two fundamental points must be emphasized. The first is that psychological or emotional factors *by themselves* do not generally cause disease, which is one misleading connotation of the terms "psychosomatic" and "psychogenic" (see Murray, 1977; Martin, 1978; Ader, 1980; Lipowski, 1984). Rather, they are one of many types of mediating factors that influence disease processes; other causal agents, such as bacteria, viruses, or tumor cells, must also be present. Stress may alter a person's likelihood of developing flu, for example, but one must also be infected with flu viruses. Immune function and disease susceptibility are influenced by many factors, such as the individual's age, sex, nutritional state, genetic makeup, past experience of

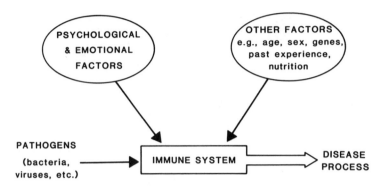

Fig. 2. Causal factors influencing immune function and disease susceptibility.

disease, and so on (Fig. 2). For instance, in humans and other species many components of immune function undergo an age-related decline starting shortly after sexual maturity. This decline is generally accompanied by an age-related increase in the incidence of viral infections, autoimmune diseases, and certain forms of cancer (Makinodan, 1976; see Cooper, 1983). Diseases are influenced by many causal factors, and cannot be divided into those that have purely "psychological" causes and those that have purely "organic" causes.

A second basic point is that the immune system, just like the central nervous system, is an enormously complex set of interacting components. Therefore, to talk of general increases or decreases in *the* immune response, as though it were a single entity, is a huge (though often convenient) oversimplification. An immune response has many steps and involves many different classes and subclasses of cells, each influencing the development and activity of other cells. Modulation of the immune response by the CNS could, in principle, occur at any of these steps. Nonetheless, here I shall follow the example of other writers and succumb to the temptation of referring to changes in *the* immune response.

III. TYPES OF EVIDENCE FOR LINKS BETWEEN BRAIN, BEHAVIOR, AND IMMUNE FUNCTION

The evidence that an organism's brain and behavior can influence its immune system comes from a variety of sources (reviewed by Rogers *et al.*, 1979; Ader, 1981b, 1983; Monjan, 1981; Plaut and Friedman, 1981; Borysenko and Borysenko, 1982; Locke, 1982; Jemmot and Locke, 1984; Goetzl, 1985; Stein *et al.*, 1985; Tecoma and Huey, 1985). The first three types of evidence (correlational, experimental, and conditioning effects) will each be discussed in more detail in later sections.

A. Clinical and Epidemiological Evidence

Correlational evidence from clinical and epidemiological studies has shown that the incidence and severity of many diseases, as well as changes in immune function, are often linked to psychological or emotional factors (see Section IV).

B. Experimental Evidence

Perhaps most compelling of all is experimental evidence that manipulations of brain or behavior—for example, experimentally induced psychological stress—can alter both disease susceptibility and immune function (see Section V).

C. Conditioning Effects

Evidence for a direct influence of the brain on immunity comes from experiments showing that changes in immune response may be subject to Pavlovian conditioning (see Section VI).

D. Anatomical Evidence

A rapidly growing body of anatomical evidence shows that the brain and immune system are potentially able to influence one another via neural connections. A variety of lymphoid tissues, such as the thymus, spleen, lymph nodes, and bone marrow, are directly innervated by neurons from the autonomic nervous system in several species of birds and mammals. At least some of these neural pathways are thought to play a role in the development and modulation of the immune system (reviewed by Bulloch, 1985; Felten *et al.*, 1985). In mice and rats, for example, the thymus gland and spleen receive autonomic innervation from the brain, and some of these neural links are thought to be functional (Bulloch and Moore, 1981; Williams and Felten, 1981; Williams *et al.*, 1981).

Furthermore, immunologically active cells such as lymphocytes carry surface receptor sites for a variety of hormones and neurotransmitters, including corticosteroids, growth hormone, estradiol, testosterone, and adrenocorticotropic hormone (ACTH). More hormone receptors are found on stimulated lymphocytes than on unstimulated lymphocytes, and receptors for some hormones—such as epinephrine and insulin—are found only on stimulated cells. Lymphocytes also carry receptors for neuropeptides such as β-endorphins and Met-enkephalin, some of which have been shown to influence immune function (Pert et al., 1985).

Nonetheless, despite the innervation of lymphoid tissue, many components of the immune response can occur in solution in the test tube. This suggests that the neural connections probably act in a regulatory manner, since they are clearly not essential for the immune response to occur. One implication of this is that the full extent of psychological and emotional influences on immune function will only emerge if immune function is measured in the intact organism, and not *in vitro* where the regulatory neural connections are absent. Regrettably, most psychoimmunological research so far has used *in vitro* rather than *in vivo* measures of immune function.

E. Brain Stimulation and Lesioning

Experimentally lesioning or electrically stimulating regions of the brain, notably the hypothalamus, can influence the immune system. For example, stimulating parts of the hypothalamus can elicit an increase in immunoglobulin levels. Conversely, when an immune response is elicited, neurons in the ventromedial hypothalamus show marked changes in their firing rates (e.g., Besedovsky et al., 1977, 1985a). Lesioning certain brain regions can produce a variety of changes in lymphocyte function (Stein et al., 1976; Brooks et al., 1982). For example, placing lesions in the anterior hypothalamus of guinea pigs has been found to reduce the responsiveness of lymphocytes to stimulation by mitogens (Keller et al., 1980; Roszman et al., 1982, 1985).

In one of a series of experiments, Besedovsky and colleagues evoked an immune response in rats by immunizing them with sheep red blood cells, and then looked at accompanying changes in their brains (Del Rey et al., 1982; Besedovsky et al., 1983). They found a reduction in norepinephrine synthesis in the hypothalamus following immunization. Analysis of individual variation showed a good correlation between the size of the reduction in norepinephrine synthesis and the strength of the immune response, as assessed by the number of plaque-forming cells in the spleen at the time of peak immune response, 4 days after immunization. Furthermore, injecting other rats with blood supernatants from immunized rats led reliably to a reduction in norepinephrine synthesis in the hypothalamus within 2 hr. The conclusion was that soluble substances, released by activated immune cells during an immune response, altered the activity of noradrenergic neurons within the brain (Besedovsky et al., 1985a). Experiments such as these support the idea of direct communication between the CNS and the immune system and, in this case, indicate that the CNS can detect changes within the immune system.

F. Other Evidence

Finally, I shall briefly mention a potpourri of circumstantial evidence that is consistent with the view that the CNS and immune system can influence one another. Changes in the levels of various hormones (which, in turn, are under neural control) can influence aspects of the immune response. Conversely, eliciting an immune response can alter the levels of various hormones. Many hormones, including corticosteroids, androgens, estrogens, and progesterone, tend to reduce immune responses while others, such as growth hormone, thyroxine, and insulin, can increase them. A variety of drugs which affect behavior are also found to influence the immune response. Conversely, some drugs that act on the immune system also produce behavioral effects.

Finally, there are some striking parallels between the pattern of development shown by the neuroendocrine and immune systems during ontogeny (Pierpaoli, 1981). There is considerable evidence that the development of lymphoid tissue and components of the immune response are subject to neuronal influences. Recent work, for example, has established that the differentiation of the developing rat's thymus is under the control of acetylcholinergic neurons from the brain.

IV. CLINICAL AND EPIDEMIOLOGICAL EVIDENCE

A. Associations between Psychological Factors and Illness

Many correlational studies of humans have uncovered associations between psychological factors and the incidence or severity of various illnesses, including infectious and autoimmune diseases and some forms of cancer (reviewed by Jemmot and Locke, 1984). Correlational evidence must, of course, be treated with caution, and causal relationships obviously cannot be inferred from correlational evidence alone.

A representative example of such work is a prospective study by Kasl *et al.* (1979), who investigated psychological factors associated with infectious mononucleosis (glandular fever) among nearly 1400 military cadets at West Point over a 4-year period. When they initially entered the academy, all cadets were screened for the antibody to Epstein–Barr virus (EBV), the pathogen responsible for infectious mononucleosis. About one third of the cadets initially lacked antibodies to EBV, showing that they had not previously been exposed to the virus and were, therefore, potentially susceptible. About 20% of the susceptible students subsequently developed antibodies to EBV and, of these, about 25% actually developed the clinical disease. Psychological measurements showed that the cadets who became ill typically had suffered high levels of academic pressure. In particular, they tended to be highly motivated to achieve but poor in their actual performance, and to have fathers who were "overachievers." These psychological factors also predicted how long they spent in hospital with the illness.

A number of observational studies of families in their normal home environments have found associations between stressful family situations and the incidence, severity, or duration of minor respiratory illnesses such as colds and flu (e.g., Boyce *et al.*, 1977). For example, Meyer and Haggerty (1962) observed 16 families over a 12-month period, and found an increased amount of family-related stress during the 2-week period immediately before children developed streptococcal infections.

The recurrence of cold sores has also been linked to psychological factors. Cold sores are caused by herpes simplex virus-I (HSV-I). An individual may carry the virus yet seldom, if ever, actually develop cold sores. A prospective study of student nurses who had HSV-I antibodies found that those who judged themselves to be more typically unhappy had significantly more recurrences of cold sores during the following year (Luborsky *et al.*, 1976).

Finally, many investigators have reported associations between psychological factors, such as stress and depression, and the incidence or prognosis of various forms of cancer (Fox, 1978, 1983; Nieburgs *et al.*, 1979; Bammer and Newberry, 1982; Cohen *et al.*, 1982; Levy, 1982; reviewed by Sklar and Anisman, 1981). For example, a 5-year prospective study of 69 women with early breast cancer found that the prognosis was significantly better for patients who had initially (3 months postoperatively) reacted with a "fighting spirit" or by denying that they had cancer, compared with patients who responded stoically or with feelings of helplessness (Greer *et al.*, 1979).

B. Associations between Life Events and Illness

Numerous studies over the past 30 years or more have investigated the associations between "life events" or "life changes" and illness (e.g., Rahe, 1972; Dohrenwend and Dohrenwend, 1974; Hurst *et al.*, 1976; Rabkin and Struening, 1976; Hull, 1977; Minter and Kimball, 1978; Rahe and Arthur, 1978; Roth and Holmes, 1985; Sarason *et al.*, 1985). Life events are changes in circumstances that force an individual to alter established patterns of behavior—

for example, bereavement, marriage, loss or change of job, moving house, and so on. The general assumption behind this research is that such changes are stressful and might, therefore, act as predisposing or precipitating factors in the development of various illnesses.

Life event stress is measured using a questionnaire, on which subjects report the life events they are currently experiencing or have experienced during a defined period in the past (usually the preceding 6–24 months). Each type of life event is assigned a rating, indicating its perceived severity. For example, death of a spouse is given the maximum score of 100, marriage is rated at 50, and birth of a baby is rated at 39. A composite score is then calculated for each individual, reflecting the overall number and severity of life events they have experienced.

The evidence generally shows that people who have experienced large amounts of life event stress subsequently tend to have slightly poorer health (Rahe, 1972; Rabkin and Struening, 1976). Several studies have found small, but statistically significant, positive correlations—typically around 0.3—between overall life event scores and the subsequent incidence, severity, and duration of various disorders, including colds, 'flu, allergies, tuberculosis, leukemia, multiple sclerosis, diabetes, and some psychiatric disorders. High life event scores are also associated with slow rates of recovery from illness. For example, in a prospective study of 313 college students, recent life change showed a small but significant (+0.33) correlation with the seriousness of subsequent illnesses (Garrity et al., 1978). Similarly, in a study of 163 Naval Academy students, negative life events in the recent past were found to be correlated with the incidence of self-reported illness (Sarason et al., 1985). Special life event scales have been developed for children; for example, Boyce et al. (1977), using such a scale, found a positive association between life event scores and the duration of respiratory illnesses in children aged 1–11 years.

Problems with Life Event Research

Life event research raises a number of methodological issues (Rabkin and Struening, 1976). Most obvious is the difficulty of disentangling cause and effect, since a correlation between stressful life events and illness clearly does not demonstrate that the former caused the latter. For instance, chronic illness or depression might facilitate the advent of stressful life events, such as divorce or losing a job, rather than the other way around. Furthermore, though large sample sizes may ensure that the correlations are statistically significant, the *strength* of the association (the "effect size") is usually rather small (see Chapter 9 in Martin and Bateson, 1986).

The reliability and validity of questionnaire-based measures of life event stress are often suspect, especially in retrospective studies. The dependent variables have also frequently been questionable, since many studies have failed to distinguish between actual illness and "care-seeking behavior," such as going to the doctor. People with high life event scores may be more likely to seek medical assistance, but this does not necessarily mean that they suffer a higher incidence of actual illness. Self-perception of illness is not the same thing as illness itself.

Another questionable assumption is that all individuals are equally exposed to sources of disease—or, at least, that variation in exposure is not correlated with life event stress. In many cases, however, it is quite likely that individuals who are subject to particular forms of life event stress may, because of concomitant differences in life-style, also be differentially exposed to pathogens. Severely depressed people, for example, may be less likely to observe conventional standards of hygiene. This problem has been overcome in a few studies by exposing volunteers to a source of infection under controlled conditions. In one experiment, colds were induced in 52 volunteers using a nasal spray containing common cold viruses (Totman et al., 1980). Prior to infection, the subjects were assessed on various measures of recent life event stress and personality. The results showed that certain aspects of life event stress were related to the intensity of the subsequent infection (see also Totman et al., 1977).

A more fundamental problem with life event research—at least, in its crudest form—is the tendency to ignore the many other causal factors that also affect illness (Rahe and Arthur, 1978; see also Fig. 2). The effects of, say, bereavement or moving house vary considerably between individuals, depending on a host of other factors. The effects of particular life events on subsequent health depend on, among other things, the individual's sex, age, marital status, education, cultural and racial background, previous experience of life events, and socioeconomic status (Masuda and Holmes, 1978). A minor irritation for one person might be devastating for another. Even physical fitness has been found to influence the impact that stressful life events have on health. In a study of 112 students, Roth and Holmes (1985) found that high levels of life event stress were associated with poorer subsequent physical health in physically unfit individuals; no relationship was found between life event stress and illness in physically fit individuals, however (see also Busse et al., 1980). A potentially important mediating factor, which is discussed later (Section IX), is the availability of social support; individuals who have high levels of social support are, in general, found to be less susceptible to the deleterious effects of stress.

C. Direct Measures of Immune Function in Humans

Compelling evidence that psychological factors affect immunity comes from research in which aspects of immune function have been directly assessed (reviewed by Palmblad, 1985). Probably the earliest work of this sort was done by Ishigami (cited by Jemmott and Locke, 1984), who reported in 1919 that during episodes of "emotional excitement," white blood cells obtained from tuberculosis patients showed a decrease in phagocytic activity. This led him to postulate that the "stress of contemporary life" could depress immune function.

1. Bereavement

One issue that is receiving increasing attention is the long-established association between bereavement and an increased incidence of illness and mortality (Parkes, 1972; Jacobs and Ostfeld, 1977; Clayton, 1979; Helsing et al., 1981; Hofer, 1984; Stroebe and Stroebe, 1986). Bereaved people are far more likely to suffer physical illness and have higher mortality rates than do comparable married individuals of the same age and sex, especially in the younger age groups. Bereaved people are more likely to seek psychiatric help and have a considerably higher risk of suicide. They are also more likely to suffer from cardiovascular disease and cancer. Interestingly, the increased health problems associated with divorce are comparable in magnitude to those of bereavement (Bloom et al., 1978; Stroebe and Stroebe, 1986).

In a pioneering study, Bartrop et al. (1977) assessed immune function in 26 people whose spouses had recently died. Their principal measure of immune function was one which is widely used: the *in vitro* responsiveness of lymphocytes to stimulation by mitogens (see Section XI). Blood samples were taken at 2 weeks and again at 8 weeks after bereavement. Lymphocytes were separated out and stimulated with one of two mitogens: phytohemagglutinin (PHA) or concanavalin A (ConA). Lymphocyte response was compared with that of control subjects, matched for age, sex, and race. Lymphocytes from the bereaved group were significantly less responsive to both mitogens than were those from controls 8 weeks after bereavement (though not 2 weeks after). The two groups did not differ in the numbers of T- or B-cell lymphocytes, or in plasma levels of cortisol, prolactin, growth hormone, or thyroxine. Thus, according to at least one measure, bereavement was associated with a significant reduction in immune function 2 months later.

Similar results have been found in a prospective, longitudinal study of 15 men whose wives had terminal breast cancer (Schleifer et al., 1983). As early as 2 weeks after their wives died, there was a significant reduction in the subjects' lymphocyte response to various mitogens (PHA, ConA, and pokeweed mitogen, or PWM), relative to prebereavement levels. A long-term follow-up showed that a moderate reduction in lymphocyte function continued for a year or more for some subjects.

Clearly, alternative interpretations must be considered for these results. For instance, the long-term stress of having a terminally ill wife might have affected immune function prior to the actual bereavement. This is supported by evidence that men whose wives die of protracted illnesses subsequently have a higher incidence of illness than do men whose wives die of acute illness (Gerber *et al.*, 1975). Changes in immune function may also be associated with depression, consequent upon bereavement. An increased incidence of illness and impairment of immune function have frequently been reported in severely depressed people (e.g., Cappel *et al.*, 1978; Shekelle *et al.*, 1981; Kronfol *et al.*, 1982; Schleifer *et al.*, 1984; reviewed by Stein, 1985). For instance, Schleifer *et al.* (1984) found that in severely depressed psychiatric patients, none of whom were receiving drugs known to affect immune function, lymphocyte responsiveness to three different mitogens (ConA, PWM, and PHA) and the total numbers of T and B lymphocytes were all significantly lower than in normal, healthy controls matched for age and sex.

A more important consideration is that bereavement may have behavioral effects on the surviving spouse that in turn alter immune function (Jacobs and Ostfeld, 1977; Jemmott and Locke, 1984). Bereavement is likely to be followed by changes in sleeping and eating patterns, exercise, and greater use of alcohol, tobacco, caffeine, tranquilizers, or other drugs. All of these factors could themselves influence immune function (see Stein, 1985). Both under- and overnutrition are associated with reduced immune function (Gross and Newberne, 1980; Chandra, 1981), for example, and chronic sleep deprivation is known to impair cell-mediated immunity (Palmblad *et al.*, 1979). However, in the study by Schleifer *et al.* (1983) discussed above, the bereaved subjects reported no major changes in any of these factors.

Future studies will have to address these issues directly, by including direct measures of the behavioral changes and other sequelae of bereavement, attempting systematically to eliminate alternative hypotheses. Ethological techniques for the direct observation and measurement of behavior (Martin and Bateson, 1986) may have an important contribution to make here, and would greatly complement self-report measures obtained from questionnaires or interviews.

2. *Commonly Occurring Social Stressors*

One of the most interesting areas being explored is the relationship between commonplace social stressors and immune function in normal, healthy people. To give one example, Kiecolt-Glaser *et al.* (1984a) studied the effects of taking an important examination on the cellular and humoral immune function of 75 medical students. They found that the cytotoxic activity of the students' natural killer cells was significantly lower on the day of the exam as compared with a baseline level measured 1 month before. In another study, the secretion of immunoglobulin IgA in the saliva of 64 students was found to be significantly lower during periods of high academic stress, as compared with periods of low academic stress (Jemmot *et al.*, 1983). Finally, in a study of medical students,

Dorian *et al.* (1981) found that the stress of an important oral examination led to a reduction in the mitogen responsiveness of lymphocytes, relative to students who were not taking the exam. Interestingly, the reduction in immune function occurred *before* the exam, suggesting that anticipation of the stressful event was involved.

Not all individuals are equally affected by a given amount of environmental stress; as emphasized earlier, factors such as the individual's experience and perception of the stressor can markedly influence his or her physiological response. This point is illustrated by the findings of a study which looked at the relations between life event stress, psychological symptoms, and natural killer cell activity in 114 healthy student volunteers (Locke *et al.*, 1984). Overall, no correlation was found between self-reported life event stress and natural killer cell activity. However, subjects who reported large amounts of life event stress together with high levels of depression and anxiety ("poor copers") had significantly lower natural killer cell activity than did subjects who reported large amounts of life event stress but few psychological symptoms ("good copers").

V. EXPERIMENTAL STUDIES OF ANIMALS

In order to unravel the complexities and ambiguities inherent in correlational studies of humans, and to analyze the mechanisms underlying these phenomena, it is usually necessary to perform controlled experiments on animals. As with human studies, these have been of two main types: those using disease susceptibility as the dependent variable, and those employing direct measures of immune function.

A. Stress and Disease Susceptibility

Since the late 1950s, many experiments have investigated the effects of various stressors on the ability of animals to resist experimentally induced diseases, including poliomyelitis, herpes simplex virus, Coxsackie B virus, tuberculosis, and a variety of parasitic infections. In most cases, stress has been found to reduce resistance to infection, although the opposite effect can also occur (see Section V.D, below).

Many early experiments looked at the effects of crowding on the resistance of mice to infection (reviewed by Plaut and Friedman, 1981). For example, Plaut *et al.* (1969) infected mice housed at different densities with *Plasmodium berghei*, a malarial parasite. They found that mortality rates were higher at higher housing densities. Other experiments found that the resistance of mice to tapeworm infection was reduced by exposure to a cat (Hamilton, 1974) or by intense fighting (Weinman and Rothman, 1967).

The majority of experiments, however, have used aversive physical stimuli, such as electric shock, as stressors. For example, Rasmussen (1969) found that mice subjected to avoidance training in a shuttlebox, with footshock as punishment, were less resistant to several viral infections, including poliomyelitis and herpes simplex. Some of the early experiments in this field—and, regrettably, some still being performed today—were crude in their use of excessive and unrealistic amounts of physical pain, and many were ethically dubious.

More recent research has shown that less drastic physical stressors can also induce significant changes in disease susceptibility. For example, Riley (1981) investigated the effects of intermittent rotation stress on the rate at which experimentally implanted tumors grew in mice. Mice were housed in their normal cages, which were rotated at 45 rpm for 10 min each hour throughout the first 6 days after the tumor was implanted. The tumors grew more rapidly in the stressed mice, compared with controls whose cages were not rotated. Riley and his colleagues have also found considerable effects of housing conditions on the susceptibility of mice to experimentally induced tumors (Riley et al., 1979; Riley, 1981). For instance, female mice infected with mammary tumor viruses developed tumors with a median latency of 358 days under normal laboratory housing conditions. However, in special "low stress" housing conditions, in which noise, vibration, odors, and disturbances were minimized, tumors developed significantly more slowly, with a median latency of 566 days.

B. Stress and Immune Function

With the advent of convenient immunological measurement techniques, a growing number of studies have employed direct measures of immune function. To give one example, Hudson (1973) used measures of immune function to study the effects of the stress of capture on Rocky Mountain bighorn sheep. During the first 20 days following capture, the response of peripheral lymphocytes to a T-cell mitogen (PHA) was markedly reduced. Riley's (1981) work on the effects of rotational stress on mice (see Section VA) also employed direct measures of immune function in addition to assessing susceptibility to implanted tumors. Intermittent rotation stress led to a 50% reduction in the number of leukocytes (mainly T lymphocytes) within 2 hr. After 24 hr of rotation stress, there was a marked reduction in thymus weight.

Numerous other studies have found changes in immune function following stress. Mice that were reared in isolation and then exposed to social stimulation showed a significantly reduced antibody response to an antigenic challenge, as compared with mice left in isolation (Edwards *et al.*, 1980). Similarly, the stress of being placed into an established group led to a significant reduction in the tuberculin reaction of immature chickens (Mohamed and Hanson, 1980), while acute immobilization of mice reduced the ability of their macrophages to kill foreign cancer cells (Pavlidis and Chirigos, 1980).

How Should the Effects of Stress Be Assessed?

Directly measuring changes in immune function, rather than inferring them from changes in disease susceptibility, is clearly desirable if one's aim is to show that the immune system is influenced by the brain. Three important points must be borne in mind, however. The first is that, despite their appearance of objectivity and precision, immunological measures are often as vulnerable to problems of reliability and validity as are behavioral measures. The responsiveness of lymphocytes to mitogenic stimulation is widely used as an index of immune function, yet it can be an unreliable, imprecise, and potentially misleading measure.

A second point is that single measures of immune function do not provide a complete picture of how the immune system is affected by stress. As emphasized in Section II, the immune system is not a simple, unified entity and cannot be adequately described by one or two measures. Just as ethologists use many different measures to describe an animal's behavior, so must psychoimmunologists use multiple measures of immune function.

A final point concerns the biological or clinical significance of changes in immune function. An experimental manipulation may cause a statistically significant change in one or more measures of immune function, but this does not necessarily mean that these changes matter for the organism concerned. Because immune responses involve many different components, it is possible that reductions in the effectiveness of some may be mitigated by increases in others. Overall, the effect of a stressor may be to cause little or no change in the animal's susceptibility to disease. The implication of this is that, ideally, studies should employ measures of both disease susceptibility and immune function. Alternatively, studies of immune function should be supported by independent evidence—as in the work on bereavement discussed in Section IV.C—showing that the stressor also causes changes in disease incidence.

C. "Control" As a Determinant of the Physiological Effects of Stress

Much confusion surrounds the meaning of the term "stress"—an issue I do not propose to become embroiled in here (see Ursin and Murison, 1983; Stein *et al.*, 1985). Nonetheless, a basic distinction must be made between the physical or social stimuli applied to the animal (sometimes referred to as "stressors") and the animal's physiological and psychological responses to those stimuli. This distinction is important because aversive and apparently "stressful" stimuli, such as electric shocks, do not always cause deleterious physiological or immunological changes ("stress responses").

A key factor in determining whether aversive stimuli such as electric shocks are stressful in a physiological sense is whether or not the animal can control or escape from the stressor by performing an appropriate behavioral response. A common finding is that uncontrollable stress has a variety of physiological and hormonal consequences that are not elicited by similar amounts of controllable stress. For example, rats that are given uncontrollable electric shocks suffer considerably greater weight loss and three times as much gastric ulceration as do rats receiving identical amounts of controllable shock (Weiss, 1968, 1971).

To investigate how the controllability of a stressor such as electric shock affects physiological stress responses, a standard yoked control procedure is used. Two animals are wired in series and therefore receive exactly the same shocks simultaneously, but differ on the controllability of shock. Each is provided with a lever; the animal given controllable shock can terminate each shock by pressing the lever, whereas the animal receiving uncontrollable shock (the yoked control) has a lever which does nothing. A third animal which receives no shocks is generally included, and this experimental procedure is often referred to as the "triplet" design.

A number of studies have investigated the differential effects of controllable and uncontrollable shock on disease susceptibility and immune function. One example is work by Visintainer et al. (1982) on the rejection of experimentally induced tumors. Adult male rats were implanted with sarcoma tumor cells. An intermediate dose was given, which would normally be rejected and not develop into a tumor in 50% of unshocked animals. The 93 subjects were randomly divided into three groups according to the standard triplet design. The shocked animals received 60 trials at random intervals and, 4 weeks later, all the animals were killed and dissected. Tumor rejection was defined as a complete absence of tumors 30 days after implantation. Only 27% of the rats given uncontrollable shock rejected the tumor, whereas 63% of those given identical amounts of controllable shock rejected the tumor. (Interestingly, the rats given controllable shock fared marginally better than the unshocked rats, 54% of which rejected the tumor.) Similar results have been found in other experiments. For example, a single session of controllable shock was found to have no significant effect on the growth rate of implanted mastocytoma tumor cells in mice, compared with no-shock controls, whereas identical amounts of uncontrollable shock led to a significantly earlier appearance of tumors, an acceleration in tumor growth, and a higher mortality rate (Sklar and Anisman, 1979, 1981; Sklar et al., 1981).

It should be said that tumor rejection and tumor growth are affected by many factors besides immune function. Peripheral blood flow and the levels of steroid hormones and prolactin all influence tumor growth, for example, and these factors are also affected by stress (Fox, 1981). However, evidence that the immune system is involved has come from experiments employing direct measures of immune function. Using the triplet design with 12 subjects in each group, Laudenslager *et al.* (1983) gave rats a single session of controllable or uncontrollable shock (or no shock). Shocked rats received 80 moderate shocks of increasing intensity. Twenty-four hours later a further brief session of five shocks was given and a blood sample was then taken. Immune function was measured in terms of the *in vitro* responsiveness of lymphocytes to a mitogen, either PHA or ConA. Uncontrollable shock produced a significant suppression of lymphocyte response to both mitogens, as compared with lymphocytes from unshocked rats. Controllable shock, however, had no effect on the response to one mitogen (PHA) and actually *increased* the response to the other mitogen (ConA).

Uncontrollable shock can also induce in animals a psychological state known as "learned helplessness," which has certain similarities to forms of human depression (Seligman, 1975). A notable behavioral symptom of learned helplessness is that the animal gives up trying to escape or control the aversive stimuli. Physiological correlates of learned helplessness include activation of the pituitary–adrenal axis, catecholamine depletion of the CNS, weight loss, and stomach ulceration. It has often been argued that perceived lack of control over events is an important factor in human depression. Both bereavement and depression are characterized by a loss of control over the environment and both are associated with an increased incidence of illness, including cancer (Fox, 1981), and impairments in immune function (e.g., Kronfol *et al.*, 1982). In a study of 114 healthy student volunteers, for example, Locke *et al.* (1984) found an association between low levels of natural killer cell activity and self-reported psychological symptoms of depression and anxiety. Not surprisingly, the parallels between the physiological effects of learned helplessness and those produced by depression and bereavement have frequently been pointed out (e.g., Seligman, 1975; Mineka and Suomi, 1978; Stein *et al.*, 1985). The extent to which stressful events are perceived to be controllable has also been implicated as an important factor mediating the effects of life event stress on health (Suls and Mullen, 1981).

D. Immunoenhancing Effects of Stress

A common impression is that stress is invariably deleterious to health. However, this is by no means always the case; depending on its nature and timing, stress can sometimes enhance aspects of immune function. Many examples of experimentally induced stress increasing animals' resistance to disease can be found in the early experimental literature. For instance, in a decidedly unpleasant experiment, Marsh et al. (1963) found that 24 hr of continuous stressful avoidance conditioning increased the resistance of adult male cynomolgus monkeys to infection with polio virus, as compared with unstressed monkeys. Seven out of 11 stressed monkeys survived, whereas only one out of 12 unstressed controls survived. The stressed monkeys also had a significantly longer incubation period for the disease and developed symptoms on average 2 days later than controls.

Several other examples can be quoted. Stressful avoidance conditioning increased the resistance of mice to passive anaphylaxis (Rasmussen et al., 1959) and Rauscher murine leukemia virus (Jensen, 1968); exposing rats to a cat increased their resistance to experimentally induced arthritic disease (Rogers et al., 1980); periodically subjecting mice to light paired with electric shock increased their survival time following injection with *Plasmodium berghei*, a malarial parasite (Friedman et al., 1973); a similar procedure also increased the resistance of mice to a form of spontaneous leukemia (Plaut et al., 1980); physically restraining rats made them more resistant to experimentally induced allergic encephalomyelitis (Levine et al., 1962); mice that were stressed by being forced to swim for long periods every day survived longer when injected with ascites tumor (Rashkis, 1952); electric shock and forced restraint both reduced the number of mammary tumors that developed in rats treated with a carcinogen (Newberry and Sengbusch, 1979), and so on.

1. Timing of Stressor

Some light was cast on the immunoenhancing effects of stress by experiments in which the relative timing of the stressor and exposure to pathogens was varied. The results suggest that an acute stressor applied just after exposure to pathogens usually reduces resistance, whereas, if the stressor precedes exposure to pathogens, resistance may be increased. For example, Jensen and Rasmussen (1963) stressed mice by exposing them to 123-dB noise for 3 hr on each of three successive days. They tested the subjects' immune responses by infecting them with vesicular stomatitis virus on the second day. Mice that were infected with the virus just after the noise ended were more resistant than were controls. However, mice infected just before the noise started were less resistant. In other words, noise stress either increased or reduced resistance, depending on whether it occurred before or after infection. Similarly, mice that were subjected to 3 days of random electric shock stress immediately before being injected with murine sarcoma virus developed smaller lesions than did unshocked controls, indicating a heightened resistance (Amkraut and Solomon, 1972). However, mice that were given 3 days of stress starting on the day of injection showed more rapid tumor development. To give a final example, rats subjected to cold stress before being injected with tumor cells developed significantly smaller tumors than did unstressed controls, again implying that the stressor had an immunoenhancing effect (Burchfield *et al.*, 1978).

2. Duration of Stressor

The duration of the stressor appears to be another important factor in determining its physiological effects. Acute stress is generally found to lower disease resistance, whereas chronic stress sometimes has the opposite effect. This was illustrated by an experiment in which mice were exposed to noise stress for different periods of time (Monjan and Collector, 1977). Mice were subjected to daily sessions of loud, uncontrollable noise (100 dB) for varying periods of up to 30 days; they were then killed and lymphocytes obtained from their spleens. Immune response was assessed in terms of two *in vitro* measures of lymphocyte function: the ability of lymphocytes to kill foreign (P815 cytoma) cells, and their proliferation following exposure to a mitogen (either bacterial lipopolysaccharides or ConA). Noise stress had a distinctly biphasic effect on lymphocyte function. In the short term, noise stress reduced lymphocyte function relative to that of unstressed controls. However, if the stress persisted for more than about a week, its effects were reversed and lymphocyte function actually increased above that of unstressed controls. This was true for both measures of lymphocyte function, and applied to both T-cell and B-cell lymphocytes. During the initial period of immune suppression there was also a reduction in the number of viable nucleated cells in the spleen and an increased level of plasma cortisol.

Thus, one appealingly simple generalization is that acute stress reduces immune function, whereas chronic stress may enhance it. Complications can, however, arise when applying this idea to naturally occurring social stressors. For instance, a prolonged and apparently stressful situation—such as having a terminally ill spouse, an unsatisfactory job, or a failing marriage—might well be described as a chronic stressor. On the other hand, it might equally well be described as a prolonged series of acute stressors. The reality will only emerge from actual measurements of the behavioral, hormonal, and immunological responses of people in these situations.

VI. CONDITIONING EFFECTS

The notion that the brain can directly influence the immune system is supported by research showing that changes in immune response are subject to

Pavlovian conditioning. Numerous experiments have demonstrated that exposing rodents to a conditioned stimulus which was previously paired with a manipulation of the immune system can elicit conditioned changes in both humoral and cell-mediated immune responses (reviewed by Ader and Cohen, 1985).

Remarkably, the first experiments on immune conditioning were performed more than 60 years ago, using the then-new Pavlovian conditioning procedure. In the 1920s, Metal'nikov and Chorine (cited by Ader, 1981c) repeatedly subjected guinea pigs to an antigenic challenge (the unconditioned stimulus, US), which was an intraperitoneal injection of either tapioca emulsion or bacterial filtrate. This US invariably elicited an immune response, which was measured as an increase in the number of circulating polymorphonuclear leukocytes (PMNs). The US was paired with an immunologically neutral conditioned stimulus (CS), namely scratching or heating an area of skin, every day for 18–25 days. Then, following a 12- to 15-day rest period, the guinea pigs were given several exposures to the CS (scratching or heating) only. Reexposure to the CS by itself elicited a significant increase in the number of circulating PMNs within a few hours. In other words, reexposing an animal to a neutral stimulus that had previously been associated with an antigenic challenge appeared to elicit a conditioned immune response.

Many comparable experiments were performed during the 1920s and 1930s, mostly with similar results. However, major disputes arose when some experimenters claimed to have demonstrated conditioned changes in specific immune responses—that is, in the production of antibodies to a particular antigen. Metal'nikov and Chorine (1928), for example, reported a conditioned increase in cholera antibodies. Most immunologists, both then and today, argued that no known mechanism could account for a conditioned change in a specific immune response, and therefore cast doubt on this type of research (Ader, 1981c).

No further work on immune conditioning or comparable phenomena appears to have been done until the 1970s, apart from a British study which looked at the effects of hypnosis on the delayed-type hypersensitivity (DTH) immune response of human subjects (Black et al., 1963). Four subjects who were particularly susceptible to hypnosis were given the Mantoux tuberculin skin test on both arms and were instructed under deep hypnosis to show a skin reaction on one arm but not the other. Three of the subjects did indeed show a greatly attenuated inflammatory response on the hypnosis-suggestion arm, although this was almost certainly due to changes in local circulation rather than to changes in cellular immune reactions. Whatever the case, this experiment did not fit with the prevailing interests of the time, and was not followed up.

Conditioned immune changes were serendipitously "rediscovered" in the 1970s by Robert Ader and colleagues in the United States during experiments on learned taste aversion (see Ader, 1981c, 1983). Rats had repeatedly been injected with a drug (cyclophosphamide) to make them nauseous, immediately after drinking saccharin-flavored water. Cyclophosphamide is also a powerful suppressor of the immune system. During the subsequent extinction phase of the experiment, when rats were given flavored water alone, a number died for no obvious reason. Mortality was highest among the rats that had drunk most flavored water. Ader conjectured that the deaths might have resulted from a conditioned immune suppression, elicited by reexposing the rats to a conditioned stimulus (flavored water) which had previously been paired with immune suppression (cyclophosphamide).

To test this hypothesis, Ader and Cohen (1975) conditioned rats by repeatedly giving them saccharin-flavored water (the CS), immediately followed by an injection of cyclophosphamide (the US). Nonconditioned animals were given plain water and cyclophosphamide, and a placebo group was given plain water and an injection of the drug vehicle. Three days later all subjects were given an antigenic challenge by inoculating them with sheep red blood cells. Some conditioned animals were then immediately reexposed to flavored water (the CS), paired with a saline injection. To control for the effects of previous conditioning, another group of conditioned animals was given plain water plus a saline injection. Finally, to control for the unconditioned immunosuppressive effects of cyclophosphamide, another group was given plain water plus an injection of cyclophosphamide. The nonconditioned animals were given flavored water plus saline injection, and the placebo group was given plain water and no injection. Six days after inoculation, blood samples were taken and immune responses measured in terms of hemagglutinating antibody activity. The principal result was that the conditioned animals that were reexposed to the CS showed a significant reduction of about 25% in their immune response as compared with the control groups (nonconditioned, and conditioned but not reexposed to the CS). This result, which was subsequently replicated in two other laboratories (Rogers et al., 1976; Wayner et al., 1978), was consistent with the hypothesis of conditioned immune suppression.

Comparable results have since been obtained in a variety of different conditioning experiments, where factors such as the nature of the unconditioned and conditioned stimuli and the relative timing of exposure to the conditioned stimuli and antigenic stimulation have been varied (e.g., Ader et al., 1982; reviewed by Ader and Cohen, 1985). In most of these experiments a conditioned reduction in antibody response of around 25% has been reliably obtained. Ader and Cohen's original experiment looked at humoral (antibody-mediated) immunity. More recent experiments have also found conditioned suppression of cellular (lymphocyte-mediated) immunity (e.g., Bovbjerg et al., 1982; Gorczynski et al., 1982). Other work has demonstrated the extinction of a conditioned immune change after repeated, unreinforced presentations of the CS, thus supporting the parallel with Pavlovian conditioning (Bovbjerg et al., 1984).

In a later experiment, Ader and Cohen (1982) assessed how a conditioned reduction in immune function influenced the development of a disease. To do this they investigated the effects of conditioning on the survival of mice suffering from an autoimmune disease, systemic lupus erythematosus (SLE). Autoimmune diseases arise when the immune system starts to attack parts of the body as though they were antigenic. Thus, for individuals suffering from an autoimmune disease, a reduction in immune response is actually beneficial. Mice with SLE develop a lethal kidney disorder and die, but the disease can be treated with an immunosuppressive drug such as cyclophosphamide. Ader and Cohen therefore set out to see whether, after conditioning, a neutral CS could substitute for an immunosuppressive drug (US). As before, saccharin-flavored water was used as the CS and cyclophosphamide as the immunosuppressive US. Some conditioned animals were subsequently reexposed to flavored water alone, in place of some regular drug injections. Development of the autoimmune disease and mortality were both significantly delayed in conditioned animals, as compared with non-conditioned animals that received the same overall amount of drug. Thus, a CS was able to act as a partial substitute for the drug. Again, the results were consistent with the hypothesis of conditioned immune suppression.

More recent experiments have also supported the general conclusion that conditioned immune suppression can markedly influence an animal's chances of surviving an illness. For example, Gorczynski *et al.* (1985) showed that reexposing animals to a CS which had previously been paired with cyclophosphamide injections caused implanted tumors to grow more rapidly, resulting in an increased mortality rate.

Immune conditioning experiments have frequently been criticized on the grounds that the unconditioned stimuli (such as cyclophosphamide or anti-lymphocyte serum) are themselves highly stressful. Thus, it is argued, the CS may merely be eliciting a conditioned physiological stress response, involving the release of immunosuppressive corticosteroid hormones (see Section VIIA, below), rather than eliciting a conditioned change in the immune response itself. However, this argument is not supported by recent experiments in which conditioned *increases* in immune function have been demonstrated. For instance, Ghanta *et al.* (1985) have established that an increase in natural killer cell activity, elicited by chemical treatment, can be successfully conditioned using a novel smell (camphor) as the CS. Results such as these, showing that increases as well as decreases in immune function can be conditioned, suggest that conditioning is a real phenomenon that acts directly on the immune system rather than arising indirectly from conditioned changes in hormone levels.

If immune conditioning does turn out to be a real and robust phenomenon, then it offers compelling evidence that the brain influences the immune system. More intriguing still are the possible medical applications of immune conditioning. Immunosuppressive drugs are widely used to prevent tissue rejection after organ transplants and in combating autoimmune diseases. These drugs are often toxic and have unpleasant side effects. Therefore, any procedure that reduced the required dose would be highly beneficial. One highly speculative possibility might be to pair a neutral CS with injections of an immunosuppressive drug, as in Ader and Cohen's experiments, and then use the CS as a partial substitute for the drug. On the other hand, the relatively small magnitude of most conditioned immune changes means that their clinical effectiveness would probably be limited. The applications of conditioned increases in immune function, such as the conditioned increase in natural killer cell activity reported by Ghanta et al. (1985; see above), are even more obvious.

Adaptive Significance: What Is the Purpose of Immune Conditioning?

In terms of their possible biological function, the immune conditioning phenomena described above are distinctly puzzling. If an organism repeatedly experiences a stimulus which reliably predicts immune suppression, then reexposure to this conditioned stimulus should, if the system is adaptive and homeostatic, elicit a compensatory *increase* in immune response. On the contrary, though, the experiments show that the CS mimics the US by eliciting an immune suppression (see Ader and Cohen, 1985).

Questions about biological function ("What is it for?") have not yet been considered in relation to immune conditioning and other aspects of psychoimmunology. Many immunologists would doubtless argue that "sociobiological storytelling" about the possible adaptive benefits of links between brain, behavior, and immunity is likely to be both vacuous and futile. Certainly, it would appear at present to be premature, given the uncertain nature of some of the empirical data. Nonetheless, there is much to be said for the general principle of thinking about biological phenomena in terms of their adaptive significance. Hypotheses about biological function ("What is it for?") can sometimes suggest interesting experiments and questions about mechanism ("How does it work?").

VII. MECHANISMS OF CNS-IMMUNE SYSTEM INTERACTIONS

As emphasized in Section II, the immune system comprises many interacting components. In principle, neural modulation of immunity could act on virtually any of these components, by influencing, for example, specific cellular components, immune modulators such as lymphokines, secondary immune reactions such as complement activation or tissue inflammation, or local blood flow. Many neurotransmitters, hormones, and neuromodulators affect blood vessels, for example, and could alter lymphocyte traffic through tissues via peripheral vascular effects.

A. The Role of Corticosteroid Hormones

Despite the plethora of theoretically possible mechanisms, explanations of how stress affects the immune system have traditionally focused on one type of mediator: the corticosteroid hormones. These hormones are released from the adrenal cortex in response to, among other things, psychological and emotional stress.

The role of psychological and emotional factors in eliciting pituitary–adrenal activation has, of course, been studied over many years (see Mason, 1968; Hennessy and Levine, 1979), starting with the pioneering work of Hans Selyé on the physiological response to stress and the "general adaptation syndrome" (Selyé, 1936, 1946, 1955; Harlow and Selyé, 1937). Psychological stressors of many types reliably elicit pituitary–adrenal activation, with a concomitant rise in plasma levels of corticosteroid hormones. Furthermore, pituitary–adrenal activation can be conditioned and is elicited by neutral stimuli which predict an aversive event.

Corticosteroid hormones have many physiological and hormonal effects, most of which are highly adaptive in enabling animals to cope with acute physical and psychological demands. These effects include mobilizing energy reserves, reducing local tissue inflammation in response to injury, increasing the secretion of stomach acids, and a variety of subtle effects on learning and perception. Of primary interest here, though, are their effects on the immune system. Though complex, the effects of corticosteroid hormones on the immune system are usually immunosuppressive (Claman, 1972; Neifeld et al., 1977; Werb, 1978; Werb et al., 1978; Fauci, 1978/9; Parillo and Fauci, 1979; Cohen and Crnic, 1982). At levels similar to those produced by stress, corticosteroids can reduce the size of the lymph nodes and thymus, reduce antibody production, and reduce the number of circulating lymphocytes. At physiological doses, corticosteroids can inhibit phagocytosis and reduce the *in vitro* responsiveness of lymphocytes (Gillis et al., 1979). Corticosteroids also alter lymphocyte traffic through the peripheral vascular system (Thompson et al., 1980) and prevent lymphokines from activating macrophages (Monjan, 1981). In general, corticosteroids have a greater effect on the traffic and distribution of immune cells than on their functional properties (Fauci, 1978/9).

In view of their generally immunosuppressive effects, a widespread assumption has been that stress reduces immune function as a secondary result of eliciting the release of corticosteroids. Consistent with this view, several studies have shown that a reduction in immune function and an increase in corticosteroid levels occur concurrently. Furthermore, chronically stressed animals show the classic symptoms, originally described by Selyé more than 50 years ago, of adrenal hypertrophy and involution of the thymus.

However, an increasing body of evidence suggests that stress-induced immune suppression is not always mediated by corticosteroid hormones. For instance, experiments by Keller et al. (1983) have shown that some measures of immune function are reduced by stress in rats with their adrenal glands removed. Adrenalectomized rats were given inescapable electric shocks over a 24-hr period, a treatment which had previously been shown to suppress lymphocyte function in a dose-dependent manner, proportional to intensity (Keller et al., 1981). The rats' lymphocyte function was assessed by measuring the *in vitro* response to a mitogen (PHA). As expected, stress caused a reduction in lymphocyte response in intact (nonoperated and sham-adrenalectomized) rats and in adrenalectomized rats given replacement corticosterone (the principal corticosteroid hormone in rats). Intact animals also showed a stress-induced reduction in the total number of lymphocytes. However, stress also reduced the response of lymphocytes from adrenalectomized rats. Thus, the stress-induced reduction in lymphocyte function (though not the reduction in lymphocyte number) occurred independently of adrenal activation. In other words, at least one aspect of a stress-induced reduction in immune function does not appear to be mediated by corticosteroid hormones.

Similarly, in the study of bereavement described in Section IV.C (Bartrop et al., 1977), reduced lymphocyte function in bereaved people was not accompanied by raised levels of cortisol, again implying that some stress-induced changes in immune function can occur independently of changes in corticosteroid concentrations. Further evidence that corticosteroids are not always involved comes from immune conditioning experiments, in which changes in immune function have been found to occur independently of changes in corticosteroid levels (see Ader and Cohen, 1985). A final point is that, although pharmacological doses of corticosteroids are usually immunosuppressive, low doses can, under certain conditions, enhance lymphocyte function (Comsa et al., 1982). Thus, it would be incorrect to assume that all stress-induced increases in corticosteroid levels are necessarily immunosuppressive.

B. Other Possible Mediators

It seems increasingly likely that there are other mediators of stress-induced immune suppression besides corticosteroid hormones. Many hormones, neurotransmitters, and neuromodulators influence immune function, and the secretion of many of these substances is also affected by stress (e.g., Maclean and Reichlin, 1981; Monjan, 1981; Jemmott and Locke, 1984). Stress can elicit changes in the levels of growth hormone, ACTH, epinephrine, norepinephrine, testosterone, insulin, prolactin, glucagon, parathyroid hormone, melanocyte-stimulating hormone, follicle-stimulating hormone, luteinizing hormone, thyroxine, thyrotropin, vasopressin, aldosterone, renin, β-endorphin, and enkephalins, and many of these substances are known to affect the immune system (Maclean and Reichlin, 1981; Monjan, 1981; Borysenko and Borysenko, 1982).

1. Catecholamines

To take but one example, the release of the catecholamines epinephrine and norepinephrine (from the adrenal medulla and sympathetic nerve endings, respectively) is stimulated by stressors similar to those that stimulate corticosteroid release (Axelrod and Reisine, 1984). Thus, increased catecholamine levels are found in humans in situations involving novelty, anticipation, and unpredictability, especially those requiring attention and vigilance (e.g., Frankenhaeuser, 1975; Dimsdale and Moss, 1980). Physical stressors, such as cold, heat, injury, vigorous exercise, surgery, and hypoxia, also elicit raised catecholamine levels (Axelrod and Reisine, 1984). Epinephrine and norepinephrine, like corticosteroids, have many physiological and behavioral effects, some of which are concerned with responding to potential threat or injury by inducing adaptive changes in metabolism, sensory, and cardiac functions.

Receptors for epinephrine and norepinephrine are found on lymphocytes and polymorphonuclear leukocytes (Pochet *et al.*, 1979; Dulis and Wilson, 1980; Brodde *et al.*, 1981) and are involved in modulating various components of the immune response (e.g., Ito *et al.*, 1977). Furthermore, stimulating β-adrenergic (epinephrine) receptors on lymphocytes and monocytes reduces their responsiveness to mitogens and has other effects, such as causing a transient increase in the number of circulating suppressor T cells and a decrease in circulating helper T cells (e.g., Yu and Clements, 1976; Crary *et al.*, 1983). β-Adrenergic drugs are immunosuppressive in experimental animals; for example, a single intraperitoneal injection of epinephrine or other β-adrenergic agonists greatly increased the proliferation of tumor cells injected into the lungs of rats (Van den Brenk *et al.*, 1976). In principle, therefore, epinephrine and norepinephrine are possible mediators of neural influences on immune function. Receptors for various other substances, including acetylcholine (Maslinski *et al.*, 1980) and histamine (Roskowski *et al.*, 1977), are also found on lymphocytes.

2. Neuropeptides

Yet another possibility which has begun to receive more attention is that substances produced by the brain might act directly on the immune system (Maclean and Reichlin, 1981). Particular attention has focused on the neuropeptides, such as β-endorphin and Met-enkephalin (Plotnikoff *et al.*, 1982; Bloom, 1985; Pert *et al.*, 1985; Wybran, 1985). The discovery of the neuropeptides has revolutionized the neurosciences by uncovering a whole new class of transmitters in addition to the classical neurotransmitters. Well over 50 different neuropeptides are known, and the number is increasing rapidly. Most, if not all, influence behavior and mood state as well as having a variety of other effects (Pert *et al.*, 1985). Some types of stress, such as uncontrollable electric shock, are known to stimulate the release of neuropeptides from the brain and from peripheral sites, and these chemicals may be involved in modifying the activity of the immune system.

Receptors for neuropeptides are found on lymphocytes and monocytes; for example, lymphocytes carry receptors for Met-enkephalin, morphine, and β-endorphins (Hazum *et al.*, 1979). Moreover, these substances have been shown to affect immune function. Met-enkephalin and morphine induce immune enhancement and immune suppression, respectively, while β-endorphins increase the mitogen response of lymphocytes in a dose-dependent manner (Gilman *et al.*, 1982). Met-enkephalin has been shown to prolong the survival of mice injected with leukemia cells (Plotnikoff *et al.*, 1982). Neuropeptides have also been shown to alter the cytotoxic activity of natural killer cells and recent work has demonstrated that another neuropeptide, Substance P, can increase the ability of macrophages to kill tumor cells.

There is great interest, therefore, in the idea that neuropeptides might be involved in neural modulation of the immune system (Hazum *et al.*, 1979; Gilman *et al.*, 1982). In a recent review of this topic, Wybran (1985) concludes that endorphins and enkephalins should be regarded as functional humoral mediators linking the central nervous system and immune system.

Circumstantial support for a role of neuropeptides, specifically the enkephalins and endorphins (or opioid peptides), comes from an experiment which looked at the effects of electric shock on the activity of natural killer cells in rats (Shavit *et al.*, 1984). The experiment was based on the finding that inescapable footshock produces quite different central effects, depending upon whether it is applied continuously or intermittently. Intermittent shock induces a state of analgesia that is mediated by opioid peptides (so-called "opioid stress"). Continuous shock, on the other hand, induces analgesia that does not depend on opioids ("nonopioid stress"). Rats were given either intermittent or continuous shock and samples of natural killer cells were then obtained. The results showed that intermittent shock ("opioid stress") led to a 26% reduction in the cytotoxic activity of natural killer cells, relative to controls. This reduction was blocked by the opioid antagonist naltrexone and could be mimicked in a dose-dependent way by morphine. Continuous shock ("nonopioid stress"), on the other hand, did not suppress natural killer cell activity. These results are consistent with the hypothesis that some forms of stress cause a central release of opioid peptides that act directly on immune cells to alter immune function. An alternative possibility, though, is that the effect is indirect, with opioid peptides merely influencing the release of other hormones. Using a similar procedure, Shavit *et al.* (1985) have also demonstrated that "opioid" (but not "nonopioid") footshock stress reduces the resistance of female rats to implanted mammary tumors.

In conclusion, then, the brain may influence the immune system in one of two principal ways: indirectly, via changes in the neuroendocrine system; or directly, via neural connections or the secretion of substances such as neuropeptides (Solomon and Amkraut, 1979, 1981; MacLean and Reichlin, 1981; Comsa *et al.*, 1982; Stein *et al.*, 1982; Besedovsky *et al.*, 1979, 1985b; Bloom, 1985; Wigzell, 1985; see also Fig. 2).

VIII. DOES THE IMMUNE SYSTEM INFLUENCE BEHAVIOR?

We have considered some of the evidence showing that the brain influences the immune system. Is there any evidence for the converse effect, namely an influence of the immune system on brain and behavior? In general, there is little solid support for this idea, but some interesting ideas and speculations are worth considering.

Some tentative and controversial evidence has suggested that some mental illnesses, notably schizophrenia, may be associated with certain immunological abnormalities, including abnormal levels of immunoglobulins, abnormalities in the structure and function of immune cells, and the presence of antibrain antibodies (e.g., Shopsin *et al.*, 1973; Johnstone and Whaley, 1975; Albrecht *et al.*, 1980; see Solomon, 1981). For example, a study of 80 men suffering from acute schizophrenic reactions found significantly elevated levels of IgA and IgM, as compared with normal controls (Amkraut *et al.*, 1973). Some studies have also reported morphological abnormalities in lymphocytes obtained from schizophrenics and their relatives, and antibrain antibodies ("autoantibodies") have reportedly been found in the brains of schizophrenics (Baron *et al.*, 1977; Knight, 1982). However, it is likely that, because of their abnormal behavior and living conditions, psychiatric patients might in any case be exposed to more immune challenges than are normal people, so results like these should be treated with caution. Another problem is that immune abnormalities might be caused by drugs given to psychiatric patients; the antipsychotic drug chlorpromazine, for example, is reported to induce changes in antibodies. Many of the ideas about immunological involvement in mental disorders are highly speculative and the empirical evidence is confused and often conflicting (see, for example, Bergen *et al.*, 1980).

Firmer evidence for links between abnormalities in brain function and immune function comes from work indicating associations between anomalous cerebral dominance, various developmental learning disorders, and abnormalities in immune function. Geschwind and Behan (1982) reported finding more than double the incidence of immune disorders (such as atopic diseases and autoimmune thyroiditis) and ten times the incidence of developmental learning disorders (such as dyslexia, stuttering, and Tourette syndrome) among strongly left-handed people as compared with strongly right-handed people. An increased incidence of immune and developmental learning disorders was also found in the relatives of left-handed people.

Geschwind and Behan suggested that these associations, which were predominantly found in males rather than females, arise because of prenatal exposure to abnormal levels of testosterone, leading to a cluster of parallel changes in the development of both brain and immune system (Geschwind, 1984; Geschwind and Behan, 1984; Behan and Geschwind, 1985). They argued that exposure to an unusually high level of testosterone *in utero* delays development of the left cerebral hemisphere, leading to right-side dominance and therefore left-handedness, and also delays development of the thymus, leading to later immunological disorders.

More recently, the notion that the links between the brain and immune system are reciprocal has been supported by work showing that the brain and immune system can communicate via signal molecules common to both systems. For instance, stimulated lymphocytes and other immunologically active tissues have been found to produce endorphinlike substances which are remarkably similar to neuroendocrine hormones in their structure and function (Smith *et al.*, 1985). The molecular basis of the two-way link between the brain and immune system is reviewed by Besedovsky *et al.* (1985b) and Blalock *et al.* (1985).

IX. A MEDIATING ROLE OF SOCIAL RELATIONSHIPS?

A promising area of research, which seems likely to flourish, concerns the role of social relationships in modifying the impact of stressors on immune function. During early life the most important social relationship for mammals is that between mother and offspring, and disruption of this relationship can influence immune function. Work on bonnet macaques, for example, found that separating infants from their mothers for 2 weeks caused a significant reduction in their lymphocyte function, relative to a preseparation baseline (Laudenslager *et al.*, 1982a,b). Lymphocyte responses returned to normal following reunion. Mother–infant separation has also been found to suppress immune function in infant squirrel monkeys (Coe et al., 1985). Seven days after separation from their mothers, 6-month-old infants showed reduced antibody responses, reduced serum complement activity, and reduced levels of serum immunoglobulin IgG, accompanied by elevated plasma cortisol levels. However, mother–infant separation for 7 days did not induce any comparable immunological changes in rhesus monkeys. In mice, daily periods of maternal separation during the first 2 weeks, followed by early weaning, had long-term effects on immune responsiveness. When tested at 7–8 weeks by immunization with sheep erythrocytes, maternally deprived mice showed a significantly reduced level of serum immunoglobulins as compared with normally reared mice, even though they did not differ in body weight or adrenal weight (Michaut et al., 1981).

The mother–infant relationship is not the only type of relationship that appears to influence immune function. Peer separation has also been found to suppress cellular immunity in pigtailed macaques. Reite et al. (1981) found that two 9-month-old infants that had been raised together since birth showed reduced lymphocyte function when separated for 2 weeks.

In humans, many epidemiological studies have found that individuals who have poor social and community relationships also tend to have a higher incidence of illness and mortality (e.g., Berkman and Symes, 1979). In particular, associations between loneliness and poor health have frequently been reported (e.g., Rubenstein and Shaver, 1980). Bereavement and divorce also appear to have a major impact on health; both are associated with increased rates of mortality, physical illness, depression, suicide, and mental illness (Bloom et al., 1978; Stroebe and Stroebe, 1986; see Section IV.C). To give one example, a prospective study of lung cancer patients found that patients whose tumors subsequently became malignant were more likely to have reported the loss of a significant relationship during the preceding 5 years, as well as less job stability and a lack of plans for the future, relative to patients whose tumors turned out to be benign (Horne and Pickard, 1979).

Several studies of the health effects of life events (see Section IVB) have found that individuals who have good social support from family and friends are generally less affected than are those who lack social support (see Vachon et al., 1982; Hofer, 1984; Stroebe and Stroebe, 1986). For instance, in a study of 163 Naval Academy students, Sarason et al. (1985) found that the association between negative life events and illness was stronger among individuals with low levels of social support as compared with those who had high levels of social support.

Direct evidence for a role of the immune system in such effects has come from recent work showing associations between loneliness and reduced immune function. For instance, high levels of self-reported loneliness were found to be associated with significantly lower levels of natural killer cell activity in a sample of 75 medical students (Kiecolt-Glaser *et al.*, 1984a). A similar association between loneliness and reduced immune function was found in a sample of 33 newly admitted psychiatric patients; those who scored above average on self-reported measures of loneliness had significantly lower levels of natural killer cell activity, a poorer T-lymphocyte response to PHA, and elevated urinary cortisol levels (Kiecolt-Glaser *et al.*, 1984b).

X. FUTURE RESEARCH: THE ROLE OF ETHOLOGY

Research into the relations between behavior, the brain, and the immune system has only begun to scratch the surface, and many potentially exciting avenues remain virtually unexplored. The time is ripe for ethologists to become involved in this work and to contribute their expertise in the recording and analysis of complex behavior, together with their capacity to deal with whole systems.

Experiments, particularly those on animals, have so far tended to be crude in terms of their behavioral measures and to rely upon unnatural forms of stress, such as inescapable electric shock, as a means of influencing immune function. There is great scope, in both animal and human work, for studying phenomena which are more biologically realistic, such as the effects of bereavement or chronic social subordination.

Research has also tended to concentrate on factors that reduce immune function. In future, more attention could be paid to factors that might have the opposite effect. Plausible candidates might include the relaxation response (e.g., Borysenko, 1982), positive social interactions (e.g., Nerem *et al.*, 1980), and conditioned increases in immune function (e.g., Ghanta *et al.*, 1985).

One specific area where psychoimmunological techniques might be applied is in the field of animal welfare, where physiological measures such as corticosteroid levels are already used for assessing the effects of stress on animals in farms, zoos, and laboratories (see Dawkins, 1980). Measures of immune function could be used as another way of assessing the impact of housing conditions, handling procedures, and transportation on well-being.

In these and in other areas of psychoimmunological research, ethology could make at least three important contributions. First, and most obvious, is the ethologist's expertise in the direct observation and description of behavior (see Martin and Bateson, 1986). A major weakness of much psychoimmunological research so far has been the relatively crude way in which behavioral variables have been measured, particularly in animal studies. Future research will need to employ multiple measures of behavior as well as multiple measures of immune function. A second contribution of ethology could be to introduce experimental models that are more realistic and biologically meaningful, moving away from the past obsession with applying electric shocks to rats.

Even evolutionary biologists might be able to make a contribution to psychoimmunology, which remains as yet a blushing virgin as far as evolutionary and functional ideas are concerned. After all, one major question that has not yet been tackled is the evolutionary significance of all these remarkable phenomena. If there truly are elaborate biological mechanisms that enable psychological and emotional factors to influence immune function, then why have they evolved and how might they help animals to survive and reproduce in their natural environments? In principle, thinking about possible biological functions can help in understanding mechanism, but whether concepts such as adaptation and optimization will contribute anything useful to psychoimmunology remains to be seen.

The benefit to ethologists of working in psychoimmunology would be the chance to contribute to an exciting and rapidly expanding field of research which is of fundamental scientific importance and may also be of great practical benefit.

XI. APPENDIX: SOME IMMUNOLOGICAL TERMS

Antibodies, or *immunoglobulins*, are blood serum proteins produced by activated B lymphocytes after they have reacted with a specific antigen. The antibody binds to the antigen and the complex is then destroyed by phagocytic white blood cells.

Antigens are foreign ("nonself") substances, such as bacteria, viruses, or grafted tissue cells, that stimulate an immune response. They react with lymphocytes carrying receptors for that particular antigen, causing the lymphocytes to produce specific antibodies.

The immune system has two main components. *Humoral immunity* involves the production of antibodies by stimulated B lymphocytes, and is elicited by bacteria and small particulate antigens. *Cell-mediated immunity* involves several components, including T lymphocytes, natural killer cells, and complement proteins, and is elicited by eukaryotic pathogens, tumor cells, foreign tissue grafts, and virally infected cells. In addition, nonspecific responses (i.e., not specific to a particular antigen) bring about tissue inflammation and the engulfing of foreign particles by phagocytic white blood cells (macrophages, monocytes, and polymorphonuclear leukocytes).

Lymphocytes are the antigen-specific cells of the immune system, and account for about 25% of white blood cells in human blood. They are of two types:

B cells and T cells. B cells produce immunoglobulins. T cells kill antigen-bearing cells after direct contact. After encountering a specific antigen, lymphocytes are stimulated to divide. Each lymphocyte has unique cell-surface receptors and will react only with one unique antigen. Control of the immune response (activation and suppression) is partly mediated by helper and suppressor T cells.

Mitogens are chemicals, such as phytohemagglutinin (PHA), pokeweed mitogen (PWM), or concanavalin A (ConA), which react with lymphocytes in a nonspecific way, stimulating them to proliferate and undergo cell division. The nonspecific responsiveness to mitogens of lymphocytes obtained from blood samples is a commonly used measure of the number of normally functioning lymphocytes in circulation.

Natural killer cells are large white blood cells which kill foreign target cells, such as tumor cells, without the need for specific antigenic recognition. They are thought to be involved in immune surveillance against tumor formation and viral infection.

XII. ACKNOWLEDGMENTS

I am grateful to Gareth Bowen, Joe Herbert, and James Serpell for commenting on an earlier draft.

XIII. REFERENCES

Ader, R. (1980). Presidential address: psychosomatic and psychoimmunologic research. *Psychosom. Med.* **42**:307–322.
Ader, R. (ed.). (1981a). *Psychoneuroimmunology*, Academic Press, New York.
Ader, R. (1981b). Behavioral influences on immune responses. In Weiss, S. M., Herd, J. A., and Fox, B. H. (eds.), *Perspectives on Behavioral Medicine*, Academic Press, New York, pp. 163–182.
Ader, R. (1981c). A historical account of conditioned immunobiologic responses. In Ader, R. (ed.), *Psychoneuroimmunology*, Academic Press, New York, pp. 321–352.
Ader, R. (1983). Developmental psychoneuroimmunology. *Dev. Psychobiol.* **16**:251–267.
Ader, R., and Cohen, N. (1975). Behaviorally conditioned immunosuppression. *Psychosom. Med.* **37**:333–340.
Ader, R., and Cohen, N. (1982). Behaviorally conditioned immunosuppression and murine systemic lupus erythematosus. *Science* **215**:1534–1536.
Ader, R., and Cohen, N. (1985). CNS–immune system interactions: conditioning phenomena. *Behav. Brain Sci.* **8**:379–394.
Ader, R., Cohen, N., and Bovbjerg, D. (1982). Conditioned suppression of humoral immunity in the rat. *J. Comp. Physiol. Psychol.* **96**:517–521.

Albrecht, P., Torrey, E. F., Boone, E., Hicks, J. T., and Daniel, N. (1980). Raised cytomegalovirus-antibody level in cerebrospinal fluid of schizophrenic patients. *Lancet* **2**:769–772.

Amkraut, A. A., and Solomon, G. F. (1972). Stress and murine sarcoma virus (Moloney) induced tumors. *Cancer Res.* **32**:1428–1433.

Amkraut, A., Solomon, G. F., Allansmith, M., McClellan, B., and Rappaport, M. (1973). Immunoglobulins and improvements in acute schizophrenic reactions. *Arch. Gen. Psychiatr.* **28**:673–677.

Axelrod, J., and Reisine, T. D. (1984). Stress hormones: their interaction and regulation. *Science* **224**:452–459.

Bammer, K., and Newberry, B. H. (eds.) (1982). *Impact of Stress on Immunity and Cancer*, C. J. Hogrefe, Toronto.

Baron, M., Stern, M., Anair, R., and Witz, I. P. (1977). Tissue-binding factor in schizophrenic sera: a clinical and genetic study. *Biol. Psychiatr.* **12**:199–212.

Bartrop, R. W., Lockhurst, E., Lazarus, L., Kiloh, L. G., and Penny, R. (1977). Depressed lymphocyte function after bereavement. *Lancet* **1**:834–836.

Behan, P. O., and Geschwind, N. (1985). Hemispheric laterality and immunity. In Guillemin, R., Cohn, M., and Melnechuk, T. (eds.), *Neural Modulation of Immunity*, Raven Press, New York, p. 73–79.

Bergen, J. R., Grinspoon, L., Pyle, H. M., Martinez, J. L., Jr., and Pennel, R. B. (1980). Immunologic studies in schizophrenic and control patients. *Biol. Psychiatr.* **15**:369–379.

Berkman, L. F., and Symes, L. S. (1979). Social networks, host resistance, and mortality: a nine year follow-up study of Alameda County residents. *Amer. J. Epidemiol.* **109**:186–204.

Besedovsky, H. O., Sorkin, E., Felix, D., and Haas, H. (1977). Hypothalamic changes during the immune response. *Eur. J. Immunol.* **7**:323–325.

Besedovsky, H. O., Del Rey, A., Sorkin, E., Da Prada, M., and Keller, H. H. (1979). Immunoregulation mediated by the sympathetic nervous system. *Cell. Immunol.* **48**:346–355.

Besedovsky, H., Del Rey, A., Sorkin, E., Da Prada, M., Burri, R., and Honegger, C. (1983). The immune response evokes changes in brain noradrenergic neurons. *Science* **221**:564–565.

Besedovsky, H. O., Del Rey, A., and Sorkin, E. (1985a). Immunological–neuroendocrine feedback circuits. In Guillemin, R., Cohn, M., and Melnechuk, T. (eds.), *Neural Modulation of Immunity*, Raven Press, New York, pp. 165–172.

Besedovsky, H. O., Del Rey, A., and Sorkin, E. (1985b). Immune–neuroendocrine interactions. *J. Immunol.* **135**:750s–754s.

Black, S., Humphrey, J. H., and Niven, J. S. F. (1963). Inhibition of Mantoux reaction by direct suggestion under hypnosis. *Br. Med. J.* **6**:1649–1652.

Blalock, J. E., Harbour-McMenamin, D., and Smith, E. M. (1985). Peptide hormones shared by the neuroendocrine and immunologic systems. *J. Immunol.* **135**:858s–861s.

Bloom, B. L., Asher, S. J., and White, S. W. (1978). Marital disruption as a stressor: a review and analysis. *Psych. Bull.* **85**:867–894.

Bloom, F. E. (1985). Neuropeptides and other mediators in the central nervous system. *J. Immunol.* **135**:743s–745s.

Borysenko, J. (1982). Behavioral-physiological factors in the development and management of cancer. *Gen. Hosp. Psychiatr.* **4**:69–74.

Borysenko, M., and Borysenko, J. (1982). Stress, behavior, and immunity: animal models and mediating mechanisms. *Gen. Hosp. Psychiatr.* **4**:59–67.

Bovbjerg, D., Ader, R., and Cohen, N. (1982). Behaviorally conditioned suppression of a graft-versus-host response. *Proc. Nat. Acad. Sci. USA* **79**:583–585.

Bovbjerg, D., Ader, R., and Cohen, N. (1984). Acquisition and extinction of conditioned suppression of a graft-vs-host response in the rat. *J. Immunol.* **132**:111–113.

Boyce, W. T., Cassel, J. C., Collier, A. M., Jensen, E. W., Ramey, C. T., and Smith, A. H.

(1977). Influence of life events and family routines on childhood respiratory tract illness. *Pediatrics* **60**:609–615.

Brodde, O.-E., Engel, G., Hoyer, D., Bock, K. D., and Weber, F. (1981). The beta-adrenergic receptor in human lymphocytes: sub-classification by the use of a new radio-ligand, (±)-^{125}-iodocyanopindolol. *Life Sci.* **29**:2189–2198.

Brooks, W. H., Cross, R. J., Roszman, T. L., and Markesbery, W. R. (1982). Neuroimmunomodulation: neuroanatomical basis for impairment and facilitation. *Ann. Neurol.* **12**:56–61.

Bulloch, K. (1985). Neuroanatomy of lymphoid tissue: a review. In Guillemin, R., Cohn, M., and Melnechuk, T. (eds.), *Neural Modulation of Immunity*, Raven Press, New York, pp. 111–140.

Bulloch, K., and Moore, R. Y. (1981). Innervation of the thymus gland by brainstem and spinal cord in mouse and rat. *Am. J. Anat.* **162**:157–166.

Burchfield, S. R., Woods, S. C., and Elich, M. S. (1978). Effects of cold stress on tumor growth. *Physiol. Behav.* **21**:537–540.

Busse, W. W., Anderson, C. L., Hanson, P. G., and Folts, J. D. (1980). The effect of exercise on the granulocyte response to isoproterenol in the trained athlete and unconditioned individual. *J. Allergy Clin. Immunol.* **65**:358–364.

Cappel, R., Gregoire, F., Thiry, L., and Sprecher, S. (1978). Antibody and cell mediated immunity to herpes simplex virus in psychotic depression. *J. Clin. Psychiatr.* **39**:266–268.

Chandra, R. K. (1981). Immunodeficiency in undernutrition and overnutrition. *Nutr. Rev.* **39**:225–231.

Claman, H. N. (1972). Corticosteroids and lymphoid cells. *N. Engl. J. Med.* **287**:388–397.

Clayton, P. J. (1979). The sequelae and nonsequelae of conjugal bereavement. *Am. J. Psychiatr.* **136**:1530–1534.

Coe, C. L., Wiener, S. G., Rosenberg, L. T., and Levine, S. (1985). Endocrine and immune responses to separation and maternal loss in nonhuman primates. In Reite, M., and Field, T. (eds.), *The Psychobiology of Attachment and Separation*, Academic Press, New York, pp. 163–199.

Cohen, J. J., and Crnic, L. S. (1982). Glucocorticosteroids, stress and the immune response. In Webb, D. R. (ed.), *Immunopharmacology and the Regulation of Leukocyte Function*, Marcel Dekker, New York, pp. 61–91.

Cohen, J., Cullen, J. W., and Martin, L. R. (eds.). *Psychosocial Aspects of Cancer*, Raven Press, New York.

Comsa, J., Leonhardt, H., and Wekerle, H. (1982). Hormonal coordination of the immune response. *Rev. Physiol. Biochem. Pharmacol.* **92**:115–189.

Cooper, E. L. (ed.) (1983). *Stress, Immunity and Aging*, Marcel Dekker, New York.

Crary, B., Borysenko, M., Sutherland, D. C., Kutz, I., Borysenko, J., and Benson, H. (1983). Decrease in mitogen responsiveness of mononuclear cells from peripheral blood after epinephrine administration in humans. *J. Immunol.* **130**:694–697.

Cunningham, A. J. (1981). Mind, body, and immune response. In Ader, R. (ed.), *Psychoneuroimmunology*, Academic Press, New York, pp. 609–617.

Dawkins, M. S. (1980). *Animal Suffering. The Science of Animal Welfare*, Chapman & Hall, London.

Del Rey, A., Besedovsky, H. O., Sorkin, E., Da Prada, M., and Bondiolotti, G. P. (1982). Sympathetic immunoregulation: difference between high- and low-responder animals. *Am. J. Physiol.* **242**:R303.

Dimsdale, J. E., and Moss, J. (1980). Short-term catecholamine response to psychological stress. *Psychosom. Med.* **42**:493–497.

Dohrenwend, B. S., and Dohrenwend, B. P. (eds.) (1974). *Stressful Life Events: Their Nature and Effects*, Wiley, New York.

Dorian, B. J., Keystone, E., Garfinkel, P. E., and Brown, G. M. (1981). Immune mechanisms in acute psychological stress. *Psychosom. Med.* **43**:84.

Dulis, B. H., and Wilson, I. B. (1980). The beta-adrenergic receptor of live human polymorphonuclear leukocytes. *J. Biol. Chem.* **255**:1043–1048.

Edwards, E. A., Rahe, R. H., Stephens, P. M., and Henry, J. P. (1980). Antibody response to bovine serum albumin in mice: the effects of psychosocial environmental change. *Proc. Soc. Exp. Biol. Med.* **164**:478–481.

Fauci, A. S. (1978/9). Mechanisms of the immunosuppressive and anti-inflammatory effects of glucocorticosteroids. *J. Immunopharmacol.* **1**:1–25.

Felten, D. L., Felten, S. Y., Carlson, S. L., Olschowka, J. A., and Livnat, S. (1985). Noradrenergic and peptidergic innervation of lymphoid tissue. *J. Immunol.* **135**:755s–765s.

Fox, B. H. (1978). Premorbid psychological factors as related to cancer incidence. *J. Behav. Med.* **1**:45–134.

Fox, B. H. (1981). Behavioral issues in cancer. In Weiss, S. M., Herd, J. A., and Fox, B. H. (eds.), *Perspectives on Behavioral Medicine*, Academic Press, New York, pp. 101–133.

Fox, B. H. (1983). Current theory of psychogenic aspects of cancer incidence and prognosis. *J. Psychosoc. Oncol.* **1**:17–31.

Frankenhaeuser, M. (1975). Experimental approaches to the study of catecholamines and emotions. In Levi, L. (ed.), *Emotions: Their Parameters and Measurement*, Raven Press, New York, pp. 209–234.

Friedman, S. B., Ader, R., and Grota, L. J. (1973). Protective effect of noxious stimulation in mice infected with rodent malaria. *Psychosom. Med.* **35**:535–537.

Garrity, T. F., Marx, M. B., and Somes, G. W. (1978). The relationship of recent life changes to seriousness of later illness. *J. Psychosom. Res.* **22**:7–12.

Gerber, I., Rusalem, R., Hannon, N., Battin, D., and Arkin, A. (1975). Anticipatory grief and aged widows and widowers. *J. Gerontol.* **30**:225–229.

Geschwind, N. (1984). Immunological associations of cerebral dominance. In Behan, P., and Spreafico, F. (eds.), *Neuroimmunology*, Raven Press, New York, pp. 451–461.

Geschwind, N., and Behan, P. (1982). Left-handedness: association with immune disease, migraine, and developmental learning disorder. *Proc. Nat. Acad. Sci. USA* **79**:5097–5100.

Geschwind, N., and Behan, P. O. (1984). Laterality, hormones, and immunity. In Geschwind, N., and Galaburda, A. M. (eds.), *Cerebral Dominance. The Biological Foundations*, Harvard University Press, Cambridge, Massachusetts, pp. 211–224.

Ghanta, V. K., Hiramoto, R. N., Solvason, H. B., and Spector, N. H. (1985). Neural and environmental influences on neoplasia and conditioning of NK activity. *J. Immunol.* **135**:848s–852s.

Gillis, B., Crabtree, G. R., and Smith, K. A. (1979). Glucocorticoid-inhibition of T cell growth factor production. I. The effect on mitogen-induced lymphocyte proliferation. *J. Immunol.* **123**:1624–1631.

Gilman, S. C., Schwartz, J. M., Milner, R. J., Bloom, F. E., and Feldman, J. D. (1982). Beta-endorphin enhances lymphocyte proliferative responses. *Proc. Nat. Acad. Sci. USA* **79**:4226–4230.

Goetzl, E. J. (ed.) (1985). Neuromodulation of immunity and hypersensitivity. *J. Immunol.* **135**(Suppl. 2):739s–863s.

Gorczynski, R. M., Macrae, S., and Kennedy, M. (1982). Conditioned immune response associated with allogeneic skin grafts in mice. *J. Immunol.* **129**:704–709.

Gorczynski, R., Kennedy, M., and Ciampi (1985). Cimetidine reverses tumor growth enhancement of plasma-cytoma tumors in mice demonstrating conditioned immunosuppression. *J. Immunol.* **134**:4261–4266.

Greer, S., Morris, T., and Pettingale, K. W. (1979). Psychological response to breast cancer: effect on outcome. *Lancet* **2**:785–787.

Gross, R. L., and Newberne, P. M. (1980). Role of nutrition in immunologic function. *Physiol. Rev.* **60:**188-302.

Hamilton, D. R. (1974). Immunosuppressive effects of predator induced stress in mice with acquired immunity to *Hymenolepsis nana. J. Psychosom. Res.* **18:**143-153.

Harlow, C. M., and Selyé, H. (1937). The blood picture in the alarm reaction. *Proc. Soc. Exp. Biol. Med.* **36:**141-144.

Hazum, E., Chang, K. J., and Cuatrecasas, P. (1979). Specific nonopiate receptors for beta-endorphin. *Science* **205:**1033-1035.

Helsing, K. J., Szklo, M., and Comstock, G. W. (1981). Factors associated with mortality after widowhood. *Am. J. Public Health* **71:**802-809.

Hennessy, J. W., and Levine, S. (1979). Stress, arousal, and the pituitary-adrenal system: a psychoendocrine hypothesis. In Sprague, J. M., and Epstein, A. N. (eds.), *Progress in Psychobiology and Physiological Psychology*, Vol. 8, Academic Press, New York, pp. 133-178.

Hofer, M. (1984). Relationships as regulators: a psychobiologic perspective on bereavement. *Psychosom. Med.* **46:**183-197.

Horne, R. L., and Picard, R. S. (1979). Psychosocial risk factors for lung cancer. *Psychosom. Med.* **41:** 503-514.

Hudson, R. J. (1973). Stress and *in vitro* lymphocyte stimulation by phytohemagglutinin in Rocky Mountain bighorn sheep. *Can. J. Zool.* **51:**479-482.

Hull, D. (1977). Life circumstances and physical illness: a cross disciplinary survey of research content and method for the decade 1965-1975. *J. Psychosom. Res.* **21:**115-139.

Hurst, M. W., Jenkins, C. D., and Rose, R. M. (1976). The relation of psychological stress to onset of medical illness. *Ann. Rev. Med.* **27:**301-312.

Ito, M., Sless, F., and Parrott, D. M. (1977). Evidence for control of complement receptor rosette-forming cells by alpha- and beta-adrenergic agents. *Nature (Lond.)* **266:**633-635.

Jacobs, S., and Ostfeld, A. (1977). An epidemiological review of the mortality of bereavement. *Psychosom. Med.* **39:**344-357.

Jemmott, J. B., and Locke, S. E. (1984). Psychosocial factors, immunologic mediation, and human susceptibility to infectious diseases: how much do we know? *Psych. Bull.* **95:**78-108.

Jemmott, J. B., Borysenko, J. Z., Borysenko, M., McClelland, D. C., Chapman, R., Meyer, D., and Benson, H. (1983). Academic stress, power motivation, and decrease in salivary secretory immunoglobulin A secretion rate. *Lancet* **1:**1400-1402.

Jensen, M. M. (1968). The influence of stress on murine leukemia virus infection. *Proc. Soc. Exp. Biol. Med.* **127:**610-614.

Jensen, M. M., and Rasmussen, A. F., Jr. (1963). Stress and susceptibility to viral infections: II. Sound stress and susceptibility to vesicular stomatitis virus. *J. Immunol.* **90:**21-23.

Johnstone, E. C., and Whaley, K. (1975). Antinuclear antibodies in psychiatric illness: their relationship to diagnosis and drug treatment. *Br. Med. J.* **2:**724-725.

Kasl, S. V., Evans, A. S., and Niederman, J. C. (1979). Psychosocial risk factors in the development of infectious mononucleosis. *Psychosom. Med.* **41:**455-466.

Keller, S. E., Stein, M., Camerino, M. S., Schleifer, S. J., and Sherman, J. (1980). Suppression of lymphocyte stimulation by anterior hypothalamic lesions in the guinea pig. *Cell Immunol.* **52:**334-340.

Keller, S. E., Weiss, J. M., Schleifer, S. J., Miller, N. E., and Stein, M. (1981). Suppression of immunity by stress: effect of a graded series of stressors on lymphocyte stimulation in the rat. *Science* **213:**1397-1400.

Keller, S. E., Weiss, J. M., Schleifer, S. J., Miller, N. E., and Stein, M. (1983). Stress-induced suppression of immunity in adrenalectomized rats. *Science* **221:**1301-1304.

Kiecolt-Glaser, J. K., Garner, W., Speicher, C., Penn, G. M., Holliday, J., and Glaser, R. (1984a). Psychosocial modifiers of immunocompetence in medical students. *Psychosom. Med.* **46:**7-14.

Kiecolt-Glaser, J. K., Ricker, D., George, J., Messick, G., Speicher, C., Garner, W., and Glaser, R. (1984b). Urinary cortisol levels, cellular immunocompetency, and loneliness in psychiatric inpatients. *Psychosom. Med.* **46**:15–23.
Knight, J. G. (1982). Dopamine-receptor-stimulating autoantibodies: a possible cause of schizophrenia. *Lancet* **2**:1073–1076.
Kronfol, Z., Silva, J., Greden, J., Dembinski, S., and Carroll, B. J. (1982). Cell-mediated immunity in melancholia. *Psychosom. Med.* **44**:304.
Laudenslager, M., Reite, M., and Harbeck, R. (1982a). Suppressed immune response in infant monkeys associated with maternal separation. *Behav. Neural Biol.* **36**:40–48.
Laudenslager, M., Reite, M., and Harbeck, R. J. (1982b). Immune status during mother–infant separation. *Psychosom. Med.* **44**:303.
Laudenslager, M. L., Ryan, S. M., Drugan, R. C., Hyson, R. L., and Maier, S. F. (1983). Coping and immunosuppression: inescapable but not escapable shock suppresses lymphocyte proliferation. *Science* **221**:568–570.
LeShan, L. L. (1959). Psychological states as factors in the development of malignant diseases: a critical review. *J. Nat. Cancer Inst.* **22**:1–18.
Levine, S., Strebel, R., Wenk, E. J., and Harman, P. J. (1962). Suppression of experimental allergic encephalomyelitis by stress. *Proc. Soc. Exp. Biol. Med.* **109**:294–296.
Levy, S. M. (ed.) (1982). *Biological Mediators of Behavior and Disease: Neoplasia*, Elsevier Biomedical, New York.
Lipowski, Z. J. (1984). What does the word "psychosomatic" really mean? A historical and semantic inquiry. *Psychosom. Med.* **46**:153–171.
Locke, S. E. (1982). Stress, adaptation, and immunity: studies in humans. *Gen. Hosp. Psychiatr.* **4**:49–58.
Locke, S. E., Kraus, L., Leserman, J., Hurst, M. W., Heisel, J. S., and Williams, R. M. (1984). Life change stress, psychiatric symptoms, and natural killer cell activity. *Psychosom. Med.* **46**:441–453.
Luborsky, L., Mintz, J., Brightman, V. J., and Katcher, A. H. (1976). Herpes simplex virus and moods: a longitudinal study. *J. Psychosom. Res.* **20**:543–548.
Maclean, D., and Reichlin, S. (1981). Neuroendocrinology and the immune process. In Ader, R. (ed.), *Psychoneuroimmunology*, Academic Press, New York, pp. 475–520.
Maddox, J. (1984). Psychoimmunology before its time. *Nature (Lond.)* **309**:400.
Makinodan, T. (1976). Immunobiology of aging. *J. Am. Geriat. Soc.* **24**:249–252.
Marsh, J. T., Lavender, J. F., Chang, S.-S., and Rasmussen, A. F. (1963). Poliomyelitis in monkeys: decreased susceptibility after avoidance stress. *Science* **140**:1414–1415.
Martin, M. J. (1978). Psychosomatic medicine: a brief history. *Psychosomatics* **19**:697–700.
Martin, P., and Bateson, P. (1986). *Measuring Behaviour. An Introductory Guide*, Cambridge University Press, Cambridge.
Maslinski, W., Grabczewska, E., and Ryzewski, J. (1980). Acetylcholine receptors of rat lymphocytes. *Biochem. Biophys. Acta* **633**:269–273.
Mason, J. W. (1968). A review of psychoendocrine research on the pituitary–adrenal cortical system. *Psychosom. Med.* **30**:576–607.
Masuda, M., and Holmes, T. H. (1978). Life events: Perceptions and frequencies. *Psychosom. Med.* **40**:236–261.
Metal'nikov, S., and Chorine, V. (1928). Rôle des réflexes conditionnels dans la formation des anticorps. *C.R. Seances Soc. Biol. Ses. Fil.* **99**:142–145.
Meyer, R. J., and Haggerty, R. J. (1962). Streptococcal infections in families. *Pediatrics* **29**:539–549.
Michaut, R.-J., Dechambre, R.-P., Doumerc, S., Lesourd, B., Devillechabrolle, A., and Moulias, R. (1981). Influence of early maternal deprivation on adult humoral immune response in mice. *Physiol. Behav.* **26**:189–191.

Mineka, S., and Suomi, S. J. (1978). Social separation in monkeys. *Psych. Bull.* **85:**1376-1400.
Minter, R. E., and Kimball, C. P. (1978). Life events and illness onset: a review. *Psychosomatics* **19:**334-339.
Mohamed, M. A., and Hanson, R. P. (1980). Effect of social stress on Newcastle disease virus (LaSota) infection. *Avian Dis.* **24:**908-915.
Monjan, A. A. (1981). Stress and immunologic competence: studies in animals. In Ader, R. (ed.), *Psychoneuroimmunology*, Academic Press, New York, pp. 185-228.
Monjan, A. A., and Collector, M. I. (1977). Stress-induced modulation of the immune response. *Science* **196:**307-308.
Murray, J. B. (1977). New trends in psychosomatic research. *Genet. Psychol. Monogr.* **96:**3-74.
Neifeld, J. P., Lippman, M. E., and Tonney, D. C. (1977). Steroid hormone receptors in normal human lymphocytes: induction of glucocorticoid receptor activity by phytohemagglutinin. *J. Biol. Chem.* **252:**2972-2977.
Nerem, R. M., Levesque, M. J., and Cornhill, J. F. (1980). Social environment as a factor in diet-induced atherosclerosis. *Science* **208:**1475-1476.
Newberry, B. H., and Sengbusch, L. (1979). Inhibiting effects of stress on experimental mammary tumors. *Cancer Detect. Prev.* **2:**225-233.
Nieburgs, H. E., Weiss, J., Navarrete, M., Strax, P., Teirstein, A., Grillione, G., and Siedlecki, B. (1979). The role of stress in human and experimental oncogenesis. *Cancer Detect. Prev.* **2:**307-336.
Palmblad, J. E. W. (1985). Stress and human immunologic competence. In Guillemin, R., Cohn, M., and Melnechuk, T. (eds.), *Neural Modulation of Immunity*, Raven Press, New York, pp. 45-52.
Palmblad, J., Petrini, B., Wasserman, J., and Akerstedt, T. (1979). Lymphocyte and granulocyte reactions during sleep deprivation. *Psychosom. Med.* **41:**273-278.
Parillo, J. E., and Fauci, A. S. (1979). Mechanisms of glucocorticoid action on immune processes. *Ann. Rev. Pharmacol. Toxicol.* **19:**179-201.
Parkes, C. M. (1972). *Bereavement: Studies of Grief in Adult Life*, International University Press, New York.
Pavlidis, N., and Chirigos, M. (1980). Stress-induced impairment of macrophage tumoricidal function. *Psychosom. Med.* **42:**47-54.
Pert, C. B., Ruff, M. R., Weber, R. J., and Herkenham, M. (1985). Neuropeptides and their receptors: a psychosomatic network. *J. Immunol.* **135:**820s-826s.
Pierpaoli, W. (1981). Integrated phylogenetic and ontogenetic evolution of neuroendocrine and identity-defense, immune functions. In Ader, R. (ed.), *Psychoneuroimmunology*, Academic Press, New York, pp. 575-606.
Plaut, S. M., and Friedman, S. B. (1981). Psychosocial factors in infectious disease. In Ader, R. (ed.), *Psychoneuroimmunology*, Academic Press, New York, pp. 3-30.
Plaut, S. M., Ader, R., Friedman, S. B., and Ritterson, A. L. (1969). Social factors and resistance to malaria in the mouse: Effects of group vs. individual housing on resistance to *Plasmodium berghei* infection. *Psychosom. Med.* **31:**536-552.
Plaut, S. M., Esterhay, R. J., Sutherland, J. C., Wareheim, L. E., Friedman, S. B., Schnaper, N., and Wiernik, P. H. (1980). Psychological effects on resistance to spontaneous AKR leukemia in mice. *Psychosom. Med.* **42:**72.
Plotnikoff, N. P., Miller, G. C., and Murgo, A. J. (1982). Enkephalins-endorphins: immunomodulators in mice. *Int. J. Immunopharmacol.* **4:**366-367.
Pochet, R., Delesperse, G., Gauseet, P. W., and Collet, H. (1979). Distribution of beta-adrenergic receptors on human lymphocyte subpopulations. *Clin. Exp. Immunol.* **38:**578-584.
Rabkin, J. G., and Struening, E. L. (1976). Life events, stress, and illness. *Science* **194:**1013-1020.
Rahe, R. H. (1972). Subjects' recent life changes and their near-future illness susceptibility. *Adv. Psychosom. Med.* **8:**2-19.

Rahe, R. H., and Arthur, R. J. (1978). Life change and illness studies: past history and future directions. *J. Human Stress* **4**:3–15.

Rashkis, H. A. (1952). Systemic stress as inhibitor of experimental tumors in Swiss mice. *Science* **116**:169–171.

Rasmussen, A. F., Jr. (1969). Emotions and immunity. *Ann. NY Acad. Sci.* **164**:458–461.

Rasmussen, A. F., Jr., Spencer, E. T., and Marsh, J. T. (1959). Decrease in susceptibility of mice to passive anaphylaxis following avoidance-learning stress. *Proc. Soc. Exp. Biol. Med.* **100**:878–879.

Reite, M., Harbeck, R., and Hoffman, A. (1981). Altered cellular immune response following peer separation. *Life Sci.* **29**:1133–1136.

Riley, V. (1981). Psychoneuroendocrine influences on immunocompetence and neoplasia. *Science* **212**:1100–1109.

Riley, V., Spackman, D. H., McClanahan, H., and Santisterban, G. A. (1979). The role of stress in malignancy. *Cancer Detect. Prev.* **2**:235–255.

Rogers, M. P., Reich, P., Strom, T. B., and Carpenter, C. B. (1976). Behaviorally conditioned immunosuppression: replication of a recent study. *Psychosom. Med.* **38**:447–451.

Rogers, M. P., Dubey, D., and Reich, P. (1979). The influence of the psyche and the brain on immunity and disease susceptibility: a critical review. *Psychosom. Med.* **41**:147–164.

Rogers, M. P., Trentham, D. E., McCune, W. J., Ginsberg, B. I., Rennke, H. G., Reich, P., and David, J. R. (1980). Effect of psychological stress on the induction of arthritis in rats. *Arthr. Rheum.* **23**:1337–1342.

Roskowski, W., Plaut, M., and Lichtenstein, L. M. (1977). Selective display of histamine receptors on lymphocytes. *Science* **195**:683–685.

Roszman, T. L., Cross, R. J., Brooks, W. H., and Markesbery, W. R. (1982). Hypothalamic–immune interactions: I. The effect of hypothalamic lesions on the ability of adherent spleen cells to limit lymphocyte blastogenesis. *Immunology* **45**:737–743.

Roszman, T. L., Cross, R. J., Brooks, W. H., and Markesbery, W. R. (1985). Neuroimmunomodulation: effects of neural lesions on cellular immunity. In Guillemin, R., Cohn, M., and Melnechuk, T. (eds.), *Neural Modulation of Immunity*, Raven Press, New York, p. 95–107.

Roth, D. L., and Holmes, D. S. (1985). Influence of physical fitness in determining the impact of stressful life events on physical and psychologic health. *Psychosom. Med.* **47**:164–173.

Rubenstein, C. M., and Shaver, P. (1980). Loneliness in two northeastern cities. In Hartog, J., and Audy, R. (eds.), *The Anatomy of Loneliness*, International University Press, New York, pp. 319–337.

Sarason, I. G., Sarason, B. R., Potter, E. H., and Antoni, M. H. (1985). Life events, social support, and illness. *Psychosom. Med.* **47**:156–163.

Schleifer, S. J., Keller, S. E., Camerino, M., Thornton, J. C., and Stein, M. (1983). Suppression of lymphocyte stimulation following bereavement. *J. Am. Med. Assoc.* **250**:374–377.

Schleifer, S. J., Keller, S. E., Meyerson, A. T., Raskin, M. J., Davis, K. L., and Stein, M. (1984). Lymphocyte function in major depressive disorder. *Arch. Gen. Psychiatr.* **41**:484–486.

Seligman, M. E. P. (1975). *Learned Helplessness: On Depression, Development and Death*, Freeman, San Francisco.

Selyé, H. (1936). A syndrome produced by diverse nocuous agents. *Nature (Lond.)* **138**:32.

Selyé, H. (1946). The general adaptation syndrome and the diseases of adaptation. *J. Clin. Endocrinol.* **6**:117–230.

Selyé, H. (1955). Stress and disease. *Science* **122**:625–631.

Shavit, Y., Lewis, J. W., Terman, G. W., Gale, R. P., and Liebeskind, J. C. (1984). Opioid peptides mediate the suppressive effect of stress on natural killer cell cytotoxicity. *Science* **223**:188–190.

Shavit, Y., Terman, G. W., Martin, F. C., Lewis, J. W., Liebeskind, J. C., and Gale, R. P. (1985). Stress, opioid peptides, the immune system, and cancer. *J. Immunol.* **135**:834s–837s.

Shekelle, R. B., Raynor, W. J., Ostfeld, A. M., Garron, D. C., Bieliauskas, L., Liu, S. C., Maliza, C., and Paul, O. (1981). Psychological depression and the 17-year risk of death from cancer. *Psychosom. Med.* **43**:117–125.

Shopsin, B., Sathananthan, G. L., Chan, T. L., Kravitz, H., and Gershon, S. (1973). Antinuclear factor in psychiatric patients. *Biol. Psychiatr.* **7**:81–86.

Sklar, L. S., and Anisman, H. (1979). Stress and coping factors influence tumor growth. *Science* **205**:513–515.

Sklar, L. S., and Anisman, H. (1981). Stress and cancer. *Psych. Bull.* **89**:369–406.

Sklar, L. S., Bruto, V., and Anisman, H. (1981). Adaptation to the tumor-enhancing effects of stress. *Psychosom. Med.* **43**:331–342.

Smith, E. M., Harbour-McMenamin, D., and Blalock, J. E. (1985). Lymphocyte production of endorphins and endorphin-mediated immunoregulatory activity. *J. Immunol.* **135**:779s–782s.

Solomon, G. F. (1981). Immunologic abnormalities in mental illness. In Ader, R. (ed.), *Psychoneuroimmunology*, Academic Press, New York, pp. 259–278.

Solomon, G. F., and Amkraut, A. A. (1979). Neuroendocrine aspects of the immune response and their implications for stress effects on tumor immunity. *Cancer Detect. Prev.* **2**:197–223.

Solomon, G. F., and Amkraut, A. A. (1981). Psychoneuroendocrinological effects of the immune response. *Ann. Rev. Microbiol.* **35**:155–184.

Stein, M. (1985). Bereavement, depression, stress, and immunity. In Guillemin, R., Cohn, M., and Melnechuk, T. (eds.), *Neural Modulation of Immunity*, Raven Press, New York, pp. 29–41.

Stein, M., Schiavi, R. C., and Camerino, M. (1976). Influence of brain and behavior on the immune system. *Science* **191**:435–440.

Stein, M., Schleifer, S., and Keller, S. (1982). The role of the brain and neuroendocrine system in immune regulation: potential links to neoplastic disease. In Levy, S. (ed.), *Biological Mediators of Behavior and Disease*, Elsevier, New York, pp. 147–168.

Stein, M., Keller, S. E., and Schleifer, S. J. (1985). Stress and immunomodulation: the role of depression and neuroendocrine function. *J. Immunol.* **135**:827s–833s.

Stroebe, W., and Stroebe, M. S. (1986). Beyond marriage: the impact of partner loss on health. In Gilmour, R., and Duck, S. (eds.), *The Emerging Field of Personal Relationships*, Erlbaum, Hillsdale, New Jersey, pp. 203–224.

Suls, J., and Mullen, B. (1981). Life events, perceived control and illness: the role of uncertainty. *J. Human Stress* **7**:30–34.

Tecoma, E. S., and Huey, L. Y. (1985). Psychic distress and the immune response. *Life Sci.* **36**:1799–1812.

Thompson, S. P., McMahon, L. J., and Nugent, C. A. (1980). Endogenous cortisol: a regulator of the number of lymphocytes in peripheral blood. *Clin. Immunol. Immunopharmacol.* **17**:506–514.

Totman, R., Reed, S. E., and Craig, J. W. (1977). Cognitive dissonance, stress and virus-induced common colds. *J. Psychosom. Res.* **21**:55–63.

Totman, R., Kiff, J., Reed, S. E., and Craig, J. W. (1980). Predicting experimental colds in volunteers from different measures of life stress. *J. Psychosom. Res.* **24**:155–163.

Ursin, H., and Murison, R. C. (1983). The stress concept. In Ursin, H., and Murison, R. (eds.), *Biological and Psychological Basis of Psychosomatic Disease*, Pergamon Press, Oxford, pp. 7–13.

Vachon, M. L. S., Sheldon, A. R., Lancee, W. J., Lyall, W. A. L., Rogers, J., and Freeman, S. J. J. (1982). Correlates of enduring distress patterns following bereavement: social network, life situation and personality. *Psychol. Med.* **12**:783–788.

Van den Brenk, H. A., Stone, M. G., Kelly, H., and Sharpington, C. (1976). Lowering of innate resistance of the lungs to the growth of blood-borne cancer cells in states of topical and systemic stress. *Br. J. Cancer* **33**:60–78.

Visintainer, M. A., Volpicelli, J. R., and Seligman, M. E. P. (1982). Tumor rejection in rats after inescapable or escapable shock. *Science* **216**:437–439.

Wayner, E. A., Flannery, G. R., and Singer, G. (1978). Effects of taste aversion conditioning on the primary antibody response to sheep red blood cells and *Brucella abortus* in the albino rat. *Physiol. Behav.* **21**:995–1000.

Weinman, L. J., and Rothman, A. H. (1967). Effects of stress upon acquired immunity to the dwarf tapeworm *Hymenolepsis nana*. *Exp. Parasitol.* **21**:61–67.

Weiss, J. M. (1968). Effects of coping responses on stress. *J. Comp. Physiol. Psychol.* **65**:251–260.

Weiss, J. M. (1971). Effects of coping behavior in different warning signal conditions on stress pathology in rats. *J. Comp. Physiol. Psychol.* **77**:1–13.

Werb, Z. (1978). Biochemical actions of glucocorticoids on macrophages in culture: specific inhibition of elastase, collagenase, and plasminogen activator secretion and effects on other metabolic functions. *J. Exp. Med.* **147**:1695–1712.

Werb, Z., Foley, R., and Munck, A. (1978). Interaction of glucocorticoids with macrophages: identification of glucocorticoid receptors in monocytes and macrophages. *J. Exp. Med.* **147**:1684–1694.

Wigzell, H. (1985). Properties of the specific immune system in relation to its possible regulation by the central nervous system. In Guillemin, R., Cohn, M., and Melnechuk, T. (eds.), *Neural Modulation of Immunity*, Raven Press, New York, pp. 197–202.

Williams, J. M., and Felten, D. L. (1981). Sympathetic innervation of murine thymus and spleen: a comparative histofluorescence study. *Anat. Rec.* **199**:531–542.

Williams, J. M., Peterson, R. G., Shea, P. A., Schmedtje, J. F., Bauer, D. C., and Felten, D. L. (1981). Sympathetic innervation of murine thymus and spleen: evidence for a functional link between the nervous and immune system. *Brain Res. Bull.* **6**:83–94.

Wybran, J. (1985). Enkephalins, endorphins, substance P, and the immune system. In Guillemin, R., Cohn, M., and Melnechuk, T. (eds.), *Neural Modulation of Immunity*, Raven Press, New York, pp. 157–160.

Yu, D. T., and Clements, P. J. (1976). Human lymphocyte subpopulations: effect of epinephrine. *Clin. Exp. Immunol.* **25**:472–479.

Chapter 8

HOW DO GENDER DIFFERENCES IN BEHAVIOR DEVELOP? A REANALYSIS OF THE ROLE OF EARLY EXPERIENCE

Lynda I. A. Birke

Animal Behaviour Research Group
Biology Department
The Open University
Milton Keynes, England

I. ABSTRACT

The predominant view of how sex differences in behavior develop is that hormones play a central, organizing role early in life. Development is then seen as proceeding along predetermined lines, to produce behavior that is assumed to be typical of one sex or the other. I argue here that this hypothesis produces a very limited view of behavioral development; by focusing on a tiny period of the animal's developmental history, the perinatal period, we have tended to ignore the many complex ways in which an animal's "typical" behavior emerges from its social relationships.

Hormones in early development may well account for a large part of the variation in behavior between adults of each sex, but this has to be proved and not assumed. They may not, moreover, account for much of the variation between individuals of the same sex, and neither do hormones necessarily predict the behavior of any one animal. To understand how behavioral differences develop in adult animals, we need to study how animals change their behavior in the course of social engagements, from infancy through to puberty and adulthood. Only in this way can we come to understand the extent to which the emergence of specific patterns of behavior in adulthood are affected, if at all, by perinatal physiology. We have yet to understand how the "typical" behavior of male and female animals develops.

II. INTRODUCTION

My concern in this paper is to question, or perhaps to qualify, the widely held view that sex-related differences in the behavior of animals develop largely as a result of determination by organizing hormones. It is, I shall argue, a view underpinned by largely unquestioned assumptions about how sexually dimorphic patterns of behavior develop. There is now an enormous research literature documenting various ways in which gonadal steroids influence sexually dimorphic patterns of behavior in a variety of species. These influences have been categorized in two ways, according to their assumed action on the central nervous system; that is, exposure to certain gonadal steroids during pre- or early postnatal life appears to produce relatively permanent, or organizational, effects on behavior, which can be contrasted with the apparently transient, or activational, effects of hormone exposure in adulthood. If the sheer magnitude of research literature is any guide, then it is reasonable to suppose that there is a kernel of truth in the claim of hormonal effects. I do not intend to cast doubt on the basic findings of this corpus of research, but rather to question some of the assumptions made within it about how those behavioral differences develop. In doing so, I want to ask whether the common assumption that we can define an animal that is typical of its sex (and hence make inferences about its development) is valid. Among other things, the notion of typicality carries questionable implications about how we view behavioral development.

The ways in which research into behavioral development is carried out, particularly in relation to hormones, beg many questions about, say, behavioral maturation or about the context of physiological effects. Indeed, so significant have the hormonal effects been seen to be that it is only relatively recently that other sources of variation in the development of sex-related differences in behavior have been studied, even in the much-studied laboratory rat. One example from the rodent literature is the demonstration that mother rats differentiate behaviorally between offspring of different sex, directing more licking of the anogenital region toward male than female pups (Moore and Morelli, 1979; Richmond and Sachs, 1984a). Not only is this differentiation itself dependent on the hormone levels of individual pups (Moore, 1981), but it has long-term consequences for sexually dimorphic patterns of behavior when those pups reach adulthood (Moore, 1984a; Birke and Sadler, 1987).

Research on mammalian species other than rodents has perhaps focused less exclusively on prenatal hormones. There is, nonetheless, a wealth of experimental data showing ways in which manipulations of prenatal hormones exert effects on later behavior in, for instance, primates, including humans (see Goy and McEwan, 1980; Donovan, 1985, for reviews). The total number of species studied, however, remains small, and the majority of primate studies have employed species (such as the rhesus monkey) having a high degree of sexual dimorphism (Mitchell, 1979). There is also considerable evidence for effects of early socialization on the development of sex-related differences; male and female infant rhesus monkeys and pigtail macaques, for example, are treated somewhat differently by their mothers (Mitchell, 1979), although the mothers' differential behavior toward their young is affected in turn by their experience of mothering and their social status (e.g., Hooley and Simpson, 1981).

In studies of primate behavior, there is undoubtedly more awareness of the importance of the early social environment than is evident in the literature on rodents. There is, too, increasing emphasis in primate studies on specifying relationships between individuals (see Hinde, 1981)—relationships which in turn can influence the expression of sexually dimorphic patterns of behavior (e.g., social play: Colvin and Tissier, 1985). Still, despite observations of this kind, the role of hormones in the differentiation of primate behavior remains primary: They are usually seen as effecting a prenatal determination, which is seen as creating an innate bias toward sex-specific behavior (e.g., Mitchell, 1979, p. 44). In a review of hormones and human behavior, for example, Donovan (1985) refers to the "organizational effect of hormones on the brain in determining future patterns of behaviour" (p. 7). Any influence of early socialization then becomes secondary, added on to the predetermined base.

It is perhaps not surprising that this bias toward assuming a central role for hormones dominates the literature: If the hypothesis tested is focused upon sex differences per se, then researchers are likely to seek explanations in antecedent causes that are similarly dimorphic. Levels of sex hormones fills the bill. Social and experiential influences are not so (apparently) clear-cut; indeed, in some studies in which social experience has been shown to be influential, the focus of the research was not on the development of sex differences at all [for instance, female kittens learn to play more if they are reared with males than if reared with females (Bateson and Young, 1979)].

Any evidence for "organizational" effects of early experience on the development of sexual dimorphism in behavior should not, however, be merely added on to the evidence for hormonal effects, as I shall argue. Nor is it sufficient to propose that the two sources of variation "interact" in some unspecified way to produce the sex differences in question. While it is apparently fashionable as an assertion, there is little evidence in the prevailing literature that positing interaction in this way helps us much, not least because it does not clearly separate "interaction" in the statistical sense (which is not necessarily interesting in behavioral terms) from the more interesting idea of an interplay between various factors (such as hormonal influences and, say, the influences of maternal differentiation by sex of pup). Certainly, asserting interaction does not seem to have had much impact on experimental research into sex-related differences, nor has it done anything to break down some of the predominant assumptions about nature versus nurture that continue to trap us.

In a review of the history of ideas about behavioral development, Oppenheim (1982) distinguished, by analogy with earlier debates about morphological development, between preformationism and epigenesis. It is, he suggests, because those interested in behavior have not been sufficiently cognizant of the history of those debates in morphology that we have become so trapped in sterile arguments about nature versus nurture. Not everyone would agree, however, with his arguably optimistic conclusion that we are now plowing more fertile ground. Oyama, for example, has recently argued (1985) that the nature/nurture dichotomy continues to sterilize thinking about development, and that the predominant mode of describing behavioral development is a fundamentally preformationist one. Perhaps not all of those studying behavioral development fall into this trap; still, I am inclined to agree with her suggestion, at least with respect to work on hormones and behavioral development. It is my contention that such notions of preformationism, of unfolding toward some "goal" of development, permeate much of the literature on the development of sex-related differences—leading us, perhaps, to overemphasize the role of endogenous hormones as determinants of later behavior.

III. HORMONAL EFFECTS ON BEHAVIORAL ORGANIZATION

The traditional concept of sexual differentiation holds that anatomical and behavioral differences between adult males and females result from differences in hormonal output during an early "sensitive period" of life. The most fundamental component of this in mammals is the presence or absence of androgens, particularly testosterone, in the embryonic testes. Testicular hormones (or their metabolites) in turn produce two major "masculinizing" effects. First, testosterone itself promotes development of the Wolffian ducts, and regression of the Müllerian ducts, while a metabolite of testosterone (dihydrotestosterone) promotes development of the urogenital sinus and the external genitalia (Schultz and Wilson, 1974). Second, androgens affect the developing neural system. Although the possibility that the brain (and hence behavior) might be sexually differentiated was mooted in the past century (in relation to human homosexuality, Ulrichs, 1898), the first experimental proof came in 1936, when Pfeiffer demonstrated that male rats that had been castrated at birth and given ovarian transplants as adults could be made to ovulate (Pfeiffer, 1935). This effect is now known to involve differentiation of nuclei in the anterior hypothalamus, such that the typically cyclic pattern of gonadotrophin secretion that characterizes female mammals persists in the absence of the aforementioned masculinization.

Permanent effects on behavior have also been demonstrated as a result of altering the hormonal milieu. Although some experimental evidence had been accumulated during the 1940s (Beach, 1981), the first paper to present formal evidence for the organizational hypothesis (that is, the hypothesis that exposure to hormones during perinatal life permanently organizes neuronal structures, thus leading to permanent behavioral alteration) appeared in 1959 (Phoenix et al., 1959). It is now widely accepted that exposure to androgenic steroids during early life (that is, prenatally in most mammalian species and immediately after birth in some species of murid rodents) has a masculinizing effect on later behavior. Thus, even if female rodents that have been neonatally androgenized are given large doses of estrogens and progestins in adulthood, they do not readily show the behavioral responses to such hormones that are typical of untreated females.

There have been various modifications of the theory since its inception. It is now held, for instance, that at least some of the masculinizing effects of testosterone result from its intracellular conversion to estrogens—the so-called "aromatization hypothesis" (see Booth, 1979; Plapinger and McEwan, 1978; Feder, 1984, for reviews). The effects of other steroids have also been studied, including the effects of hormones produced by the embryonic ovary (e.g., Döhler, 1978; Dunlap et al., 1973; White et al., 1979). "Masculinization" of the brain of males is now thought to be a different—and potentially separable—process from that of "defeminization" (see Goy and McEwan, 1980; Davis et al., 1979). In addition, comparisons of processes of sexual differentiation in different species of vertebrates have shed some light on similarities and differences among them. For instance, adult sexual behavior of some species of fish, such as the cichlid, *Sarotherodon mossambicus*, can be influenced, as it is in mammals, by exposure to androgens during early development (Billy and Liley, 1985). On the other hand, in some vertebrate species, particularly those in which it is the female that is the heterogametic sex, it seems to be females whose behavior appears to be more strongly influenced by prenatal hormones (see Adkins-Regan, 1981, for review). The role of prenatal hormones even extends, in some species of reptile, to influencing sex itself: Gutzke and Bull (1986) found that injections of either estradiol or testosterone into eggs of snapping and painted turtles caused all embryos to develop anatomically as females.

Slight modifications aside, however, the basic tenets of the traditional hypothesis have remained unchallenged. That is, hormonally based processes of masculinization and/or feminization are assumed to have acted during a critical period of early life to organize neural structures in some permanent fashion, and thus cause differences between males and females in patterns of adult behavior.

This is believed to be the primary process, although it may be subject to other, less obviously permanent effects in later life (such as the influence of other animals, sexual experience, or changes in the physical environment). My intention in this paper is primarily to question some of the assumptions made about behavioral development by the traditional hypothesis, but first, I want briefly to consider two areas of research that appear to challenge some parts of the hypothesis. They both do this by suggesting that there are more sources of variation in the individual expression of sexually dimorphic patterns of behavior than the organizational hypothesis implies.

IV. BLURRING THE DICHOTOMY: ORGANIZATION IN CONTEXT

First, the rather rigid dichotomy between early organizational (i.e., permanent) and later activational (i.e., impermanent) effects of hormones on brain and behavior has been challenged by Arnold and Breedlove (1985), who argue instead for labile effects of hormones on brain structure throughout life, and for the possibility of more permanent, structural changes occurring in adulthood. The second source of evidence concerns the effects of varying social experience during early life, as noted in the Introduction.

Although it is the second of these areas that concerns me here, the notion that organization and activation are not quite as mutually exclusive—or as universal—as is generally supposed merits further attention. Petersen (1986), for example, found no evidence for androgenic organization of the brain in voles (*Microtus canicaudus*) and queried the assumption that in mammals, neural differentiation *always* proceeds by masculinization of "neural substrates. . . that are inherently feminine." (See also Döhler, 1978, for a rejection of the idea of "inherently feminine" structures.) Arnold and Breedlove (1985) point out numerous "exceptions to the rules," and conclude from these that, rather than pointing to a strict two-process theory of steroid influences, "the evidence suggests. . . that various neural processes (e.g., growth, synaptogenesis, regulation of receptors, perhaps even neurogenesis) are influenced by steroids throughout life." They also stress the difficulty of distinguishing in practice between permanent and impermanent effects: How long does the latter have to exist before it becomes the former? One interesting example involves the seasonally breeding mouse *Peromyscus leucopus*, which shows seasonal changes in size of motor neurons involved in copulatory behavior (neurons in the spinal nucleus of the bulbocavernosus muscle, which innervates penile musculature). In rats, the size of these motor neurons seems to be organized permanently by prenatal androgens (Sachs and Thomas, 1985). The seasonal changes in *Peromyscus*, by contrast, are structural changes that might be classified as "permanent" by the organizational model—yet they are clearly impermanent over months. Hormones can provoke structural, apparently permanent, changes in adult as well as prenatal animals: Testosterone stimulates an increase in volume in preoptic nuclei when given to adult Mongolian gerbils (Commins and Yahr, 1984), for example, and the sexual dimorphism of some components of play behavior in the ferret, *Mustela furo*, can be influenced by androgens throughout adolescence (Stockman et al., 1986). Even the assumption that mating behavior itself is activated by adult hormones is far from universally true; data on gonadal activity and mating behavior in various species of vertebrates suggest that the two are temporally dissociated in many species, including vespertilionid bats and several species of snakes (Crews, 1984). Sexual behavior is thus not directly activated by hormones in these species.

At its simplest, the traditional hypothesis assumes that exposure to particular steroids at specific periods of life leads to behavioral differentiation. Within this framework, differences between individuals, even those of the same genetic sex, are given little attention, and are assumed to arise from differences in the levels of the relevant hormones (or perhaps in the density of the appropriate steroid receptors). Thus, male rats born to mothers that were stressed during pregnancy tend to show "masculine" behavior, such as mounting, less readily during adulthood (Ward and Weisz, 1980). The explanation of this phenomenon was that the mothers responded to stress by increasing adrenal production of steroids, including the weakly androgenic androstenedione, which was assumed to cross the placentas into the circulation of the infants and there compete with endogenous testosterone in the male pups for binding to androgen receptors. Here, interindividual differences were ascribed to different levels of available testosterone resulting from the treatment given to the mothers.

Various studies have focused upon experimental treatments that result in variations in the levels of available steroids in mammals, although the explicit concern of these is rarely to investigate individual differences per se. Some of these are summarized in Table I.

The significance of the blurring of the organization/activation dichotomy described by Arnold and Breedlove (1985) is that it allows for greater flexibility, for other influences besides individual hormones to act throughout life, and for possible hormonal effects on, say, neurogenesis even in adult life. One important source of variation in the expression of adult behavioral sexual dimorphism is early experience, particularly during the preweaning period. Mother rats spend much time licking and handling pups, but, as noted above, they discriminate between the sexes, directing more anogenital licking toward males. This behavior appears to be stimulated by chemical cues: Urine from male pups is more effective than that from females in eliciting a licking response from maternal dams (Moore, 1981), although different rates of responding by male and female

Table I. Hormonal Factors in the Development of Sex-Related Differences in the Behavior of Mammals

Factor	References
Absolute levels of testosterone and other steroids that might compete for receptor sites	Phoenix et al. (1959) Barraclough and Gorski (1962) Gerall et al. (1973)
Levels of aromatizable androgens	Naftolin et al. (1972) DeBold and Clemens (1978) Whalen and Olsen (1981)
Number of androgen, estrogen, and progestin receptors in target tissues, particularly brain, during development or in adulthood	White et al. (1979) Olsen and Whalen (1980)
Plasma binding; e.g., the role of α-fetoprotein in "mopping up" of androgens in neonatal female rats	Raynaud (1972) Plapinger and McEwan (1978)
Sensitivity of target tissues—this is known to decline after, for example, castration or after neonatal androgen treatment of females, presumably because of receptor and/or enzyme availability	Gerall et al. (1972)
Sensitivity of target tissues also increases around puberty	Dudley (1981)
Proximity in utero to males, especially those located caudally; androgens cross between placentas in animals having multiple young, such as rats, and have some masculinizing effects on adjacent female fetuses	Vom Saal (1981) Richmond and Sachs (1984b)
"Defeminization" of females by estrogens secreted by the maternal ovary	Witcher and Clemens (1987)
Enzyme availability; physical and behavioral masculinization is, for example, incomplete in rats having the *Testicular Feminization Mutation*, because they are deficient in 5-α-reductase	Olsen (1979)
Maternal stress, which leads to an increase in adrenal steroids	Ward and Weisz (1980) Moore and Power (1986) Fride et al. (1986)
Pubertal changes in hormone levels after infancy	Slob et al. (1986) Stockman et al. (1986)

pups might also be important (Moore, 1986). Moreover, hormonal treatment of pups can also alter maternal licking: Neonatal exposure to androgens (Moore, 1982) and progestins (Birke and Sadler, 1985), for example, has been shown to result in increased rates of anogenital licking. Whether or not increased maternal licking in turn alters pups' endogenous hormone levels has yet to be determined.

This difference in tactile experience during early life appears to have long-term consequences for the development of sex-related behavioral differences in the rat. Most significantly, maternal licking of the anogenital region has effects on the temporal patterning of masculine behavior in adulthood, such that animals receiving less licking as a result of interference with maternal olfaction tend to display longer intermount intervals (Moore, 1984a). The opposite effect has occurred in females that had received more anogenital stimulation (artificially) than did controls; they showed shorter intermount intervals when tested for masculine behavior as adults (Moore, 1984b). One of our own studies looked at various types of behavior as animals grew up, following manipulations to alter maternal licking rates: Animals that had received less licking as pups (because their body odor was altered by perfume) played more during the juvenile period, and males showed longer intermount intervals when tested for sexual behavior as adults (Birke and Sadler, 1987).

Both play behavior and sexual behavior are patterns that are sexually dimorphic in most species studied. And both have been subject to a number of investigations of hormonal determinants. Play behavior in rats, for instance, occurs mainly during the juvenile, peripubertal period; males tend to show higher levels of play and of play initiation (pouncing on another animal: Olioff and Stewart, 1978). Play behavior is also affected by neonatal hormone treatments; androgens given to female rats increased juvenile play-fighting, and castration of males reduced it (Meaney and Stewart, 1981). On the other hand, neonatal exposure to a synthetic progestin, medroxyprogesterone acetate, resulted in reduced levels of juvenile play in both sexes (Birke and Sadler, 1983).

These studies certainly suggest that there is an effect of early hormone exposure on the development of play in this species. What is not always clear is how the effect is mediated. That hormonal treatment of pups also alters maternal behavior toward those pups means that the results of experiments in which hormone levels are artificially manipulated cannot be unequivocally interpreted. A standard technique in rodent studies of hormonal effects in development is to use split litters, such that half the litter receives hormone, and half are controls. One problem with this approach is that controls may differ between litters because maternal behavior toward controls may change as a function of the hormonal treatment given to the other half-litter; indeed, inspection of data from control animals in play behavior studies in this laboratory suggests that this may well be the case, since controls seemed to differ from litters receiving different hormonal treatments (Birke and Sadler, 1985). This would tend to reduce differences between control and experimental groups, while increasing variance.

Hormonal treatments do appear to act upon the developing brain in the rat, but in most experiments these effects cannot readily be dissociated from possible effects on maternal behavior. Studies using brain implants of hormones provide the clearest evidence for direct action on the brain: Females given implants of crystalline testosterone directly into the amygdala between days 1 and 5 showed higher levels of juvenile play than even control males (Meaney *et al.*, 1985). Still, under normal conditions it seems likely that maternal differentiation acts in concert with differences in endogenous hormones to generate differences in later

Table II. Social/Environmental Variables Affecting the Development of Sexually Dimorphic Behavior Patterns in Adult Rodents

Variable	References
Rates of licking by the mother directed toward the anogenital region, as in rats	Birke and Sadler (1987) Moore (1984a) Moore and Morelli (1979)
Odors associated with suckling, which affect the development of adult male sexual behavior in rats	Fillion and Blass (1986)
The number of littermates in infancy, affecting the expression of adult male sexual behavior in rats	Hård and Larsson (1968)
Isolation of male rats after weaning, which alters the expression of adult sexual behavior	Thor (1980)

social behavior. And, of course, there is always the possibility that even when implants are used, mother rats manage to differentiate between pups in some way!

Maternal behavior, then, is one source of variation that has, to date, been given insufficient regard in work on the development of differences in behavior between individuals, at least in studies of sex-related differences using rodents. The role of other environmental factors in the development of social organization in rodent species, such as the effects of other siblings, including older ones, or the status of the mother in the social organization of the group, is also little known. Table II lists a few of the studies that have been done that indicate some of the range of effects.

V. ASSUMPTIONS OF THE HYPOTHESIS: ORGANIZING TYPICAL ANIMALS

At least one reason why these factors have been little studied is the prevailing emphasis on the traditional organizational theory. Endogenous hormones are assumed to act within the individual pup during the critical period (see Fig. 1B). But that would take the individual pup out of its social context: One thing that is strikingly clear about behavior of infant rats is that the mother and her litter are highly coordinated. Only slowly do individual pups appear to separate out behaviorally and the appearance of coordination begin to break down. Even as this occurs, the rats' behavior remains highly directed toward the mother for some time as they orient preferentially toward her odor (Leon and Moltz, 1972). Yet the organizational hypothesis assumes the individual pup and its specific physiology to be the primary unit.

Why should this be so? Why is it that the *context* in which, for instance, the rat pup grows up has been so ignored? Even in studies of primates, with their differing and complex social organizations, there seems to have been a tendency to underemphasize variability and social–environmental context in the generation of gender differences. For instance, while they acknowledge that there are many primatologists who do emphasize individual differences and variability, Baldwin and Baldwin (1979) conclude that the "primate literature is full of descriptions that tie behavior to anatomy." One result of this, they suggest, is that the degree of variance in sexually dimorphic behavior to be found between primate species has been underestimated. Part of the problem lies in the assumptions that are made both about the processes of development, and about sex-related differences per se. There is no doubt that, on average, there are behavioral differences between males and females in most species. But that is a general statement about differences between subsets of individuals within a group or population; it should not be the basis for predictions about the behavior of specific individuals. Problems arise precisely when we extrapolate from state-

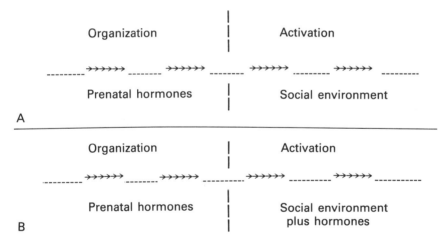

Fig. 1. (A) The assumed additive relationship between hormones and (social) environment in, for example, primates. Prenatal hormones organize the brain (and hence behavior), and variables in later social environment (in, for example, maternal behavior) reinforce the previously organized behavioral dimorphism. Adult hormones may modulate behavior but do not activate as strongly in primates as they do in rodents. Developmental time is represented by the arrows running from left to right. (B) The hormonal–environmental interaction, as exemplified by rat development. In rodents, adult hormones are usually thought necessary to activate behavior.

ments about the group to statements about individuals. In work on development and hormonal effects, it is assumed that, because of the existence of sex-related differences at the *population* level, we can posit the existence of "typical" male behavior, or "typical" female behavior within individuals. Only by so doing can we begin to ask questions about the role of hormones within the developmental history of any individual animal: The effect of the hormones is then interpreted as modifying those typical patterns.

There are three dangers in assuming the existence of typical individuals. The first is that, by focusing on "typical" individuals, we tend to lose sight of the fact that "typical" individuals may not always be the *same* animals in all cases; that is, some individuals may be "typical" (i.e., approximating the population mean) males on measures of, say, aggressive behavior, but may not be at all typical on measures of some other behavior. In short, it is rarely possible to specify exactly what "the" typical individual is like. Moreover, saying that there are sex-related differences in a population does not mean that it is always the same individuals occupying the same points in the distribution.

Perhaps the clearest example of this is provided by some species of lizard that reproduce parthenogenetically, such as *Cnemidophorus uniparens*. Although all the animals are female, both "malelike" (by comparison with other *Cnemidophorus* species) and "femalelike" behavior can be found. "Pseudocopulatory" behavior between two females apparently serves as a primer of reproductive function, and animals engage in each of the two roles at different stages of the ovarian cycle (Moore *et al.*, 1985). Whether or not any individual exhibits, say, malelike behavior depends partly upon hormonal changes associated with that cycle, and partly upon the social situation (for example, the presence of others acting in similar fashion). For these parthenogenetic lizards at least, "typical" behavior is hard to define.

The second danger in assuming typicality is that, while it is obviously necessary for sexually dimorphic behavior patterns such as, say, intromission or lordosis to occur if reproductive success is to be achieved, the strategies adopted by animals to achieve this may vary, even within any one species. Assuming "typicality" thus underestimates an important question: Under what conditions—social or environmental—is an animal's behavior typical of its sex? In their detailed studies of the sexual behavior of Norway rats living in groups, McClintock and Adler (1979) pointed out that "typical" sexual behavior for a female rat had been seen predominantly in terms of the lordosis response; the active role taken by females in soliciting and perhaps choosing between males, and in pacing copulatory behavior (McClintock and Adler, 1979; see also Beach, 1976) had been overlooked, they suggested, partly because of assumptions about typicality. Partly, too, it was overlooked because the small cages in which animals were tested did not allow females to display their full range of sexual responses. Wallen and Winston (1984) similarly point out that, for rhesus monkeys, the effects of ovarian hormones on female sexual behavior depend upon the social conditions of testing: Hormonal effects are more clearly demonstrated when animals are tested in multifemale social groups than when they are tested (as is more common in laboratory studies) in pair tests. "Typicality" in these experiments, then, is a function of the testing conditions.

The third danger is that development becomes reduced to a process of maturation, of unfolding toward some specified end point—the "typical" individual.[1] This in turn has two consequences: Conceptualizing development as unfolding renders the organism itself passive, a product of preformation, rather than being an active, engaging part of the developmental process (Oyama, 1982; Gollin, 1984). Focusing on maturation toward a specified endpoint also implicitly assumes a constancy of process itself. (In relation to the emergence of sex differences, for example, the processes of masculinization and defeminization, when applied to behavior, are seen essentially as unitary processes.) That is, maturation is nothing more than unfolding, and it is assumed that the form or function of a behavior at any one age can be fairly easily mapped in relation to that at another age. So, not only is "typical" assumed to be applicable to all the behavior of a (hypothetical) individual at any one time, but typicality is also assumed to apply across developmental time. "Typical" individuals have thus had "typical" developmental histories with respect to particular behaviors.

For example, the set of behavioral responses called "play-fighting" in, say, juvenile rats includes components such as boxing, wrestling, and "pinning" of one individual by another, all of which appear to be similar in form to those occurring in many adult aggressive encounters. A common assumption is that the whole complex represents a precursor of adult aggression (see discussion in Meaney et al., 1985; Martin and Caro, 1985; Fagen, 1981), as though juvenile play is simply an immature form of adult aggression. No doubt some of the components do develop into components of adult aggressive sequences, but many do not; as Meaney et al. (1985) note, the patterns and form of these components do change with age, even during the juvenile period itself. Moreover, there is considerable similarity between some of the later forms of these behaviors and adult sexual behaviors, such as mounting (Moore, 1984b). Precisely because of the tendency to see development as maturation (Oyama, 1982), studies of development of, say, sex difference focus only on a limited number of stages of the organism's life—typically its perinatal life, including effects of hormone treatments, and its adult responses. As a result of this, we know remarkably little about the ways in which different patterns of behavior, and

---

[1] Although I am concerned here with the development of sex-related differences in behavior, we do, of course, tend sometimes to make similar assumptions about species typicality. When this is done, it is all too easy to slip into preformationist assumptions (Bateson, 1982; Oyama, 1985) about the origins, genetic or other, of that typicality.

changing relationships between components, *emerge* and crystallize out of previous patterns and relationships.

At any rate, there seems to be no good reason to assume that the motivational and structural organization of behavior in adult life can necessarily be directly extrapolated from that in infancy or pubertal life, unless we have direct empirical evidence to show the continuity between such arbitrary points. As Bateson (1981) pointed out, there may be ecological changes during an organism's development that can be accompanied by marked behavioral discontinuity. Indeed, a number of recent studies of sex-related differences in behavior in juveniles of some primate species suggest that, contrary to previous expectations, sex-related differences occurring prepubertally are not necessarily predictive of adult sex differences (e.g., in talapoins: Wolfheim, 1978; in vervet monkeys: Raleigh *et al.*, 1979; in androgenized female rhesus monkeys: Phoenix and Chambers, 1982).

One period of a mammal's life in which we might expect to find such reorganization of the structural relationships between behavioral components is puberty. This is a period of rapid and often dramatic physiological change, during which patterns of endocrine secretion begin to shift towards the adult pattern. Sensory feedback may be altered during, say, mounting behavior, as a result of pubertal changes in genital size. Social relationships, too, may change; some animals may leave the natal group at puberty, for instance, or become seasonally territorial. Sexual dimorphism of body size might increase, which itself may alter social interactions. Some pharmacological studies have focused on neurochemical changes in adolescent rodents, in order to explain short-term changes in behavior that occur at this time (such as temporary increases in activity, or changes in learning performance: see review by Spear and Brake, 1983); in addition, it has been suggested that pubertal changes in levels of androgens affect later behavior of rats in open-field tests (Slob *et al.*, 1986), or the adult expression of masculine sexual behavior or aggression in mice (Shrenker and Maxson, 1983). Yet the changes in the organization of behavior that occur at puberty have been remarkably little studied. Even in primates—which have been subjected to more longitudinal studies than have rodents—there have been very few genuinely developmental investigations during the period of puberty (Mitchell, 1979).

But the biggest problem with assuming "typical" individuals is that we are projecting backward within the life of that typical individual and seeking causes of that typical behavior in events internal to that individual. Rather than investigating the richness and complexity of differences between varied individuals, we tend to reduce the range before we start. So, having defined male-typical and female-typical behavior for a species, we look for physiological antecedents that are similarly dichotomous (see Birke, 1982, for further discussion of this point); the most obvious of these is early differences between the sexes in levels of particular steroids. Put another way, the assumption seems to be that the same dichotomous categories apply across all levels of biological organization. Thus, even though it is well known that it is only quantitative, not qualitative, differences in hormone secretion that distinguish males and females, it is still commonplace to talk of "male" and "female" hormones as though there is a straightforward dichotomy.[2]

VI. THE DICHOTOMY REVISITED: THE ROLE OF HORMONES IN DEVELOPMENTAL PATHWAYS

Yet the same dichotomous categories do *not* apply across all levels of biological organization. The most important discrepancy lies in the unquestioned assumption of typical individuals. As noted, this is actually a statement about populations. There is undoubtedly some validity to it at that level. If we specify exactly what kind of behavior pattern we expect then we may well be able to identify the mean frequency of that behavior pattern for each sex: This is the "typical" frequency for that sex, although it is rarely completely dichotomous but overlaps. But it is a big jump to go from statements about one specific behavior at the population level to statements about how that behavior maps onto underlying physiology or biochemistry, and a still bigger jump to make causal assumptions that link the two.

It is, of course, precisely this latter kind of approach that led to the well-established finding that androgen levels are critical in "masculinizing" areas of the brain involved with patterns of gonadotrophin release, or in masculinizing certain behavioral potentials. Androgens are quintessentially "male" hormones, so it was inevitable that they became the focus of the search for physiological antecedents of male behavior. Yet just because we are seeking gender-typical patterns, the variability between individuals both in behavior and in hormone levels has been largely ignored. Moreover, it does not necessarily follow that any clear dimorphism observed does result from hormonal causes: It may emerge from other constraints operating during social development. One example is the increased mounting behavior coinciding with the group breeding season in juvenile male macaques (*Macaca fuscata*). This occurs despite the fact that young macaques do not show seasonal rises in testosterone until they are much older; the behavior, then, appears to be socially, not hormonally, activated (Eaton *et al.*, 1986).

It should be evident by now that I am partly making a plea for greater focus

---

[2]Attempts are sometimes made to excuse this categorization by, for example, asserting that "male" hormones are those functionally associated with characteristically male physiology. They can, therefore, be defined as male. Apart from the obvious circularity, this type of assertion is an instance of precisely the kind of dichotomizing downward that I am discussing. For how else is a "male physiology" to be defined?

on individual differences, and the processes of their development, in the study of sex-related differences—although I do not intend that merely to encourage a search for yet more sources of variation. Of course, hormonal influences *may* account for a large part of the variation *between* males and females. But that needs to be shown experimentally, not assumed a priori; it is, moreover, unlikely to account for much of the variation within a sex, which has often been neglected. In discussing the nature/nurture problem in the field of human developmental psychology, McCall (1981) pointed out that there are "two realms of developmental psychology." One, older, realm was primarily concerned with individual differences between children, while the newer, more "scientific," realm brings from animal research experimental methods that focus upon groups and the experimental manipulation of variables. McCall considers that these remain essentially two distinct sects, the "two realms," and that this in turn has implications for our understanding of behavioral development.

These two realms can also be discerned in ethology to some extent. The original, traditional naturalist, who perhaps paid more heed to individual differences between animals, has largely given way to the scientific approach, with its emphasis on groups and experimental control. In some ways this is, of course, desirable: We stand to gain replicability and predictability to aid our theorizing. But the cost is that, although we know a great deal about sources of variation between individuals, we know little about how individuals got to be the way they are, about how their behavior has developed through their own individual histories (see also Magnusson, 1986). Indeed, studies of behavioral development in animals seem similar to the trends that McCall claims are dominant in human studies: "Contemporary developmental psychology," he claims, "is mainly the parametric study of immature organisms; it is not primarily the study of longitudinal changes" within individuals themselves (McCall, 1981, p. 2). That is just as true of studies of the development of sex-related differences.

The emphasis on maturational unfolding, alluded to above, lends itself to the age-old nature/nurture dichotomy. The picture that emerges from the literature relating hormones to behavior over developmental time is one in which hormones are determining in early life; later, there is much greater lability and potential for environmental input. Primates are often cited as examples of mammals whose behavior is much more emancipated from endocrine control than, say, the laboratory rat. But, as Phoenix (1974, p. 30) pointed out, this evidence "relates to the role of gonadal hormones in regulating adult sexual behavior, and not to the importance of the steroidal hormones (especially testosterone) during prenatal life." For primates, then, nature (the hormones) determines the basic potentialities, while nurture (e.g., social interactions, physical environment) acts later in life to enhance (or possibly reduce) those basic tendencies. The relationship is, however, essentially additive. That is, over developmental time, the hormones act first, and are seen as primary, and *then* experiential factors are added on to that primary base.

The traditional hypothesis can be summarized as a diagram looking like a 2 × 2 table (see Fig. 1). Permanent, organizing effects occur early in life (on the left-hand side of the arrow representing the trajectory through developmental time); transient, activational effects occur later. For data from primate studies, this dichotomy corresponds approximately to a division between hormonal and environmental (including social) determinants of behavioral differences, as shown in Fig. 1A, while for data from studies of rodents, the dichotomy is more like that in Fig. 1B. In the latter, organizational effects of hormones are represented on the left, and the combined action of experience and adult, activational effects of hormones are shown on the right. However it is conceptualized, the relationship has generally been interpreted as additive.

Most students of animal behavior are, of course, now more sophisticated than to advocate a simple opposition or addition between nature and nurture. Fashions prevail in behavioral science as elsewhere, and a current fashion is to espouse *interactionism*, i.e., a belief in the interplay between the effects of causal factors internal to the organism (e.g., the genotype, hormones) and of environment. In the case of sex-related differences, this might be an interplay between perinatal hormones and environment. If we are to move away from the notion that hormones determine behavior, then it is, of course, certainly very necessary to emphasize interaction in this way, pointing, for example, to the existence of *reciprocal causality* (behavior affecting hormones, as well as vice versa).

The trouble with such a fashion, however, is that it is not always clear just what is "interacting." Obviously the genes themselves are not interacting directly with, say, social behavior. The implication is that the *effects* of each of the supposed factors interact to produce an outcome; that is, both internal factors (genes/hormones) and environmental factors influence outcome causally. In the case of perinatal, "organizing" effects of hormones, the posited interaction appears to mean that the effect of the hormones in, say, provoking structural changes in the brain acts in concert with the effects of environmental variables, and thus modulates the development of behavior patterns affected by those changed brain structures. What is not always clear is the time scale involved. There is little problem with ideas of interaction if the two influences are contemporary, such as maternal discrimination of pups by sex and postnatal changes in hormone profiles of rodents. But if later social experience is considered, the structural change, and its perinatal hormonal antecedent, have taken place long before other social/experiential changes may occur. In this case, it is much less clear what form the interaction is, in practice, supposed to take.

There seem to be two problems with this view of interaction. The first problem is that there is really rather little evidence for us to go on. What we do have is a body of empirical knowledge about the ways in which animals behave that shows us some of the sources of variance within the population: but that does not necessarily tell us anything about the *interplay* of those sources of variance.

The second problem is that the organism itself remains passive, just as it was in the additive, nature/nurture viewpoint. Its behavior is the product of interaction in its past, but cannot directly be part of the current interaction. Gollin (1984) stressed the need for more dynamic approaches than that implied by simple interactionism; as he pointed out, drawing on some of Lehrman's earlier insights:

> ... the transactions are between an organism and an environment, not between heredity and environment, or genes and environment. The process is not one of unfolding but of building. The building is guided by a dynamic scaffold that provides both opportunities for change and constraints upon its direction. (Gollin, 1984, p. 2)

In the literature on the development of sex-related behavioral differences, especially that on rodents, this central transaction (between an organism and its environment) is all but forgotten. There is now an extensive literature documenting effects of perinatal hormones on the etiology of sexually dimorphic behavior, but very little on how the individual organism engages with its environment (social or otherwise) and how sexually dimorphic patterns of behavior emerge from that engagement. One result of this dearth is that hormonal influences become seen as determining, almost by default. Admittedly, there is now increasing interest in the contributory role of social factors during early life, including, for example, some of our own experiments with rat social development (e.g., Birke and Sadler, 1985, 1987). While this research trend is heartening, in that it removes the spotlight from hormonal determination, it still fails to address the organism as actively engaging with its environment. As Moore (1986) has recently shown, for example, it is not just that mother rats differentiate between offspring on the basis of olfactory cues, but the pups themselves behave in slightly, subtly different ways: Male pups are fractionally faster than females in initiating and completing the leg extension that follows the tactile stimulation received from the mother.

The need to focus on process and developmental systems has often been stated, and there is no doubt that it is preferable to think of perinatal hormones as being part of a complex and evolving system, including the whole organism engaging with its environment at all stages of its life, rather than simply as determinants. Gandelman, in a review of the influence of gonadal hormones on sensory function, wrote of the problems of ignoring the ways in which the environment may act in relation to hormonal effects:

> Statements that hormones stimulate or activate behavior . . . while satisfying a reductionistic approach to the specification of hormone action, can be viewed as less than satisfactory explanations for they do not take the environment into account. . . . Given that environmental stimuli are evocators of behavior, an understanding of the hormone–behavior relationship may best be achieved by viewing hormones as modulators of afferent input and, as such, modulators of the environment. (Gandelman, 1983)

Although Gandelman was reviewing hormonal effects on adult behavior, his assertion is equally valid for the role of hormones in development. We have to understand that role as part of a process, transacting with the organism's environment.

This is not intended to be an argument for the kind of "interactionism" that typically treats hormones and environment as principal causes, and the organism itself as a dependent variable (see Buss, 1979; Oyama, 1985, for discussion of this form of interactionism). Gollin (1984) argued similarly against this use of interactionism, commenting:

> Nor is the issue of causality resolved by assigning causal function to sets of events or properties because of temporal priority. This is tantamount to confusing development with history. Rather, what appears to be involved at every period in development is a configuration or pattern of structural and functional components constrained by organismic and experiential transactions. (Gollin, 1984, p. 3).

While these two sets of factors—"hormones" and "environment"—are seen in this way as alternative prime causes, it is likely that any stability, any constancy of outcome, will be attributed to the hormones; concomitantly, any lability will be attributed to the environment.

Indeed, this is just what has happened in the study of social play in rats. Because several investigators found evidence of sex differences, they sought hormonal bases (see Meaney *et al.*, 1985, for review). Yet many studies have reported *no* evidence of sex differences in this behavior (e.g., Panksepp, 1981; see also Thor and Holloway, 1984). The emergence of any sex difference may, therefore, depend upon an interplay between many factors, not only hormones. For instance, Janus (1987) reported an effect of age of weaning on the pattern of juvenile play in rats, while Panksepp *et al.* (1985) noted how previously established social learning altered the extent to which opiate drugs could affect play behavior. Neither of these studies was concerned primarily with sex differences (indeed, Janus found none), but they do serve to remind us that the structure of play is a function of earlier experience at least as much as it is of hormones.

If the problem with focusing excessively on hormones is that it too easily becomes deterministic and narrowly causal, the trouble with focusing on the plethora of possible social inputs during early life is that what emerges seems to be an impossible array of outcomes. The organism becomes simply a tabula rasa. Bateson (1976) suggested that this is treating the animal ". . . as though it were a billiard ball whose path is the resultant of the various external forces that acted upon it" (Bateson, 1976, p. 418). As he points out, the problem then is that the development of behavior patterns that are relatively buffered against environmental changes is underplayed, and that it is not clear how we proceed to investigate all the forces impinging upon the metaphorical billiard ball.

While I would argue that we do indeed need to draw attention to the variety of inputs to the development of behavioral systems that are, or become, sexually dimorphic, it is also the case that such development is relatively buffered against environmental change. The fact remains that we *can* roughly specify a "typical" male or female rat or chimpanzee, even given the caveats outlined above. Whatever the developmental pathways by which these end points are reached—and they may be multiple—there is, for the emergence of sex-related behavioral differences at least, *some* degree of equifinality (Bateson, 1976).

All too often, writing concerned with sex-related differences and their development slips into the assumption that such buffering results from perinatal hormonal determination. The next step is to examine this by manipulating perinatal hormone levels, and then to investigate the behavioral differences that occur in adulthood. But the extent to which a perturbation to the system is effective in provoking change in the developmental system, or is ineffective (i.e., the system is buffered), depends on what we look at. In his 1976 paper, Bateson cites as an example of buffering the result when growing animals suffer temporary nutritional deficiency. Within limits, the developing system catches up as soon as adequate nutrition is restored, but if the limits are exceeded, the system fails to recover and stunting results. Behavioral outcomes are not always so (apparently) simple, however. As an example, consider the fact that, within limits, we can alter the environment of a developing male rat and it will still be perfectly capable of mounting, intromitting, and ejaculating as an adult—that is, of behaving in "typical" male ways sexually. The limits are set by things like being reared in isolation (the presence of littermates in early infancy seems to be necessary for the development of adult male mounting behavior: Hård and Larsson, 1968), or by fairly drastic hormonal changes. The presence or absence of these behaviors seems to be fairly well buffered. What is much more susceptible, however, to environmental perturbation is the fine detail, such as the temporal organization of such behavior. As I've noted, altering the extent to which mother rats direct anogenital licking toward male pups, even in the absence of hormonal changes, results in a change in temporal patterning: Males receiving less anogenital licking as pups showed longer intermount intervals (Moore, 1984a; Birke and Sadler, 1987). Rearing males in isolation after weaning results in similarly extended intermount intervals (Thor, 1980). Moreover, Morali *et al.* (1985) suggest that the major effect of neonatal treatment with androgens seems to be on sexual *motivation* rather than on the patterns of behavior as such; patterns such as mounting, they suggest, are basically very similar for both sexes. So, whether or not the observer perceives extensive buffering in the developmental systems depends upon what the outcome is perceived to be. This might be said to constitute a second form of projecting backward from the perceived outcome, the "typical" animal.

Very few people working in the field of hormones and behavior today would claim that hormones actually *determine* behavioral outcomes in any straightforward way. There is increasing evidence to the contrary. But what I would argue is that the existence of *apparently* well-buffered systems such as the differentiation of two genders[3] leads to interpretations that remain deterministic at heart. Furthermore, the stress on "typical" females and males itself results in overemphasis on the dichotomy, implying greater "buffering" or divergence of the developmental pathways than may, in fact, exist. Still, there is some buffering: Males and females rarely grow up to behave exactly the same. How is this to be explained if we are not to retreat back into determinism?

VII. THE IMPORTANCE OF BEING SOCIAL: SOME IMPLICATIONS FOR RESEARCH

What we do need to do is to avoid looking simply at very immature animals and then correlating their behavior with that of adults (or, worse, assuming causal and functional continuity); the organism, as Lehrman (1953) stressed, is different at every stage of its development. What we need to do instead is to establish the rules of gender development (or the rules by which sex-related differences in behavior develop, if you prefer). Under what conditions, for example, do particular characteristics emerge?

It is, of course, quite a daunting task to undertake research on a longitudinal basis, looking at detailed changes in the organization of behavior and social interactions throughout the various stages of an animal's life. Even in small laboratory rodents there is much work involved. Not surprisingly, longitudinal studies of this kind are rare; Chalmers (1980) describes changes in behavioral organization during ontogeny in olive baboons, and points out the dangers of assuming adult organization to be simply a maturation of juvenile behavior. As he points out, even if continuity of form can be demonstrated between two types of behavior, it does not necessarily follow that they serve the same function at different times during ontogeny. It is this assumption, Chalmers suggests, which bedevils many longitudinal studies of behavioral development.

Still, I would argue that, however difficult, it is precisely a longitudinal approach that is needed. I suggest that it is only by discovering how patterns of

---

[3]The distinction between "sex" and "gender" is forcefully maintained in the literature on human development; for this reason, I have (largely) used the term "sex-related differences" throughout the present paper to refer to the behavior of nonhuman animals. It is always slightly foolhardy to extrapolate from animals to humans, and given that I have always been critical of hormonal determinism in relation to human behavior (e.g., Birke, 1982), the risks should be evident. Still, the literature on human gender development rests on just the same kind of assumptions and problems that I have outlined here. "Sex," in this literature, is the primary division to which children are allocated at birth. This is held to be the biological basis, onto which the environmental (cultural) edifice of "gender" is built. The relationship between the two is essentially additive and dichotomous, rather than genuinely developmental. For criticism of this approach to gender development in humans, and an attempt to outline an alternative approach, see Birke and Vines (1987).

sexually dimorphic behavior actually emerge from the transactions in which individuals engage, rather than assuming hormonal determinism, that we will begin to understand the rules of gender development. This cannot be gained by focusing on discrete physiological changes in, say, the first few days of life and correlating these with adult behavior. That is not to say that hormones are unimportant: what I am suggesting is that they are necessary, but not sufficient, for the eventual emergence of sex-related differences in behavior. To understand how sex differences in behavior develop will require longitudinal studies of patterns of behavior and how they become transformed over time. Chalmers (1988) has argued for the existence of different kinds of rules in behavioral development, distinguishing between *directing rules*, which govern the developmental pathways as the frequency of that behavior changes with age, and *stopping rules*, which govern when that phase of behavioral development is to come to an end. It is quite possible that such putative rules could differ in their effects on the development of specific patterns of behavior in each sex. But this can only be studied empirically by focusing on longitudinal changes in the behavior of both sexes.

These changes, moreover, have to be studied relationally. Most studies focus upon behavioral changes *within* any one individual. But individuals change their behavior as a function of present and past transactions, both with the physical world and with other individuals: Biben (1986), for example, noted squirrel monkeys that had learned that they were likely to win in play fights were more likely to initiate bouts of play. What is needed, then, is greater attention to the behavior of animals in those transactions. Kortmulder (1983), in a somewhat controversial discussion of patterns of play behavior, suggested that it might be helpful to think about two animals interacting socially by using the analogy of an electromagnetic field that "shapes" and constrains the patterns of interaction. However we choose to conceptualize it, there is no doubt that to understand the development of sex-related differences in behavior, we need to put greater emphasis on changing relationships *between* animals (and the effect of these on actual behavior) than on behavioral development within certain individuals.[4] After all, the most biologically salient sex-related differences are those involved in sexual behavior itself—and what is that if it is not a social activity?

One critical question about development of sex-related differences in behavior that needs to be addressed is a functional one—not to what extent a behavior

---

[4]This need applies also to theories about human gender development (Birke and Vines, 1987). Biologically based theories tend to see the behavior of individuals as emerging from internal causes, as though social context is irrelevant, while theories based on ideas of social learning view the infant as a tabula rasa, acted upon by social pressures, but itself lacking agency. It is these drawbacks that have led some theorists to turn again to psychoanalytic theories of gender development which, whatever else their limitations may be, do emphasize the emergence of gender-differentiated behavior from *relationships* (e.g., with parents).

within any given species is determined by hormones, but at what age, or stage in the life cycle, would it be functionally necessary to buffer the system (if at all) to achieve reproductive success?[5] Obviously, not all species will be the same. If some degree of sexual dimorphism is functionally necessary—in order that copulation can occur effectively, for example—then a species which is normally solitary might need a stronger influence from prenatal hormones. Highly social species, by contrast, might be expected to have greater flexibility, with any buffering that occurs emerging from later behavioral transactions. Unfortunately, comparisons between species with regard to how sex-related differences develop are rare; more commonly, what is compared are variations in hormonal determination.

There are certainly constraints on these developmental systems, i.e., some degree of buffering; the interesting question is how those constraints emerge developmentally. The task is to analyze the rules by which sexual dimorphism of any particular behavior pattern emerges during social transactions, including the role of hormones in those developing systems. What I do not believe is that sexual dimorphism is caused exclusively by hormonal differences.

VIII. ACKNOWLEDGMENTS

I am very grateful to Pat Bateson, Neil Chalmers, and Gail Vines for helpful comments on earlier drafts of this paper.

IX. REFERENCES

Adkins-Regan, E. (1981). Early organizational effects of hormones: an evolutionary perspective. In Adler, N. T. (ed.), *Neuroendocrinology of Reproduction: Physiology and Behavior*, Plenum Press, New York, pp. 159–228.
Arnold, A. P., and Breedlove, S. M. (1985). Organizational and activational effects of sex steroids on brain and behavior: a reanalysis. *Horm. Behav.* **19**:469–498.

---

[5]The dominance of the prevailing belief in well-buffered, determined systems by which sex-related differences develop stems also, I suspect, from a belief in optimal adaptation. Of course, the behavior of animals subjected to massive perturbations in development is likely to be maladaptive—the inability of monkeys reared in isolation to organize or direct their sexual behavior is one example. But not all animals are "typical," and not all show behavior that is optimally adapted. Most animals will exhibit apparently adaptive sexually dimorphic behavior, such as exhibiting lordosis or mounting. On the other hand, slight differences between individuals in the patterning, the choreography, of sexual behavior might well have consequences for their reproductive fitness that have been largely ignored.

Baldwin, J. D., and Baldwin, J. I. (1979). The phylogenetic and ontogenetic variables that shape behavior and social organisation. In Bernstein, I. S., and Smith, E. O. (eds.), *Primate Ecology and Human Origins: Ecological Influences on Social Organization*, Garland STPM Press, London, pp. 89–116.

Barraclough, C. A., and Gorski, R. A. (1962). Studies on mating behavior in the androgen-sterilized rat and their relation to the hypothalamic regulation of sexual behavior in the female rat. *J. Endocrinol.* **25**:175–182.

Bateson, P. (1976). Rules and reciprocity in behavioral development. In Bateson, P. P. G., and Hinde, R. A. (eds.), *Growing Points in Ethology*, Cambridge University Press, Cambridge, pp. 401–422.

Bateson, P. (1981). Ontogeny of behavior. *Br. Med. Bull.* **37**:159–164.

Bateson, P. (1982). Behavioral development and evolutionary processes. In King's College Sociobiology Group (eds.), *Current Problems in Sociobiology*, Cambridge University Press, Cambridge, pp. 131–151.

Bateson, P., and Young, M. (1979). The influence of male kittens on the object play of their female siblings. *Behav. Neur. Biol.* **27**:374–378.

Beach, F. A. (1976). Sexual attractivity, proceptivity and receptivity in female mammals. *Horm. Behav.* **7**:105–138.

Beach, F. A. (1981). Historical origins of modern research on hormones and behavior. *Horm. Behav.* **15**:325–376.

Biben, M. (1986). Individual and sex-related strategies of wrestling play in captive squirrel monkeys. *Ethology* **71**:229–241.

Billy, A. J., and Liley, N. R. (1985). The effects of early and late androgen treatments on the behavior of *Sarotherodon mossambicus* (Pisces: Cichlidae). *Horm. Behav.* **19**:311–330.

Birke, L. I. A. (1982). Cleaving the mind: speculations on conceptual dichotomies. In Dialectics of Biology Group (ed.), *Against Biological Determinism*, Allison and Busby, London, pp. 60–78.

Birke, L. I. A., and Sadler, D. (1983). Progestin-induced changes in the play behaviour of the prepubertal rat. *Physiol. Behav.* **30**:341–347.

Birke, L. I. A., and Sadler, D. (1985). Maternal behavior in rats and the effects of neonatal progestins given to the pups. *Dev. Psychobiol.* **18**:467–475.

Birke, L. I. A., and Sadler, D. (1987). Maternal behavior of rats and its effects on the sociosexual development of offspring. *Dev. Psychobiol.* **20**:627–640.

Birke, L., and Vines, G. (1987). Beyond nature vs. nurture: process and biology in the development of gender. *Wom. Stud. Int. For.* **10**:555–570.

Booth, J. E. (1979). Sexual differentiation of the brain. In Finn, C. A. (ed.), *Oxford Reviews of Reproductive Biology*, Oxford University Press, Oxford, pp. 74–158.

Buss, A. R. (1979). *A Dialectical Psychology*, Irvington, New York.

Chalmers, N. R. (1980). Developmental relationships among social, manipulatory, postural and locomotor behaviors in olive baboons, *Papio anubis*. *Behaviour* **74**:22–37.

Chalmers, N. (1988). Developmental pathways in behavior. *Anim. Behav.* **35**:659–674.

Colvin, J., and Tissier, G. (1985). Affiliation and reciprocity in sibling and peer relationships among free-ranging immature male rhesus monkeys. *Anim. Behav.* **33**:959–977.

Commins, D., and Yahr, P. (1984). Acetylcholinesterase activity in the sexually dimorphic area of the gerbil brain: sex differences and influences of adult gonadal steroids. *J. Comp. Neurol.* **224**:132–140.

Crews, D. (1984). Gamete production, sex hormone secretion, and mating behavior uncoupled. *Horm. Behav.* **18**:22–28.

Davis, P. G., Chaptal, C. V., and McEwan, B. S. (1979). Independence of the differentiation of masculine and feminine behavior in rats. *Horm. Behav.* **12**:12–19.

DeBold, J. F., and Clemens, L. G. (1978). Aromatization and the induction of male sexual behavior in male, female and androgenized female hamsters. *Horm. Behav.* **11**:401–413.
Döhler, D.-D. (1978). Is female sexual differentiation hormone-mediated? *Trends in Neurosci.* November:138–140.
Donovan, B. T. (1985). *Hormones and Human Behavior*, Cambridge University Press, Cambridge.
Dörner, G. (1976). *Hormones and Brain Differentiation*, Elsevier, Amsterdam.
Dudley, S. D. (1981). Prepubertal ontogeny of responsiveness to estradiol in female rat central nervous system. *Neurosc. Biobehav. Rev.* **5**:421–435.
Dunlap, J. L., Gerall, A. A., and McLean, L. D. (1973). Enhancement of female receptivity in neonatally castrated males by prepubertal ovarian transplants. *Physiol. Behav.* **10**:1087–1094.
Eaton, G. G., Johnson, D. F., Glick, B. B., and Worlein, J. M. (1986). Japanese macaques' (*Macaca fuscata*) social development: sex differences in juvenile behavior. *Primates* **27**:141–150.
Fagen, R. (1981). *Animal Play Behavior*, Oxford University Press, Oxford.
Feder, H. H. (1984). Hormones and sexual behavior. *Ann. Rev. Psychol.* **35**:165–200.
Fillion, T. J., and Blass, E. M. (1986). Infantile experience with suckling odors determines adult sexual behavior in male rats. *Science* **231**:729–731.
Fride, E., Dan Y., Feldon, J., Halevy, G., and Weinstock, M. (1986). Effects of prenatal stress on vulnerability to stress in prepubertal and adult rats. *Physiol. Behav.* **37**:681–687.
Gandelman, R. (1983). Gonadal hormones and sensory function. *Neurosci. Biobehav. Rev.* **7**:1–17.
Gerall, A. A., Stone, L. S., and Hitt, J. C. (1972). Neonatal androgen depresses female responsiveness to estrogen. *Physiol. Behav.* **8**:17–20.
Gerall, A. A., Dunlap, J. L., and Hendricks, S. E. (1973). Effect of ovarian secretions on female behavioral potentiality in the rat. *J. Comp. Physiol. Psychol.* **82**:449–465.
Gollin, E. S. (1984). Ontogeny, phylogeny and causality. In Gollin, E. S. (ed.), *The Comparative Development of Adaptive Skills: Evolutionary Implications*, Erlbaum, Hillsdale, New Jersey, pp. 1–17.
Goy, R. W., and McEwan, B. S. (1980). *Sexual Differentiation of the Brain*, MIT Press, Cambridge, Massachusetts.
Gutzke, W. H. N., and Bull, J. J. (1986). Steroid hormones reverse sex in turtles. *Gen. Comp. Endocrinol.* **64**:368–372.
Hård, E., and Larsson, K. (1968). Dependence of adult mating behavior in male rats on the presence of littermates in infancy. *Brain Behav. Evol.* **1**:405–419.
Hinde, R. A. (1981). The bases of a science of interpersonal relationships. In Duck, S. W., and Gilmour, R. (eds.), *Personal Relationships*, Vol. 1, *Studying Personal Relationships*, Academic Press, London, pp. 1–22.
Hooley, J. M., and Simpson, M. J. A. (1981). A comparison of primiparous and multiparous mother–infant dyads in *Macaca mulatta*. *Primates* **22**:379–392.
Kortmulder, K. (1983). Play-like behavior: an essay in speculative ethology. *Acta Biotheoret.* **32**:145–166.
Lehrman, D. S. (1953). A critique of Konrad Lorenz's theory of instinctive behavior. *Quart. Rev. Biol.* **28**:337–363.
Leon, M., and Moltz, H. (1972). The development of the pheromonal bond in the albino rat. *Physiol. Behav.* **8**:683–686.
Martin, P., and Caro, T. M. (1985). On the functions of play and its role in behavioral development. In Rosenblatt, J. S., Beer, C., Busnel, M.-C., and Slater, P. J. B. (eds.), *Advances in the Study of Behavior*, Vol. 15, Academic Press, New York, pp. 59–104.
McCall, R. B. (1981). Nature–nurture and the two realms of development: a proposed integration with respect to mental development. *Child Devel.* **52**:1–12.

Janus, K. (1987). Early separation of young rats from the mother and the development of play fighting. *Physiol. Behav.* **39**:471–476.

Magnusson, D. (1987). Implications of an interactional paradigm for research on human development. *Int. J. Behav. Dev.*

McClintock, M., and Adler, N. T. (1979). The role of the female during copulation in wild and domestic rats (*Rattus norvegicus*). *Behaviour* **67**:67–96.

Meaney, M. J., and Stewart, J. (1981). Neonatal androgens influence the social play of prepubescent male and female rats. *Horm. Behav.* **15**:197–213.

Meaney, M. J., Stewart, J., and Beatty, W. W. (1985). Sex differences in social play: the socialization of sex roles. In Rosenblatt, J. S., Beer, C., Busnel, M.-C., and Slater, P. J. B. (eds.), *Advances in the Study of Behavior*, Vol. 15, Academic Press, New York, pp. 1–58.

Mitchell, G. (1979). *Behavioral Sex Differences in Nonhuman Primates*, Van Nostrand Reinhold, New York.

Moore, C. L. (1981). An olfactory basis for maternal discrimination of sex of offspring in rats (*Rattus norvegicus*). *Anim. Behav.* **29**:383–386.

Moore, C. L. (1982). Maternal behavior is affected by hormonal condition of pups. *J. Comp. Physiol. Psychol.* **96**:123–129.

Moore, C. L. (1984a). Maternal contributions to the development of masculine sexual behavior in laboratory rats. *Dev. Psychobiol.* **17**:347–356.

Moore, C. L. (1984b). Development of mammalian sexual behavior. In Gollin, E. S. (ed.), *The Comparative Development of Adaptive Skills: Evolutionary Implications*, Erlbaum, Hillsdale, New Jersey, pp. 19–56.

Moore, C. L. (1986). Sex differences in self-grooming of rats: effects of gonadal hormones and context. *Physiol. Behav.* **36**:451–455.

Moore, C. L., and Morelli, G. A. (1979). Mother rats interact differently with male and female offspring. *J. Comp. Physiol. Psychol.* **93**:677–684.

Moore, C. L., and Power, K. L. (1986). Prenatal stress affects mother–infant interaction in Norway rats. *Dev. Psychobiol.* **19**:235–245.

Moore, M. C., Whittier, J. M., Billy, A. J., and Crews, D. (1985). Male-like behaviour in an all-female lizard: relationship to ovarian cycle. *Anim. Behav.* **33**:284–289.

Morali, G., Carrillo, L., and Beyer, C. (1985). Neonatal androgen influences sexual motivation but not the masculine copulatory motor pattern in the rat. *Physiol. Behav.* **34**:267–275.

Naftolin, F., Ryan, K. J., and Petro, Z. (1971). Aromatization of androstenedione by the diencephalon. *J. Clin. Endocrinol. Metab.* **33**:368–370.

Olioff, M., and Stewart, J. (1978). Sex differences in the play behavior of prepubescent rats. *Physiol. Behav.* **20**:113–115.

Olsen, K. (1979). Androgen-insensitive rats are defeminised by their testes. *Nature (London)* **279**:238–239.

Olsen, K., and Whalen, R. E. (1980). Sexual differentiation of the brain: effects on mating behavior and [^{3}H]-estradiol binding by hypothalamic chromatin in rats. *Biol. Repro.* **22**:1068–1072.

Oppenheim, R. W. (1982). Preformation and epigenesis in the origins of the nervous system and behavior: issues, concepts, and their history. In Bateson, P. P. G., and Klopfer, P. H. (eds.), *Perspectives in Ethology*, Vol. 5, Plenum Press, New York, pp. 1–100.

Oyama, S. (1982). A reformulation of the idea of maturation. In Bateson, P. P. G., and Klopfer, P. H. (eds.), *Perspectives in Ethology*, Vol. 5, Plenum Press, New York, pp. 101–131.

Oyama, S. (1985). *The Ontogeny of Information: Developmental Systems and Evolution*, Cambridge University Press, Cambridge.

Panksepp, J. (1981). The ontogeny of play in rats. *Dev. Psychobiol.* **14**:327–332.

Panksepp, J., Jalowiec, J., De Eskinazi, F. G., and Bishop, P. (1985). Opiates and play dominance in juvenile rats. *Behav. Neurosci.* **99**:441–453.

Petersen, S. L. (1986). Perinatal androgen manipulations do not affect feminine behavioral potentials in voles. *Physiol. Behav.* **36**:527–531.

Pfeiffer, C. A. (1935). Origin of functional differences between male and female hypophyses. *Proc. Soc. Exp. Biol. Med.* **32**:603–605.

Phoenix, C. H. (1974). Prenatal testosterone in the nonhuman primate and its consequences for behavior. In Friedman, R. C., Richart, R. M., and van der Wiele, R. L. (eds.), *Sex Differences in Behavior*, Wiley, New York, p. 30.

Phoenix, C. H., and Chambers, K. C. (1982). Sexual behavior in adult gonadectomized female pseudohermaphrodite, female, and male rhesus macaques (*Macaca mulatta*) treated with estradiol benzoate and testosterone propionate. *J. Comp. Physiol. Psychol.* **96**:823–833.

Phoenix, C. H., Goy, R. W., Gerall, A. A., and Young, W. C. (1959). Organizing action of prenatally administered testosterone propionate on the tissues mediating mating behavior in the female guinea pig. *Endocrinology* **65**:369–382.

Plapinger, L., and McEwan, B. S. (1978). Gonadal steroid–brain interactions in sexual differentiation. In Hutchinson, J. B. (ed.), *Biological Determinants of Sexual Behaviour*, Wiley, Chichester, pp. 153–218.

Raleigh, M. J., Flannery, J. W., and Ervin, F. R. (1979). Sex differences in behavior among juvenile vervet monkeys (*Cercopithecus aethiops sabaeus*). *Behav. Neur. Biol.* **26**:455–465.

Raynaud, J. P. (1973). Influence of rat estradiol binding plasma protein (EBP) on uterotrophic activity. *Steroids* **21**:249–258.

Richmond, G., and Sachs, B. D. (1984a). Maternal discrimination of pup sex in rats. *Dev. Psychobiol.* **17**:87–89.

Richmond, G., and Sachs, B. D. (1984b). Further evidence for masculinization of female rats by males located caudally *in utero*. *Horm. Behav.* **18**:484–490.

Sachs, B. D., and Thomas, D. A. (1985). Differential effects of perinatal androgen treatment on sexually dimorphic characteristics in rats. *Physiol. Behav.* **34**:735–742.

Schultz, F. M., and Wilson, J. D. (1974). Virilization of the Wolffian duct in the rat fetus by various androgens. *Endocrinology* **94**:979–986.

Shrenker, P., and Maxson, S. C. (1983). The genetics of hormonal influences on male sexual behavior of mice and rats. *Neurosci. Biobehav. Rev.* **7**:349–359.

Slob, A. K., Huizer, T., and van der Werff ten Bosch, J. J. (1986). Ontogeny of sex differences in open-field ambulation in the rat. *Physiol. Behav.* **37**:313–315.

Spear, L. P., and Brake, S. C. (1983). Periadolescence: age-dependent behavior and psychopharmacological responsivity in rats. *Dev. Psychobiol.* **16**:83–109.

Stockman, G. R., Callaghan, R. S., Gallagher, C. A., and Baum, M. J. (1986). Sexual differentiation of play in the ferret. *Behav. Neurosci.* **100**:563–568.

Thor, D. H. (1980). Isolation and copulatory behavior of the male laboratory rat. *Physiol. Behav.* **25**:63–67.

Thor, D. H., and Holloway, W. R. (1984). Social play in juvenile rats: a decade of methodological and experimental research. *Neurosci. Biobehav. Revs.* **8**:455–464.

Ulrichs, K. H. (1898). *Forschungen über das Rätsel der männlichen Liebe*, Republished by Arno Press, New York, 1975.

vom Saal, F. S. (1981). Variation in phenotype due to random intrauterine positioning of male and female fetuses in rodents. *J. Repro. Fert.* **62**:633–650.

Wallen, K., and Winston, L. A. (1984). Social complexity and hormonal influences on sexual behavior in rhesus monkeys (*Macaca mulatta*). *Physiol. Behav.* **32**:629–637.

Ward, I. L., and Weisz, J. (1980). Maternal stress alters plasma testosterone in fetal males. *Science* **207**:328–329.

Whalen, R. E., and Olsen, K. L. (1981). Role of aromatization in sexual differentiation: effects of prenatal ATD treatment and neonatal castration. *Horm. Behav.* **15**:107–122.

White, J. O., Hall, C., and Lim, L. (1979). Developmental changes in the content of oestrogen receptors in the hypothalamus of the female rat. *Biochem. J.* **184**:465–468.

Witcher, J. A. and Clemens, L. G. (1987). A prenatal source for defeminization of female rats is the maternal ovary. *Horm. Behav.* **21**:36–43.

Wolfheim, J. H. (1978). Sex differences in behavior in a group of captive juvenile talapoin monkeys (*Miopithecus talapoin*). *Behaviour* **63**:110–128.

Chapter 9

BEHAVIORAL DEVELOPMENT: TOWARD UNDERSTANDING PROCESSES

Carel ten Cate

Sub-Department of Animal Behaviour
University of Cambridge
Madingley, Cambridge CB3 8AA, England
and
Zoological Laboratory
University of Groningen
Postbus 14, 9750AA Haren, The Netherlands

I. ABSTRACT

Behavior develops out of an interplay of various internal and external factors. Therefore, to understand development it is necessary not only to identify these factors, but also to analyze the processes by which they exert their influence and how different factors interact with one another. In this chapter I shall use the phenomena of "imprinting" and "song learning" in birds to illustrate how such a process-oriented, interactive approach toward development can lead to new insight in, and interpretations of, several aspects of these phenomena.

The Introduction outlines what I mean by a "process-oriented" approach to development. This approach raises a number of new questions about imprinting and song learning, the discussion of which forms the remaining part of the chapter. First, it is examined how individuals may build up an internal representation of their parents' appearance or song as a consequence of having had experience with these. Both imprinting and song learning are characterized by a gradual decrease in sensitivity to external stimulation. Which processes may lead to the ending of this sensitivity is the second question examined. Next, it is shown how the integration of various findings about the acquisition process provides a general, descriptive model that may help to explain what at first sight seem to be anomalous findings in studies on song learning in the white-crowned sparrow. In the next section, attention is drawn to the active role that an individual itself may play in the acquisition of knowledge. Finally, in the Discussion section, the findings are summarized and some attention is given to a functional perspective.

II. INTRODUCTION

If we took some bird eggs, from a domestic chicken for instance, and put half of them in an incubator, leaving the other half out, it would not be a surprise to discover that only the eggs put in the incubator will hatch. From this simple experiment we could correctly conclude that "warmth" is necessary for the transformation of eggs into chicks. However, the experiment tells us no more than that we have identified only one of the participants in a complex process. Of course, we all know that the warmth interacts with other factors outside and inside the egg. And although we can trace these factors, such as genes, yolk, etc., in the same way as the effect of warmth is assessed, a full understanding of how they assemble a chick is only obtained by studying their interactions: the embryological process.

The trivial example above illustrates two principles which are just as important in the study of behavioral development as they are in the study of physical development. The first principle is that identifying (or failure to identify) one factor contributing to the multifactorial and interactive process of development does not allow a conclusion about the importance of other factors: If it is shown that a particular environmental factor has an influence on the development of a specific behavior pattern, it does not exclude contributions of other factors. On the other hand, if this environmental factor was shown to have no influence on the emergence of the behavior, it would be wrong to accept this as proof of the influence of some other specific factor like, for instance, genes. A direct, positive assessment of the contribution of each of these factors is needed for such acceptance.

The second principle arising from the "egg and chick" example is the distinction between *identifying* the various factors influencing behavioral development and studying *how* these factors exert their effect. An experiment in which a change in adult behavior occurs as a consequence of a manipulation earlier in life need not tell us something about the process by which this manipulation produced the change.

The aim of this paper is to demonstrate the implications of these principles for understanding behavioral development. This enterprise is not a new one. Other people have put forward similar ideas and contributed to their empirical foundation (e.g., Lehrman, 1953, 1970; Kruijt, 1964, 1966; Bateson, 1976, 1983). Recently, Oyama (1985) has provided a powerful conceptual reassessment of these principles. The present paper should be seen as a case study, illustrating the use of these principles in developmental research. To this end, I will focus mainly on two subjects deeply rooted in traditional ethology: the phenomenon of "imprinting," and in particular "sexual imprinting," and that of song learning in birds. Basically, what I want to advocate is a change in attitude toward the study of behavioral development and, in particular, toward imprinting and related processes, like song learning. This change in attitude parallels the one which has occurred among learning theorists (Dickinson, 1980). The traditional behaviorist focused on the behavioral change per se, brought about by some manipulation. The limitations of the explanatory value of this approach gave rise to modern learning theory, in which the conditioning experiment is used as an analytical tool for studying *how* animals acquire knowledge through experience (see Dickinson, 1980).

After a short introduction to the subject of imprinting, I shall discuss some studies showing how application of the above-mentioned principles leads to a number of new questions and interpretations of this well-examined phenomenon. This will lead me to discuss research on another traditional topic among developmental ethologists: that of song learning, which shows many resemblances to imprinting. Finally, a short paragraph will give some attention to the "so what?" question: What, if any, is the importance of these ideas in a broader context, for instance for understanding evolutionary processes?

III. IMPRINTING AS A PHENOMENON

The young of many precocial birds, such as goslings or chicks, have a strong tendency to form a social preference for the first moving, conspicuous object they encounter after hatching. In doing so, they are fairly unselective, and thus exposure to human beings or artificial stimuli such as a rotating light can lead to a preference for these objects over conspecifics. This preference develops without the presence of any obvious external reinforcement, it may be achieved within a couple of hours (which gave rise to the notion of a so-called "sensitive phase"), and it usually shows great stability. These special characteristics gave the process its name: *filial imprinting*. The publication and translation of Lorenz's paper *Der Kumpan in der Umwelt des Vogels* (Lorenz, 1935, 1937), and his claim for the uniqueness of the process (Lorenz, 1935), are especially seen as the source of inspiration for many further studies. These studies focused on the characteristics of imprinting, its relation with other learning processes, and also on the attractiveness of different stimuli (for reviews, see Bateson, 1966; Sluckin, 1972). Lorenz also pointed to a long-term effect of early experience: Some birds that he had hand-raised, or that were reared by a foster species,

Table I. Mating Preference of Zebra Finch Males Raised by Mixed Pairs[a]

Preference[b]	Zebra finch males raised by mixed pairs	
	Immelmann (1972a)	ten Cate (1984)
Z	17	30
B	4	1

[a]Mixed pairs: Male Zebra finch and female Bengalese finch or vice versa.
[b]Z, conspecific females; B, Bengalese finch females.

later showed a sexual preference for humans or the foster species, respectively (Lorenz, 1935). This so-called "sexual imprinting" has since been shown to occur not only in precocial birds such as ducks (Schutz, 1965), snow geese (Cooke and McNally, 1975), and quail (Gallagher, 1976), but also in altricial birds such as doves and pigeons (Warriner *et al.*, 1963; Brosset, 1971) and zebra finches (Immelmann, 1969a; Kruijt *et al.*, 1983).

At present, the evidence suggests that filial imprinting and sexual imprinting are more likely to be two separate (though partially overlapping) processes, rather than, as Lorenz (1935) thought, two expressions of one underlying process (Bateson, 1979, 1981). For instance, in a number of species showing filial imprinting during the first days of life, the sensitive period for sexual imprinting is known to occur at a later age and may involve a different stimulus object (see Bateson, 1979). The questions that guided research on sexual imprinting have been very similar to those concerning filial imprinting. The important pioneering work of Immelmann on the zebra finch, for instance, addressed questions such as the timing of the sensitive period and the stability of the preference (Immelmann, 1972a, b; Immelmann and Suomi, 1981). He also addressed the question of whether a preference for an individual's own species was more easily obtained than one for a foster species. One such study involved the rearing of zebra finch males by mixed pairs consisting of one zebra finch parent and one Bengalese finch parent (Immelmann, 1972a,b). Males raised in this way usually preferred conspecific females over Bengalese finch females (Table I), a finding which I have confirmed (ten Cate, 1982, 1984; Table I).

IV. IMPRINTING AS A DEVELOPMENTAL PROCESS

At first sight, the extensive studies on sexual imprinting may suggest it to be one of the best-understood developmental processes. They have shown that sexual preferences in zebra finch males are influenced by experience with the parents, that imprinting occurs within a limited time span, that the preference is relatively stable, and also that a preference for one's own species is established more easily than one for another species. But, however important in their own right, what do these findings say about the development of species recognition in zebra finches viewed as a developmental *process*? That we can make further progress here, using the principles outlined in the Introduction, becomes clear after a closer examination of the interpretation of the data in Table I. These data have been interpreted as showing that zebra finches not only *learn* about parental characteristics, but also possess an *unlearned* preference for characteristics of their own species (e.g., Immelmann, 1972a; Bischof, 1979). To hypothesize the presence of such a preference is certainly valid, and initial preferences for specific stimuli have been shown for filial imprinting (e.g., Bateson and Jaeckel, 1976; Bateson, 1978a). However, in the present case, it should be noted that the evidence is obtained indirectly. The finding that the final preference of zebra finch males seems incompletely explained by imprinting is taken as evidence for the influence of another specific factor. But there could be many ways by which an adult male might become biased toward its own species. It has, for instance, been shown that in cross-fostered zebra finch males not only parents, but also the conspecific siblings, may exert an influence (Kruijt *et al.*, 1983). Another potential source for the learning of conspecific characteristics is a bird's own body. Evidence for such autoimprinting is present from fowl (Kruijt, 1962, 1985; Vidal, 1975, 1982). So, insofar as external participants in the process are concerned, there may be more than just parental influence. If we could control these alternative ways of learning conspecific characteristics, and a preference for the individual's own species persisted, that finding would of course be of interest. However, the question would still remain: Who are the participants in that process, and, for that matter, what is the nature of the process itself?

The point I have raised above supports the view that the influence of parents cannot, on its own, provide a satisfactory explanation for data such as those in Table I. But can it? This brings me to the developmental questions about the process of sexual imprinting. Identifying the importance of parental influence or the time of exposure, for instance, was an important step toward understanding the acquisition of a sexual preference by zebra finch males. The next step is to examine the processes bringing about these characteristics: *How* does exposure to a stimulus influence the acquisition? *How* is the timing achieved? *What* is the nature of the stored information, i.e., the "internal representation," resulting from the process? *How* does this representation exert its control over the behavioral output?, and so on.

In the following, I shall discuss in some detail three issues illustrating some of the progress that has been, or could be, made on these questions. First, I address the question of how exposure to one's parents may lead an individual to build up a representation of their appearance or their song. The second problem concerns the timing of the sensitive period, and, in particular, the mechanism underlying the ending of the sensitivity. The third issue is whether a developing individual can itself exert some influence on the progress of the acquisition process. The connection between these issues will, I hope, become clear later on.

A. The Acquisition Process

Immelmann (1972a) showed that young zebra finches developed a preference for the parental species, even if they were surrounded by a majority of birds of other species. This pointed to the importance of the social bond between parents and young for the imprinting process. Therefore, to tackle the first question, a logical start seemed to be the examination of what actually goes on between (foster) parents and young. To this aim, I carried out observations on various types of pairs (zebra finch, Bengalese finch, and mixed) from hatching until some 55 days of age [an extensive report of these studies is given in ten Cate (1982, 1984)]. Over this period, the relationships between young zebra finches and conspecific parents changed quite dramatically. Initially, the main interactions consisted of the young begging and the parents feeding them and brooding. After fledging, begging and feeding continued until about 35 days of age. Young and parents also spent considerable time clumping together, during which mutual preening (allopreening) occurred. A decrease in feeding coincided with the emergence of, and a steady increase in, aggression from both parents toward the juveniles. Although roughly similar changes in interactions between Bengalese parents and zebra finch young occurred, some of these interactions differed systematically in their timing and intensity from those with conspecific parents. This was examined in more detail in the mixed pairs. It appeared that in this situation parental care (feeding and brooding the young) and aggression were shown significantly more by zebra finch than by Bengalese finch parents (an example—feeding in male zebra finch/female Bengalese finch pairs—is given in Fig. 1A). Toward the end of the exposure, more clumping and allopreening was shown with the Bengalese finch parent, but most clumping occurred with the (conspecific) sibling. These differences were independent of the type of mixed pair used (male zebra finch/female Bengalese finch or vice versa), which led to the conclusion that overall, young zebra finches had been exposed to more interactions with conspecifics (parents and/or siblings), than with Bengalese finches. So, could it be that the quantity of such interactions was linked with the learning process, i.e., did males raised by mixed pairs become biased toward the zebra finch (Table I) as a consequence of more interactions with their conspecifics earlier on?

To test this hypothesis, two separate sets of experiments were set up in which young zebra finches remained visually exposed to conspecifics and

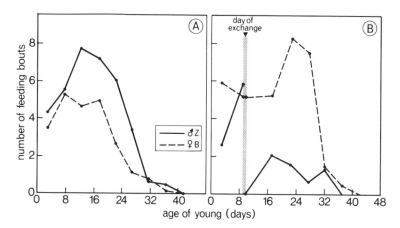

Fig. 1. (A) Feeding frequency in mixed pairs in relation to age of young. Indicated is the mean number of feeding bouts/4-hr observation period per 5-day interval received by individual male juveniles. The zebra finch parent fed significantly more frequently than the Bengalese finch parent ($p < 0.05$; Wilcoxon matched-pairs signed-ranks test (WMP); N=8). (B). Feeding frequency in mixed pairs in which the zebra finch parent was replaced by a nonbreeding zebra finch at around day 10. After exchange most feeding was done by the Bengalese finch parent ($p < 0.01$; WMP; N=8).

Bengalese finches, but in which the interactions with these species were manipulated. The first set of experiments consisted of a *successive* exposure to each species (ten Cate et al., 1984). After being reared by their own parents for 1 month, young males were transferred to a group of Bengalese finches for another month. For one series of males, the interactions between the Bengalese finches and the males were recorded. The later preference of these males showed a highly significant correlation with the amount of interactions with the Bengalese finches earlier on: The more aggressive was the behavior and clumping shown by the Bengalese finches, the more Bengalese-finch-directed was the later preference. The tentative conclusion, that the initiatives of the Bengalese finches caused the shift in preference toward them, was supported by the finding that exposure to Bengalese finches behind a wire screen (which limited the behavior of the Bengalese finches toward the young, but maintained visual exposure to them) led to a smaller shift (Fig. 2). A further reduction of the preference for Bengalese finches was obtained by exposure to stuffed Bengalese finches. This occurred in spite of the fact that the young males themselves showed a variety of social behavior, including precocial sexual behavior, toward the stuffed models.

A second set of experiments was carried out by manipulating the behavior of the zebra finch parent in mixed pairs, i.e., with young zebra finches subjected to *simultaneous* exposure to zebra finch and Bengalese finch (ten Cate, 1984). One such manipulation was, for instance, to replace the zebra finch parent by a bird of the same sex from a nonbreeding group when the young were around 10

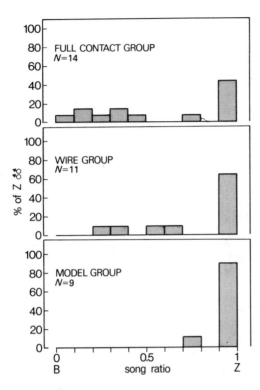

Fig. 2. Sexual preferences shown by zebra finch males that had individually been placed in a group of Bengalese finches (full contact group); in a group of Bengalese finches but separated from them by a wire screen (wire group); and in a cage containing a number of stuffed Bengalese finches (model group). The number of zebra finches raised in each condition is indicated. Sexual preference is measured by the relative frequency of singing to zebra finch and Bengalese finch females in a choice test (see ten Cate, 1985c, for details of this method and its validation); a song ratio of 1.0 indicates an exclusive preference for the zebra finch female, a ratio of 0.0 indicates an exclusive preference for the Bengalese finch female.

days of age. This new zebra finch only occasionally fed the young (and if this increased it was replaced). The Bengalese finch parent usually responded with an increase in its feeding so that, in contrast to the situation in nonmanipulated pairs (Fig. 1A), most feeding was now done by the Bengalese finch parent (Fig. 1B). None of the other behavioral interactions between the new zebra finch parent or the Bengalese finch parent and the young showed any significant change compared with the nonmanipulated pairs. So, any difference in later preference of the young must be due to this reversal of parental feeding. The effect of this and other manipulations is presented in Fig. 3. A significant decrease in preference for zebra finch occurred as a consequence of manipulating feeding behavior of the male zebra finch parent. Figure 3 also shows the effect on the preference of other manipulations, in which the possibility of the young males to interact with the zebra finch parent was further decreased. This resulted in a shift in preference from zebra finch to Bengalese finch. Since in all experiments the visual exposure to zebra finches was maintained, these results also strongly suggest that social interactions are necessary to build up a representation of the appearance of the individual with which a young male interacts.

So the above-mentioned experiments bring us a step closer to the mechanism underlying the acquisition of a preference. Nevertheless, it could be that the larger amount of interactions with conspecific parents in normal mixed pairs arose because the young zebra finches showed more responsiveness toward the conspecific parent. This may have induced the parent to show more feeding or aggression toward them. However, a more detailed study, on the difference between the parental species in mixed pairs raising mixed broods of both species, suggested the opposite (ten Cate, 1985a). Parents of both species preferentially fed conspecific young and some observations suggested that this initially arose from a greater reactivity of the parents to conspecific stimuli. This seems to add another step to the process outlined above: Parents are more responsive to conspecific young, this leads to more interactions with these young, and this in turn results in a stronger preference of these young for conspecifics later on.

The bias toward the individual's own species in Table I can now be interpreted as being a consequence of the way in which the imprinting process

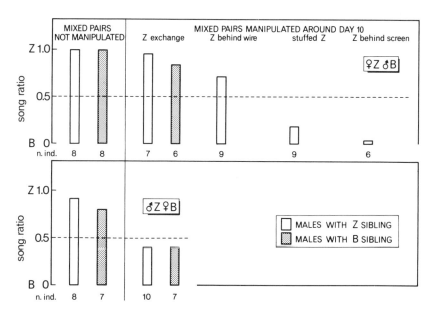

Fig. 3. Influence of various manipulations of parental behavior on the later sexual preference of zebra finch males raised by mixed pairs. Columns indicate the mean preference (measured as the song ratio—see Fig. 2) over the indicated number of males. *Z exchange*, replacement of the zebra finch parent by a same-sexed nonbreeding bird; *Z behind wire*, the new bird was placed behind a wire screen; *stuffed Z*, a stuffed zebra finch was placed in the cage; *Z behind screen*, the new zebra finch was placed behind an opaque screen (thus it could be heard, but not seen). See ten Cate (1984) for further details.

operates. If those birds with which most interactions take place are acting as imprinting stimuli, then young zebra finch males should indeed be more likely to prefer conspecific females later on. No "unlearned" preference for species-specific characteristics has to be assumed to explain these data: *It is the way a stimulus bird behaves, rather than its appearance, which makes it a suitable imprinting object.*

Of course, this raises a new question: What gives behavioral interactions their impact? It is as though the initiatives of parents or other birds somehow act as an unconditioned stimulus, making parental appearance become the conditioned one. One way to obtain this effect could be that each behavioral initiative that brings a parent very close to a young, or makes it touch it (aggression, feeding, clumping, and allopreening all produce this effect), leaves an impact. The cumulative effect of these experiences would be the development of an internal representation. In addition to the mechanism above, an interesting possibility is that a contingency between a bird's own behavior and that of others may have an impact. This hypothesis originates from work by Watson (1972, 1981) and Watson and Ramey (1972). Experiments with human babies suggested that a guiding factor for directing social behavior might be the contingency between the baby's behavior and some response to it. When movement of a mobile was linked with head movements of the baby, this not only stimulated the baby to show these movements, but the baby also started to direct social behavior to it. So, translated to the imprinting situation, a bird might become imprinted on the individual that responds to its behavior. An attempt to test this hypothesis in a filial imprinting experiment (ten Cate, 1986a) has been undertaken, but although it showed that for filial imprinting, as well, the learning process was indeed most enhanced when chicks were exposed to a contingent moving ("behaving") stimulus, the question of whether or not this enhancement was due to a contingency effect could not be answered conclusively.

A developmental process very similar to the one suggested here may also be important in song learning. Here, too, the learning process seems enhanced by giving a young bird the opportunity to interact with a tutor, as compared with a situation in which such interaction is absent, or where recorded songs are played to an isolated individual [e.g., in zebra finches (Immelmann, 1969b; Price, 1979; Böhner, 1983), in canaries (Waser and Marler, 1977), in nightingales (Todt *et al.*, 1979), in Bengalese finches (Dietrich, 1980), in indigo buntings (Payne, 1981), in chaffinches (Slater, 1983), and in white-crowned sparrows (Baptista and Petrinovich, 1984)]. In addition, when two potential live tutors are available, zebra finches seem to choose the one with which most aggressive interactions occur (Clayton, 1987). Pepperberg (1985), reviewing the enhancement of song learning by social interactions, draws attention to what might be a human analog of the stimulating effect of social interactions on achieving behavioral modifications: Live, social interactions with other humans are necessary to overcome existing strong inhibitions toward specific learning tasks.

It must be noted that demonstrating the enhancement of filial and sexual imprinting and song learning by interaction with the stimulus object (or by exposure to a moving rather than a nonmoving stimulus), does not exclude the possibility that some stimuli possess physical characteristics (e.g., color, shape, sound) that make them more attractive than others. Social interaction and physical characteristics should be seen as two factors which both may influence the outcome of the learning process. I will discuss the interactive effects of these two factors in more detail later on.

B. The Timing Process

An important feature of both the imprinting and the song-learning process is that learning seems limited to a specific period early in life. For sexual imprinting, a number of studies have concentrated on the delineation of the sensitive phase, e.g., in quail (Gallagher, 1977, 1978), mallard (Schutz, 1965), domestic cockerels (Vidal, 1975), and also in the zebra finch (Immelmann, 1972a, b; Immelmann and Suomi, 1981). However, such delineations need to be treated carefully, not because of the methodological difficulties facing the experimenter (although they are often underestimated: see Bateson, 1987b), but because their application is necessarily restricted until we know the process that underlies them. Are onset and offset of a phase simply a consequence of some internal clock (such as the physiological state at different ages), or are they influenced by rearing conditions, and if so, how are they influenced? If the latter were the case, any delineation would only be valid within the conditions of the experiment showing it. Only knowledge about the process would allow a prediction concerning the outcome of subjecting an animal to various treatments at different ages.

That delineation of a sensitive phase is a problem quite different from examining its cause is well demonstrated by work of Eales (1985) on song learning in zebra finches. Zebra finch song consists of one phrase, composed of smaller elements, which is repeated several times to form a song bout. The characteristics of this phrase are highly constant for a given individual, but phrases of different males may vary considerably. Young males usually copy the song of their (foster) father, even if this is a Bengalese finch male (Immelmann, 1969b). Since young males become independent of their parents at around 35 days of age (Immelmann, 1962; ten Cate, 1982), Eales (1985) was interested in whether song acquisition occurred before or after this age. To examine this, she carried out a number of experiments (see Eales, 1985, for details). In a first series, young males were removed from their parents and denied further interaction with adult males when they were either 35, 50, 65, or 120 days old. It appeared that the songs of males removed at 35 days of age shared no characteristics with those of their father, whereas those removed at 65 or 120 days were nearly identical to those of their fathers (see Fig. 4 for an example), with intermediate copying from males removed at 50 days. In another experiment, males were transferred from their own father to a different male at, again, 35, 50, or 65 days. As shown in Fig. 5, these males, too, seem to have learned their song after day 35 and before day 65. It should be noted that, strictly speaking, such an experiment varies two factors—not only age but also the length of the exposure, which also may influence from whom song is copied. Nevertheless, the data suggest a sensitive phase for song learning somewhere between 35 and 65 days of age. This was confirmed by later experiments of Clayton (1987), in which the length of exposure to successive tutors was kept constant. In addition, Clayton (1987) showed that introduction to a new tutor after 80 days of age did not alter the song acquired from previous tutors. Thus a sensitive phase was neatly delineated. But it should be noticed that the experiments do not demonstrate *how* the sensitive phase is brought to an end, a question that will be central to the remainder of this section.

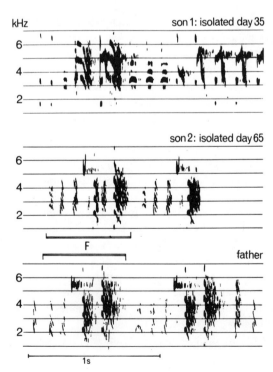

Fig. 4. Sonograms of the song of one zebra finch father and two of its sons, which were removed from the father at different ages. Son 2 copied its father's phrase (F), while the song of son 1 showed no resemblance to its father's song (Eales, 1985).

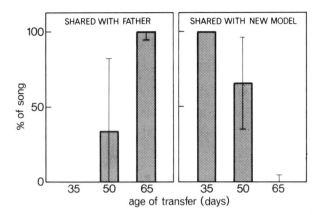

Fig. 5. Percentage of song elements shared by zebra finch males with either their father or a new model, in relation to the age of transfer from father to the new model (modified from Eales, 1985). Median value and interquartile ranges are given.

A number of hypotheses have been proposed to explain how sensitive phases may come to an end (see Bateson 1979, 1987a,b). One possible hypothesis is the traditional "clock" model, which assumes that some physiological factor, intrinsic to the animal but extrinsic to the learning system, ends the receptivity to external stimulation, i.e., stops the learning process. So, for song learning in zebra finches, physiological changes at about 65 days of age might make further learning impossible. Another hypothesis is the "self-termination" model, which assumes that learning stops when a certain specified amount of input has been received. So, the cumulative input of 65 days of exposure to adult song may have been enough to reach the limits of the zebra finches' capacity to store or process information about songs. Further stimulation would then be incapable of getting access to the song-learning system, and thus no further learning would occur. To make a black and white distinction between these hypotheses: The clock model assumes an infinite capacity of the learning system but a limited time span available for learning; the self-termination model assumes a limited capacity of the learning system but an infinite time span available for learning. Formulated in this way, the models lead to different predictions. A testable difference is that in the absence of relevant stimulation (e.g., singing males) until well after the age of 65 days, the clock model predicts that no learning of tutor songs will take place later on, whereas the self-termination model predicts that, since little or no learning could have occurred previously, the capacity for song learning will be extended to later ages. For the zebra finch, relevant information with respect to both models emerged when Eales (1985) had a closer look at males that were removed from their parents at 35 days of age and thereafter not exposed to other males. As mentioned, their song shared no characteristic with their father's song. Most of these males developed highly abnormal and unpatterned songs, showing no phrase structure (Fig. 6A). The interesting finding occurred when these males were placed with adult males at the age of 6 months. All males then modified their songs into the characteristic phrase structure, and thus apparently learned their song well after the assumed sensitive phase (Fig. 6B). This indicates that, lacking adequate tutoring, sensitivity is maintained until such a tutor becomes available. This supports the model that song learning is, at least to some degree, *self-terminating*; i.e., *the progress of the learning process itself may influence the ending of the sensitive phase.*

The notion of self-terminating sensitivity has also been put forward for filial imprinting. Bateson (1987a,b) showed that the puzzling variation in the timing of sensitivity among studies of filial imprinting becomes understandable if here, too, the termination of the learning process is dependent on its progress.

Although similar detailed experiments on the timing underlying the process of sexual imprinting are lacking, some of the present findings would fit in well with the self-termination mechanism. For instance, Immelmann (1972a; Immelmann and Suomi, 1981) reported that the preference of zebra finch males could be more easily altered when they had been exposed first to Bengalese finches and next to zebra finches than when the transfer occurred in the opposite direction. As we have seen in the previous section, the acquisition of a preference is enhanced by interactions between (foster) parents and young. The quantity of these interactions is lower between Bengalese finch parents and zebra

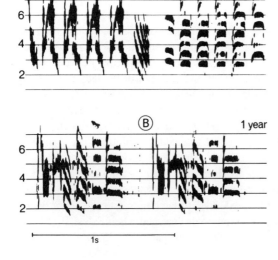

Fig. 6. Sonagrams of a zebra finch male removed from its father at 35 days of age. (A) Sonagram at 4 months of age, i.e., when adult, showing an abnormal, unpatterned song. (B) Sonagram of the same male at 1 year of age. The male had been placed in a group of adult males at 6 months of age, after which the characteristic phrase structure developed (Eales, 1985).

finch young than between zebra finch parents and young. Given this, the learning process may proceed more slowly with Bengalese finch parents. If imprinting is self-terminating, i.e., the learning stops after a specific amount of experience, then an earlier ending of sensitivity should be expected with conspecific parents (but see ten Cate et al., 1984, for alternative explanations).

As mentioned above, the concept of self-termination requires that there be a limited capacity somewhere in the imprinting (or song-learning) system. This might be a limitation in memory space, or in access from the memory to some behavioral system responsible for the later behavioral output. The plausibility of this assumption, as well as an extensive overview of the evidence and full implications of the various models for the termination of sensitive phases, are given by Bateson (1979, 1981, 1987a, and, in particular, 1987b). The notion of a self-terminating developmental process is not unique to imprinting and song learning. In a recent paper, Chalmers (1987) discusses several "rules" that may underlie the predictable, characteristic changes that are often seen in behavior during development. He argues that one of the ways such changes are brought about is that the behavioral performance reaches a set point that is independent of age, but dependent on quality or quantity of the performance; in other words, the progress of the developmental process brings itself to an end.

An Interactive Model for Imprinting and Song Learning

The data of Eales (1985, 1987) indicate another phenomenon, mentioned in the previous section: Exposure to the juvenile songs of age-mates only or to the vocalizations (calls) of females produces very slow, if any, song learning in zebra finch males. This in spite of opportunities to interact with age-mates or females. It indicates that, although female calls can be used as song elements (Eales, 1987), the physical qualities of the acoustic stimulation may influence song learning: Male song is of higher quality (leads to more rapid learning, hence earlier termination of sensitivity) than female vocalizations or the juvenile songs of age-mates. A similar preference to copy some song types rather than others was obtained by Marler (1970), who showed that tape-tutored white-crowned sparrows were more likely to accept conspecific than heterospecific songs as models. Marler postulated "a sensory 'template', somewhere in the auditory pathway, which focuses attention on sound patterns of a certain type" (Marler, 1970). Conclusive evidence for such phenomena for sexual imprinting is still lacking (see previous section and ten Cate, 1984, 1985b, 1988; ten Cate and Mug, 1984), although the possibility is still open. For filial imprinting, ample evidence is available that naive chicks prefer some physical characteristics over others (see Sluckin, 1972). As mentioned in the previous section, these findings show that imprinting and song-learning processes can be influenced in two ways. Frequent interactions with a stimulus (or exposure to a moving one) will enhance the acquisition of knowledge of its appearance or song type. At the same time, the physical qualities (e.g., appearance or sound) of the stimulus are important.

The next step is to relate these findings to the self-terminating character of imprinting and song-learning processes. Since ending of the learning process, i.e., ending of the sensitive phase, seems dependent on the progress of the process, which in turn depends on both physical qualities and behavior of the stimulus, the developmental process is an *interactive* one. A descriptive model showing these interactive relationships is given in Fig. 7. The two factors influencing the learning process are indicated along the two axes. As we have seen, learning speed increases with increasing amount of social interaction (or stimulus behavior), which is indicated along the y-axis. The zero level of social interactions would be exposure to a taped tutor in song learning, or a nonmoving stimulus such as a stuffed bird in sexual or filial imprinting. Intermediate stimulation might be provided by a stimulus bird behind a wire screen, and optimal stimulation might involve tutor and young bird in the same cage. The x-axis indicates that learning is also enhanced by physical characteristics of the stimulus: the color of the light used for filial imprinting, the acoustic qualities of conspecific songs over other song types or female calls, etc. It is the combination of both factors which influences the progress of the learning process: It will proceed slowly with a stimulus of poor physical quality and limited possibilities of enhancing its attraction due to constraints on its behavior. The notion of self-termination brings an important extra dimension in the model. It suggests that the cumulative effects of experience are asymptotic. As a result, novel experience exerts a smaller and smaller influence on the bird's behavior. The model implies in the top right part of the figure that this state can be reached very quickly, and

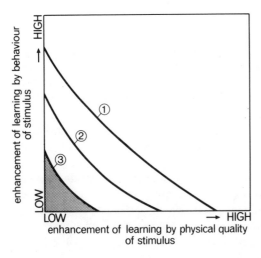

Fig. 7. Graphical representation of the dependence of sensitive-phase duration on both physical quality of a stimulus and the enhancement of its attraction by movement or behavior shown by this stimulus. 1, 2, and 3 are isoclines, each connecting points of equal learning speed, i.e., giving sensitive phases of equal length. Line 1 connects points at which the combination of stimulus quality and behavior leads to a rapid learning (a short sensitive phase). The shaded area below line 3 indicates the condition in which learning proceeds too slowly to produce a noticeable effect (the sensitive phase never ends). Line 2 represents an intermediate learning speed (a sensitive phase of intermediate length).

in the bottom left part that it will only be reached very slowly. In fact, it can well be imagined that in the bottom left corner a threshold occurs, below which the stimulation is so limited that learning is never complete; i.e., even an infinite exposure to that condition never leads to the end of the sensitivity. Also, the model implies that an equally long sensitive phase (i.e., identical learning speed) can be obtained by a rearing condition involving limited possibilities of interacting with a high-quality stimulus, and by a condition involving excellent opportunities to interact with a stimulus of poor physical quality. Now some hypothetical isoclines can be drawn, connecting points for which the combination of both factors requires identical periods of exposure to produce self-termination. The completed figure now captures the connections between the main influences on imprinting and song learning. The shaded area indicates the threshold at which learning only occurs after infinite exposure. The different lines indicate the different levels of learning speed, with high learning speed implying a rapid learning and thus a shorter period of sensitivity.

Does this model have any use, in the sense that it can clarify findings for which other hypotheses could not account? There is no better way of illustrating this than with an example, which is provided by some at-first-sight paradoxical findings in studies of song learning in the white-crowned sparrow.

Marler (1970) demonstrated two things: that tape-tutored white-crowned males would copy conspecific songs only, and that these songs were learned before 50 days of age. More recently, Baptista and colleagues (Baptista and Morton, 1981; Baptista and Petrinovich, 1984, 1986; Baptista, 1985; Petrinovich and Baptista, 1987) demonstrated that hand-reared birds could learn after 50 days of age when exposed to a live tutor, but not when exposed to a taped one. Also, not only conspecific songs could be learned from a live tutor, but heterospecific songs could be learned when birds were exposed to members of these species or to conspecifics singing heterospecific songs. That some filtering occurs in tape-tutored birds, similar to Marler's findings, is not disputed. But in addition, Baptista and Petrinovich (1986) state that "social interaction can override any auditory gating mechanism that would prevent white crowned sparrows from learning allospecific songs" (see also Baptista, 1985; Pepperberg, 1985). About the observation of learning occurring after the 50-day limit observed by Marler (1970), Baptista and Petrinovich (1986) state: "Perhaps there is no sensitive phase at all when a live tutor is presented after a period of isolation from song as existed in the present study: the lack of acoustic stimulation could delay the period of sensitivity." They continue by saying:

> Two facts mediate against this interpretation. First, tape-tutored birds do not learn after 50 days of age. If isolation merely delays the onset of the sensitive phase these birds should learn also. Second, two male and two female students were held in group isolation until they were 100 days old and then exposed to a live tutor. None of the birds copied the tutor song. . . . Both of the male students were singing an isolate song prior to tutoring, and this isolate song persisted. (Baptista and Petrinovich, 1986)

The quotations above illustrate that although the traditional clock-model approach of sensitivity ending at around 50 days of age is not capable of explaining the data, Baptista and Petrinovich also reject the possibility of delayed sensitivity resulting from isolation. In a recent paper (Petrinovich and Baptista, 1987), a modified clock-model is proposed, with the suggestion that sensitivity to strong stimulation (i.e., to live tutors) is maintained for a longer period than is sensitivity to weaker forms of stimulation. Although indeed the findings can be interpreted in this way, I suggest that they also fit a self-termination model if the effects of social stimulation or sensitivity to certain song types over others are not seen as independent, dichotomous categories but as interactive ones, as presented in Fig. 7. Using this model, the results obtained by Marler (1970) and Baptista and Petrinovich (1984, 1986; Petrinovich and Baptista, 1987) fit very well together, and the findings which are presented as anomalies with respect to the possibility of delayed sensitivity in fact support the interactive model.

So, how can we fit the various findings in the model outlined in Fig. 7? Marler exposed his birds to a high-quality stimulus (conspecific song). Baptista and Petrinovich first isolated white-crowned sparrows for 50 days. This condition does not exclude all learning: As has been demonstrated by deafening males, males may learn from their own auditory feedback (Marler, 1970). However, it seems plausible that a bird's own juvenile isolate song is a lower-quality stimulus than full adult song (surprisingly, no experiments have been reported to examine whether isolate song is indeed the less attractive stimulus of the two). Assuming that the learning proceeds very slowly with the bird having feedback from its own vocalizations only, both anomalies observed by Baptista and Petrinovich become understandable. First, the learning may proceed so slowly that at 50 days of isolation the process is not even halfway to the required amount of stimulation. When the birds are next subjected to full contact with heterospecific stimulus birds, this creates exposure to a better acoustical stimulus, the effect of which can be enhanced by interaction with it. This experience may be in time to override the preemptive effect that a bird's own isolate song had, thus explaining why isolate birds may learn from a live tutor after 50 days.

Assuming that conspecific song provides a higher-quality stimulation than isolate songs, the preemptive effect of conspecific song may have been sufficient to prevent an effect of exposure to a live tutor in birds that were exposed to taped, conspecific tutor song for the first 50 days. Thus the self-termination model explains the first observation with which Baptista and Petrinovich have difficulties.

Another thing the interactive model would predict is that *given enough time* a bird's own isolate song, although of poor quality, may eventually produce enough cumulative stimulation to terminate the learning process. This state may have been reached in the white-crowned sparrows that were isolated (i.e., exposed to their own song and that of age-mates, since the birds were reported to be

Behavioral Development

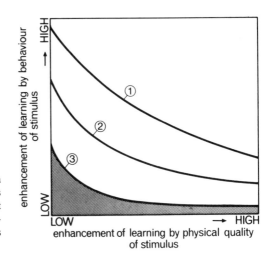

Fig. 8. Like Fig. 7, but representing a situation in which, even for a stimulus of high physical quality, enhancement of the stimulation by some form of behavior (or movement) of the stimulus is essential for learning to occur.

group-isolated) for 100 days, before they were introduced to the live tutor. So again the interactive model explains a seeming anomaly in the data obtained by Baptista and Petrinovich. Thus, at the moment, the interactive self-termination model seems a possible alternative to the clock model for understanding song learning in the white-crowned sparrow.

The framework presented here might also be of use to examine interspecific differences. Zebra finches, for instance, seem not to be capable of learning from a taped tutor, whether or not a conspecific song is played (Price, 1979). This suggests that the enhancement of learning the characteristics of a stimulus by some more intensive exposure to it is essential for learning to occur. This is in contrast with a species like the white-crowned sparrow. It means that Fig. 7 would look different for the zebra finch—the shaded area representing no learning would extend along the x-axis, as shown in Fig. 8.

C. An Active Role for the Developing Individual?

As shown above, the developmental process underlying imprinting and song learning is an interactive one, in which progress of learning is influenced by several factors. Presumably, further research will show that the process is more complex. For instance, the amount of experience required to reach self-termination might vary to some degree with developmental age, so reality may be in between the clock and self-termination models.

In this section I want to point to an implicit assumption made in previous sections, which is that the developing bird itself is supposed to be passively "waiting" for the moment that the cumulative experience terminates the learning process. But why should it do so? It would be most adaptive if, in conditions leading to very slow learning, the young bird itself could to some extent compensate for this. This would prevent it, for instance, from being forced toward independence without sufficient knowledge of what its mates should look like, or what its song should sound like. The example I will use to illustrate this point is, I admit, a speculative one. However, its aim is to stimulate research on, rather than to prove, this possibility.

A number of estrildid finch species (the group of species to which the zebra finch belongs) have been reported to show "listening" or, as it sometimes has been called, "peering" behavior (Morris, 1958; Immelmann, 1962; Immelmann and Immelmann, 1967). Immelmann (1962) describes listening as follows: "during the singing of one male, another male perches beside him, stretches his neck until his head is close to the bill of the singing male and freezes in this posture." Although studies on the singing and song learning of zebra finches never reported its occurrence, the behavior can be shown by this species (ten Cate, 1986b). However, the conditions under which this occurs seem quite special. It was observed predominantly in juvenile males, and in particular in juvenile males having a Bengalese finch foster father, rather than a zebra finch father (Table II). The age at which the behavior was observed corresponded with the period for which, under similar conditions, the data of Eales (1985) suggested that young males learn their father's song, i.e., after day 35 of age. This coincidence suggests a link between listening and song learning, as has also been suggested for listening behavior shown by juvenile Bengalese finches (Dietrich, 1980). But if such a link exists, why should listening be shown with Bengalese finch fathers only, and not with conspecific ones? One possible suggestion (see ten Cate, 1986b) arises when the singing behavior of the Bengalese and zebra finch fathers is compared. Bengalese finch fathers sing less than do zebra finch fathers (Fig. 9). Also, the song of Bengalese finches is usually longer, more variable, softer, and has different acoustic features as compared to zebra finch

Table II. Occurrence of Listening Behavior in Juvenile Zebra Finch Males, with Respect to the Species of Their Foster Fathers

Species of foster father	Number of juvenile zebra finch males	
	Observed	Listening
Zebra finch	38	—
Bengalese finch	56	14

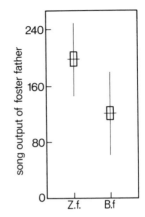

Fig. 9. Song output by Bengalese and zebra finch fathers raising zebra finch offspring. Song output was measured when the offspring were between 30 and 55 days of age. The numbers refer to the frequency of scores indicating that a father sang within a 1.5 min period during a total of 20 hr of observation/father. Mean, S.D., and S.E.M. are given; zebra finch fathers have a significantly higher song output than do Bengalese finch fathers ($p < 0.001$; Mann-Whitney U-test).

song. Some, or the combination, of these factors might mean that the stimulation offered by a singing Bengalese finch male drops below a certain threshold needed by a young zebra finch male to enable accurate copying. By listening behavior, the young male may actively increase its exposure to Bengalese finch songs, thus enabling adequate copying. So, it seems plausible that, *in case of low stimulation either in quantity or in quality, a developing individual might behave in such a way that it actively increases the exposure to that stimulation.*

For imprinting processes, too, there is some evidence suggesting an active role of the young bird in acquiring information. In a filial imprinting experiment, chicks and ducklings will actively work (e.g., by pressing a pedal) to present themselves with the most effective form of stimulation (Bateson, 1971).

The example of the listening behavior also illustrates another point. Had only the outcome of the learning process been studied, i.e., whether zebra finch males copied the song of either species, we might well have concluded that young zebra finches are just as capable of copying Bengalese finch song as they are of copying conspecific song. The observation of listening to Bengalese finch fathers suggests that such a conclusion might well have been wrong. Only observation of the behavior of the young birds during their development can bring to light whether active acquisition of knowledge by a juvenile bird contributes to the learning process.

V. DISCUSSION

The examples discussed above illustrate a number of things. The sexual-imprinting studies demonstrate that awareness of behavioral development as a multifactorial process leads to new questions about how observed behavioral phenomena arise and how such questions can be tackled. The song-learning studies show how the interaction of several factors may operate to produce a unifying explanation for phenomena previously treated as being independent and only partially explainable. Finally, the examples emphasize a methodological point: the importance of observing the actual course of development in an individual and the experience to which it is subjected, not only to get a handle on the processes underlying behavioral change, but also, as the listening example illustrates, as the only way to discover a process.

Of course, the examples were selectively chosen to illustrate various points. Other questions about certain aspects of song-learning and imprinting processes can be asked. An important one concerns the way in which information about a stimulus is stored: What is the way in which a stimulus is internally represented? What determines the apparently limited capacity of a "memory store"? Can previously acquired knowledge of a stimulus be "erased" by a later one? Can previously acquired knowledge be overridden but still maintain an influence over behavioral output, or can such existing knowledge be "updated" under certain conditions, thus modifying a previously acquired representation and resulting in "mixed" songs or preferences? There is, for instance, some evidence that successive exposure of young zebra finch males to conspecific parents and next to Bengalese finches may produce not only males singing songs consisting of elements of both species (Clayton, 1987), but also males with a sexual preference for individuals combining elements of both species in their appearance (ten Cate, 1986c, 1987). This suggests that these males possess an internal representation consisting of a mixture of characteristics of both species. Knowledge of such phenomena is at present very limited, but nevertheless necessary to predict when and how experience terminates the learning process.

Better knowledge of the processes of song learning and imprinting raises other questions. First, whether the way in which both processes operate makes sense against the context in which they occur in nature. Second, whether the better knowledge does tell more about what these processes are for, i.e., about the function of song learning and imprinting.

The experiments discussed above have shown that both physical qualities of, and interaction with, a stimulus may enhance the learning of its characteristics. Under the conditions occurring in nature, it seems likely that these factors will in particular promote the learning of characteristics of parents and siblings. Other individuals may lack the appropriate stimulus qualities (e.g., by singing heterospecific songs) or may have no interactions with the juveniles. In addition to this, the learning processes are highly flexible; rapid progress is made when the conditions for learning are "right," e.g., when parents are nearby and active, and learning is very limited when the conditions are "wrong," e.g., when parents are absent. The ending of learning after a specific amount of experiences has been received, rather than after a specific time period has elapsed, also seems a mechanism to produce effective learning. At the same time, an active role of the juvenile bird may prevent the sensitivity getting extended to situations in which no additional adequate learning may occur. So, the combination of the different factors suggests that, under normal circumstances, they will promote learning of a particular type of stimuli.

This conclusion highlights that, apparently, it is very important to limit learning to the characteristics of specific individuals. Why should this be so? For sexual imprinting, the coincidence in the timing of imprinting and the development of adult plumage characteristics in young birds led Bateson (1979) to suggest that imprinting is primarily a mechanism to learn the characteristics of close kin. As shown above, the "design" of the process fits with this suggestion. Learning the characteristics of kin may enable an individual to achieve a balance between inbreeding and outbreeding (Bateson, 1987b, 1980, 1982). Better knowledge of how the imprinting process operates also enabled a better assessment of how sex-related differences in mate choice arise (ten Cate, 1985b, 1988).

As for the function of song learning, this is at the moment a subject of discussion. Three different hypotheses have been suggested: Song learning may primarily have evolved as a mechanism to transfer a complex behavior pattern (a specific song type) to the next generation, to allow individuals to match their songs to that of neighbors, which may have a number of advantages, or to allow identification of birds from genetically different populations, which may have an advantage in choosing mates (see Baker and Cunningham, 1985, and accompanying commentaries, for the various views on this subject). A species on which the discussion focuses is the white-crowned sparrow, and the timing of song learning in this species is used as a way to discriminate between the various hypotheses. As we have seen, the timing of the sensitivity in the white-crowned sparrow is probably more dependent on rearing conditions than was previously realized. It can therefore be doubted whether present knowledge allows a firm conclusion about when songs are learned in nature, and from what tutor. This leaves the question about the function of song learning in the white-crowned sparrow still open. Nevertheless, study of the song-learning process has helped to identify the factors which may, under natural circumstances, contribute to song learning and its timing. Measuring parameters such as exposure to song and interactions with other individuals in free-living birds seems necessary for translating the existing findings to natural conditions.

To conclude, the aim of this chapter has been to demonstrate that, however interesting and useful it is to identify participants in the developmental process, it must be seen as a first step in studying behavioral development. Phenomena such as an own-species bias in imprinting, or a sensitive phase for song learning, may at first seem self-explanatory and well understood; they are, in fact, descriptions of the end product of dynamic processes and open to further investigation. Finally, I hope to have demonstrated that an interactive approach to development is more than just stating that life is complicated, and that it does lead to testable hypotheses and unifying explanations concerning behavioral development.

VI. ACKNOWLEDGMENTS

This chapter has been awarded the first Niko Tinbergen-Förderpreis by the German Ethologische Gesellschaft e.V. (the association of German ethologists). The text arose out of a talk given at the Association for the Study of Animal Behaviour 50th Anniversary meeting (June 1986). I am very grateful to Pat Bateson and Joan Stevenson-Hinde for inviting me to present that talk.

An important influence on my own development as an ethologist was exerted by Jaap Kruijt and Pat Bateson, and the ideas presented here owe much to their guidance and support over the years. They, as well as Lucy Eales, Karen Hollis, and Peter Slater, commented on the manuscript. Lucy Eales also kindly provided photographs of the sonagrams. Dick Visser prepared the figures. Writing was done while I held a Science and Engineering Research Council fellowship at the Sub-Department of Animal Behaviour at Madingly. Preparation of the final manuscript has been made possible by a fellowship of the Royal Netherlands Academy of Arts and Sciences. Hil Lochorn-Hulsebos typed the final version of the manuscript.

VII. REFERENCES

Baker, M. C., and Cunningham, M. A. (1985). The biology of bird-song dialects. *Behav. Brain Sci.* **8:**85–133.
Baptista, L. F. (1985). The functional significance of song sharing in the white crowned sparrow. *Can. J. Zool.* **63:**1741–1752.
Baptista, L. F., and Morton, M. L. (1981). Interspecific song acquisition by a white crowned sparrow. *Auk* **98:**383–385.
Baptista, L. F., and Petrinovich, L. (1984). Social interaction, sensitive phases and the song template hypothesis in the white crowned sparrow. *Anim. Behav.* **32:**172–181.
Baptista, L. F., and Petrinovich, L. (1986). Song development in the white-crowned sparrow: social factors and sex differences. *Anim. Behav.* **34:**1359–1371.
Bateson, P. P. G. (1966). The characteristics and context of imprinting. *Biol. Rev.* **41:**177–220.
Bateson, P. P. G. (1971). Imprinting. In Moltz, H. (ed.), *Ontogeny of Vertebrate Behavior*, Academic Press, New York, pp. 369–378.
Bateson, P. P. G. (1976). Specificity and the origins of behavior. *Adv. Stud. Behav.* **6:**1–20.
Bateson, P. P. G. (1978a). Early experience and sexual preferences. In Hutchison, J. B. (ed.), *Biological Determinants of Sexual Behaviour*, Wiley, London, pp. 29–53.

Bateson, P. (1978b). Sexual imprinting and optimal outbreeding. *Nature* **273**:659–660.
Bateson, P. (1979). How do sensitive periods arise and what are they for? *Anim. Behav.* **27**:470–486.
Bateson, P. (1980). Optimal outbreeding and the development of sexual preferences in Japanese quail. *Z. Tierpsychol.* **53**:231–244.
Bateson, P. (1981). Control of sensitivity to the environment during development. In Immelmann, K., Barlow, G. W., Petrinovich, L., and Main, M. (eds.), *Behavioral Development*, Cambridge University Press, Cambridge, pp. 432–453.
Bateson, P. (1982). Preferences for cousins in Japanese quail. *Nature* **295**:236–237.
Bateson, P. (1983). Genes, environment and the development of behavior. In Halliday, T. R., and Slater, P. J. B. (eds.), *Genes, Development and Learning*, Blackwell Scientific, Oxford, pp. 52–81.
Bateson, P. (1987a). Biological approaches to the study of behavioral development. *Int. J. Behav. Devel.* **10**:1–22.
Bateson, P. (1987b). Imprinting as a process of competitive exclusion. In Rauschecker, J. P., and Marler, P. (eds.), *Imprinting and Cortical Plasticity*, John Wiley, New York, pp. 151–168.
Bateson, P. P. G., and Jaeckel, J. B. (1976). Chicks' preference for familiar and novel conspicuous objects after different periods of exposure. *Anim. Behav.* **24**:386–390.
Bischof, H. J. (1979). A model of imprinting evolved from neurophysiological concepts. *Z. Tierpsychol.* **51**:126–137.
Böhner, J. (1983). Song learning in the zebra finch (*Taeniopygia guttata*): selectivity in the choice of a tutor and accuracy of song copies. *Anim. Behav.* **31**:231–237.
Brosset, A. (1971). l'"Imprinting", chez les Columbidés—Etude des modifications comporte mentales au cours du vieillisement. *Z. Tierpsychol.* **29**:279–300.
Chalmers, N. R. (1987). Developmental pathways in behavior. *Anim. Behav.* **35**:659–674.
Clayton, N. S. (1987). Song tutor choice in zebra finches. *Anim. Behav.* **35**:714–721.
Cooke, F., and McNally, C. M. (1975). Mate selection and colour preferences in lesser snow geese. *Behaviour* **53**:151–170.
Dickinson, A. (1980). *Contemporary Animal Learning Theory*, Cambridge University Press, Cambridge.
Dietrich, K. (1980). Vorbildwahl in der Gesangentwicklung beim Japanischen Mövchen (*Lonchura striata* var. *domestica*, Estrildidae). *Z. Tierpsychol.* **52**:57–76.
Eales, L. A. (1985). Song learning in zebra finches: some effects of song model availability on what is learned and when. *Anim. Behav.* **33**:1293–1300.
Eales, L. A. (1987). Song learning in female-raised zebra finches: another look at the sensitive phase. *Anim. Behav.* **35**:1356–1365.
Gallagher, J. (1976). Sexual imprinting: effects of various regimens of social experience on mate preference in Japanese quail (*Coturnix coturnix japonica*). *Behaviour* **57**:91–115.
Gallagher, J. (1977). Sexual imprinting: a sensitive period in Japanese quail (*Coturnix coturnix japonica*). *J. Comp. Physiol. Psychol.* **91**:72–78.
Gallagher, J. (1978). Sexual imprinting: variations in the persistence of mate preference due to difference in stimulus quality in Japanese quail (*Coturnix coturnix japonica*). *Behav. Biol.* **22**:559–564.
Immelmann, K. (1962). Beiträge zu einer vergleichende Biologie australischer Prachtfinken (*Sperm.*). *Zool. Jb. Syst.* **90**:1–196.
Immelmann, K. (1969a). Ueber den Einfluss frühkindlicher Erfahrungen auf die geschlechtliche Objektfixierung bei Estrildiden. *Z. Tierpsychol.* **26**:677–691.
Immelmann, K. (1969b). Song development in the zebra finch and other estrildid finches. In Hinde, R. (ed.), *Bird Vocalisations*, Cambridge University Press, Cambridge, pp. 61–81.
Immelmann, K. (1972a). The influence of early experience upon the development of social behavior in estrildine finches. *Proc. XV Int. Ornithol. Congr.*, Den Haag 1970, pp. 316–338.

Immelmann, K. (1972b). Sexual and other long-term aspects of imprinting in birds and other species. *Adv. Stud. Behav.* **4**:147–174.
Immelmann, K., and Immelmann, G. (1967). Verhaltensökologische Studien an afrikanischen und australischen Estrildiden. *Zool. Jb. Syst.* **94**:609–686.
Immelmann, K., and Suomi, J. (1981). Sensitive phases in development. In Immelmann, K., Barlow, G. W., Petrinovich, L., and Main, M. (eds.), *Behavioral Development*, Cambridge University Press, Cambridge, pp. 395–421.
Kruijt, J. P. (1962). Imprinting in relation to drive interactions in Burmese red junglefowl. *Symp. Zool. Soc. London* **8**:219–226.
Kruijt, J. P. (1964). Ontogeny of social behavior in Burmese red junglefowl (Gallus gallus spadiceus) Bonnaterre. *Behaviour Suppl. XII.* 1–201.
Kruijt, J. P. (1966). The development of ritualized displays in junglefowl. *Phil. Trans. Royal Soc. London B* **251**:479–484.
Kruijt, J. P. (1985). On the development of social attachments in birds. *Neth. J. Zool.* **35**:45–62.
Kruijt, J. P., ten Cate, C. J., and Meeuwissen, G. B. (1983). The influence of siblings on the development of sexual preferences in zebra finches. *Dev. Psychobiol.* **16**:233–239.
Lehrman, D. S. (1953). A critique of Konrad Lorenz's theory of instinctive behavior. *Quart. Rev. Biol.* **28**:337–363.
Lehrman, D. S. (1970). Semantic and conceptual issues in the nature–nurture problem. In Aronson, L. R., Tobach, E., Lehrman, D. S., and Rosenblatt, J. S. (eds.), *Development and Evolution of Behavior*, Freeman, San Francisco, pp. 17–52.
Lorenz, K. (1935). Der Kumpan in der Umwelt des Vogels. *J. Ornithol.* **83**:137–213, 289–413.
Lorenz, K. (1937). The companion in the bird's world. *Auk* **54**:245–273.
Marler, P. (1970). A comparative approach to vocal learning: song development in white crowned sparrows. *J. Comp. Physiol. Psychol. Monogr.* **71**:1–25.
Morris, D. (1958). The comparative ethology of grassfinches (*Erythrurae*) and Mannikins (*Amadinae*). *Proc. Zool. Soc. Lond.* **131**:389–439.
Oyama, S. (1985). *The Ontogeny of Information*, Cambridge University Press, Cambridge.
Payne, R. B. (1981). Song learning and social interaction in indigo buntings. *Anim. Behav.* **29**:688–697.
Pepperberg, J. M. (1985). Social modelling theory: a possible framework for understanding avian learning. *Auk* **102**:854–864.
Petrinovich, L., and Baptista, L. (1987). Song development in the white-crowned sparrow: modification of learned song. *Anim. Behav.* **35**:961–974.
Price, P. H. (1979). Developmental determinants of structure in zebra finch song. *J. Comp. Physiol. Psychol.* **93**:260–277.
Schutz, F. (1965). Sexuelle Prägung bei Anatiden. *Z. Tierpsychol.* **22**:50–103.
Slater, P. J. B. (1983). Chaffinch imitates canary song elements and aspects of organization. *Auk* **100**:493–495.
Sluckin, W. (1972). *Imprinting and Early Learning*, 2nd ed., Methuen & Co Ltd., London.
ten Cate, C. (1982). Behavioral differences between zebra finch and Bengalese finch (foster) parents raising zebra finch offspring. *Behaviour* **81**:152–172.
ten Cate, C. (1984). The influence of social relations on the development of species recognition in zebra finch males. *Behaviour* **91**:263–285.
ten Cate, C. (1985a). Differences in the interactions between zebra finch and Bengalese finch parents with conspecific versus heterospecific young. *Z. Tierpsychol.* **67**:58–68.
ten Cate, C. (1985b). On sex differences in sexual imprinting. *Anim. Behav.* **33**:1310–1317.
ten Cate, C. (1985c). Directed song of male zebra finches as a predictor of subsequent intra- and interspecific social behavior and pair formation. *Behav. Proc.* **10**:369–374.

ten Cate, C. (1986a). Does behavior contingent stimulus movement enhance filial imprinting in Japanese quail? *Dev. Psychobiol.* **19**:607–614.
ten Cate, C. (1986b). Listening behaviour and song learning in zebra finches. *Anim. Behav.* **34**:1267–1268.
ten Cate, C. (1986c). Sexual preferences in zebra finch males raised by two species: I. A case of double imprinting. *J. Comp. Psychol.* **100**:248–252.
ten Cate, C. (1987). Sexual preferences in zebra finch males raised by two species. II. The internal representation resulting from double imprinting. *Anim. Behav.* **35**:321–330.
ten Cate, C. (1988). The causation and development of sex differences in partner preferences. *Proc. XIX Int. Ornithol. Congr. (1986)* Ottawa (in press).
ten Cate, C., and Mug, G. (1984). The development of mate choice in zebra finch females. *Behaviour* **90**:125–150.
ten Cate, C., Los, L., and Schilperoord, L. (1984). The influence of differences in social experience on the development of species recognition in zebra finch males. *Anim. Behav.* **32**:852–860.
Todt, D., Hultsch, H., and Heike, D. (1979). Conditions affecting song acquisition in nightingales (*Luscinia megarhynchos* L.). *Z. Tierpsychol.* **51**:23–35.
Vidal, J.-M. (1975). Influence de la privation sociale et de 'l'autoperception' sur le comportement sexuel du coq domestique. *Behaviour* **52**:57–83.
Vidal, J.-M. (1982). 'Auto-imprinting': effects of prolonged isolation on domestic cocks. *J. Comp. Physiol. Psychol.* **96**:256–267.
Warriner, C. C., Lemon, W. B., and Ray, T. S. (1963). Early experience as a variable in mate selection. *Anim. Behav.* **11**:221–224.
Waser, M. S., and Marler, P. (1977). Song learning in canaries. *J. Comp. Physiol. Psychol.* **91**:1–7.
Watson, J. S. (1972). Smiling, cooing, and 'the Game.' *Merrill-Palmer Quart.* **18**:323–339.
Watson, J. S. (1981). Contingency experience in behavioral development. In Immelmann, K., Barlow, G. W., Petrinovich, L., and Main, M. (eds.), *Behavioral Development*, Cambridge University Press, Cambridge, pp. 83–89.
Watson, J. S., and Ramey, C. T. (1972). Reactions to response-contingent stimulation in early infancy. *Merrill-Palmer Quart.* **18**:217–227.

INDEX

Adaptive behavior, 123, 133
Adaptive significance of behavior, 47, 48, 51, 53
Adaptiveness of behavior, 3, 10, 14, 15, 18, 50, 55, 57, 75
Allee, Warder Clyde, 98, 101
Allopreening, 248, 252
Altmann, Margaret, 102
Animal behavior, history of study in North America, 85–117
   1890–1917, 89–100
   1917–1945, 100–110
   1945–present, 100–117
Animal cognition, 123, 126–128
Animal models, 129–132
Animal psychology, 123–133
   animal models, 129–132
   history, 124–129
Animal welfare, 52, 203
Animals as models for people, 123, 124
Ape language, 126–128
Appeasement behavior, 4, 11
Appetitive behavior, 8, 12, 91
Applied ethology, 52, 53
Aromatization hypothesis, 219
Audubon, John James, 97

Autoimmune disease
   and conditioned immune response, 194
   psychological factors insusceptibility, 173, 179
Autoimprinting, 247
Avoidance-image formation, 154, 155

Baerends, Gerard, 111, 112
Baldwin, James Mark, 96, 99
Bascom, John, 88, 89
Bateson, cake-bake theory, 7, 8
Beach, Frank A., 105, 111
Behavior genetics, 107
Behavioral development, processes, 253–266
   environmental factors, 244
   imprinting, 246–263
      acquisition process, 248–253
      active role of developing individual, 261–263
      timing process, 253–261
   process-oriented, interactive process, 243, 244
Behavioral ecology, 50–52, 107
Behavioral ontogeny, 97, 98

Bereavement
  and illness, 181, 202
  and immune function, 183, 184, 197
Bill pecking, 140
Body size, sexual dimorphism, 62, 63, 72
Breeding system, 62, 63, 74
Buffering of sexually dimorphic behavior development, 233, 234

Cake-bake model, 7, 8
Cancer, psychological factors in susceptibility, 173, 179, 180
Carmichael, Leonard, 107
Carpenter, C. Ray, 105, 109, 111
Causal analysis of behavior, 10, 11
Causal systems, 4, 17
Chains of behavior, 11, 12
Chick recognition, 159, 160
Clark University, 99, 100
Clumping and development of imprinting, 248, 252
Communication, ethology versus sociobiology, 19–21
Comparative approach in ethology, 55–76
  aim of comparative approach, 59–63
    generation of hypothesis, 61
    identifying form of relationship, 60, 61
    testing hypotheses, 62, 63
  future directions, 73–76
  limitations, 64–73
    bias in distribution of data, 66, 67
    data, 64–66
    establishing causation, 71–73
    intraspecific variation, 67–69
    methodological decisions in data usage, 69–71
  when is it appropriate, 56–58

Comparative physiology, 109, 110, 117
Conditioned immune suppression, 193–195
Conditioning effects, 191–195
  adaptive significance, 195
  immune conditioning, 192
Comsummatory behavior, 8, 12, 30, 49, 91
Corticosteroid hormones, and CNS–immune system interactions, 196, 197
Costs and benefits of behavior, 18
Craig, Wallace, 90, 91, 98, 99
Cross-species comparisons, 56
Cross-species difference in behavior, 55
Crozier, William J., 99, 102

Darwin, Charles, 4, 5, 87, 125
Defeminization of brain, 219, 222
Development of animal behavior, 48
Development of behavior, 5, 17; *see also* Gender differences in behavior
Distribution of behavioral or ecological traits among species, 73, 74
Dominance relationships, chickens, 101
Donaldson, H. H., 92, 98–100

Early experience and development of sexually dimorphic behavior, 217, 221
Ecology and ethology, 17, 18
Egg recognition, 159
Encephalization quotients, 59, 60
Endorphins and immune function, 199
Energetics, 26

Enkephalins and immune function, 199
Environmental variables in development of sexually dimorphic behavior, 224, 225, 230, 231
Epinephrine and stress, 198
Ethogram, 12–14; *see also* Modal action patterns (MAPs)
Ethological attitude, 98, 109
Ethology: *see* Comparative approach in ethology
Ethology, future of, 47–53
  applied ethology, 52, 53
  neuroethology, 49, 50
  and sociobiology and behavioral ecology, 50–52
Ethology and sociobiology, 1–37
  contrasting ethology and sociobiology, 13–20
  definitions and introduction, 1–5
  ethology in retrospect, 5–13
    causal analysis, 10, 11
    experimental line, 5–8
    hierarchical structure and chains of behavior, 11, 12
    homology/analogy, comparative approach, 8–10
    survival value, 10
  evolving continuity between ethology and osciobiology, 21–31
    proximate explanations, 21–23
    role of ethology in behavioral research, 25–31
    ultimate causation, 25
    whither ethology, 31–35
European ethology, influence on North America, 111–116
  Americans in Europe, 113, 114, 116
  effects of interaction, 115
  Europeans in North America, 114, 115

European ethology (*continued*)
  international conferences, 113–115
  publications, 113
Evolution of behavior, 4, 15, 19, 23, 53
Evolution and comparative psychology, 95, 106, 107
Evolution of mammalian brain size, 59–61

Feeding behavior, 141
Feeding frequency, mixed pairs finches, 249, 252
Field study research, primates, 105
Filial imprinting, 7, 137, 157–159, 245, 246, 252, 253, 256, 247, 263
Fisher, R. A., 4
Fixed action patterns, 49
Food-begging vocalizations, 139, 140
Food-recognition learning, 137
  acquisition, 138–146
    from parent–offspring interactions, 139–142
    imitation, 143, 144
    local enhancement, 143
    social facilitation, 142, 143
    without reinforcement, 144–146
  released-image recognition, 146–152
    comparison with Pavlovian conditioning, 150–152

Game theory, 14, 17, 18
  applied to combat, 26, 27
Gender differences in behavior, development of, 215–237
  hormonal effects on behavioral organization, 218–220
  importance of being social, 235–237
  organization in context, 220–224

Gender differences in behavior, development of (*cont.*)
  organizing the typical animal, 224–229
  role of hormones in developmental pathways, 229–235
Genetic determinism, 6, 30, 32
Genetics and individuality, 16, 17
Growth rate, herbivores/frugivores versus carnivores, 61

Hall, G. Stanley, 96, 99, 100
Harvard University, 98–100
Herrick, C. Judson, 92, 98
Herrick, C. L., 92, 93
Hess, Eckhard, 112, 113
Heterogeneous summation, 33, 34
Hierarchical structure of behavior, 11, 12
History: *see* Animal behavior
Holmes, Samuel Jackson, 92, 98
Home range movements in carnivores, 60, 61, 63, 73
Hormonal effects on behavioral organization, 218–220, 222, 225
Hormone–behavior interactions, 102, 103
Hormones
  prenatal, role in behavior development, 216
  role in developmental pathway, 229–235
Hull, Clark L., 125

Imitation and food recognition, 142, 144, 146, 152, 153
Immune function, 203
  and bereavement, 183
  enhancement by stress, 189–191
  and stress, 186, 184
Immune suppression, conditioned, 193

Immune system, 174, 176
  and cancer, 175
  link with brain and behavior, 200, 201
  mechanisms of CNS–immune system interaction, 195–200
  role in behavior, 200, 201
Immunoenhancing effects of stress, 189–191
  duration, 191
  timing, 190, 191
Imprinting, 4, 7, 156–163, 243, 245, 246, 264; *see also* Filial imprinting; Sexual imprinting
  developmental process, 246–263
    acquisition process, 248–253
    active role of developing individual, 261–263
    timing process, 253–261
  and social interactions, 250
Inclusive fitness, 48, 53
Individual conspecific recognition, 161–163
Individual variability in sexually dimorphic behavior, 229, 230
Infectious disease, psychological factors in susceptibility, 173, 179
Interactive model for imprinting and song learning, 257–261
Interbirth interval, 69
  and female bodyweight, 67
Interspecific allometry, 59
Intraspecific variation, 67–69
Instinct–learning intercalation, 6, 7
Instrumental conditioning, 137, 139, 145

Jennings, Herbert Spencer, 91, 99
Job change or loss, and illness, 181
Johns Hopkins University, 98, 99

Index

Kin recognition, 1, 30
Kin selection, 29, 30
Kuo, Zing-Yang, 6, 107

Lashley, Karl S., 105, 106
Learned helplessness, 189
 model for human depression, 130, 131, 133
Learning theory, 7, 95, 103, 104
 process oriented, 109
Lekking, 22, 27
Lek mating, 57
Life events or changes
 and illness, 180–182
 problems with research, 181, 182
"Listening" behavior, finches, 262, 263
Local enhancement and food recognition, 142, 145, 146
Loeb, Jacques, 91, 98
Lorenz, Konrad, 1–6, 8, 9, 11, 14, 16, 111, 112, 141
Lymphokines and corticosteroids, 196

Manipulation of behavior and communication, 20
MAPs: *see* Modal action patterns
Mate choice, 28, 30, 37
Maternal behavior, differences towards sexes, 216, 218, 222–224, 231
Mating behavior, 27–29, 221
Mating preferences, 246
McDougall, 107
Merriam, C. Hart, 88
Modal action patterns (MAPs), 12, 13, 19, 26, 30, 34, 35
Morgan, L. H., 88
Morton, Thomas, 8, 17
Motivation, 126

Motivational models, 32
Motivational studies, 11

Natural selection, 18
Nature/nurture debate, 8
Nature/nurture dichotomy, 218
Neuropeptides and immune system, 198–200
Nice, Margaret Morse, 101
Noble, G. Kingsley, 102
Norepinephrine and stress, 198

Observational learning of food recognition, 152, 153
One-trial learning, 6, 8
Operant conditioning, 125
Optimal foraging, 1, 8, 31
 theory, 51
Organizing hormones, and development of sexually dimorphic behavior, 216

Parental behavior and development of imprinting, 248
Pavlovian conditioning, 137, 139, 145, 148, 153, 154, 159, 163
 comparison with released-image recognition model, 150–152
Peckham, Elizabeth G. and George W., 89, 90
Phylogenetic explanations, 58, 72, 75
Plasticity of behavior, 30, 31
Play behavior, 57
 play-fighting, 227
 sexual dimorphism, 221, 223, 233
Postcopulatory display, 24, 25
Problem solving, 127
Predator recognition, 155, 156
Proximate mechanisms, 21–23

Psychoimmunology, 173–205
  clinical and epidemiological evidence, 179–185
    direct measure of immune function in humans, 182–185
    life events and illness, 180–182
  conditioning effects, 191–195
  evidence for links between brain, behavior, and immune function, 176–179
  experimental studies in animals, 185–191
    control as a determinant of psychological effects of stress, 187–189
    stress and disease susceptibility, 185, 186
    stress and immune function, 186, 187
  future research, 203, 204
  immune system influence on behavior, 200, 201
  mechanisms of CNS–immune system interactions, 195–200
  mediating role of social relationships, 201–203
Psychological factors and susceptibility to disease, 173, 174, 179

Range of species, 97
Recognition learning in birds, 137–164
  acquisition of food recognition, 138–146
    food recognition without reinforcement, 144–146
    from parent–offspring interactions, 139–142
    social transmission, 142–144
  recognition learning phenomena, 152–163

Recognition learning in birds (*cont.*)
    avoidance image formation, 154, 155
    imitation, 152, 153
    imprinting, 156–163
    predator recognition, 155, 156
    search image formation, 153, 154
    released-image recognition, 146–152
      comparison with Pavlovian conditioning, 150–152
Recognition learning phenomena, 152–163
  avoidance image formation, 154, 155
  imitation, 152, 153
  imprinting, 156–163
    chick recognition, 160, 161
    egg recognition, 159
    filial, 157–159
    individual conspecific recognition, 161–163
    sexual, 160, 161
  predator recognition, 155, 156
  search image formation, 153, 154
Released-image recognition learning, 137
  food recognition, 138, 142
Released-image recognition model, 147–152, 163, 164
  comparison with Pavlovian conditioning, 150–152
Released inhibitory processes, 141
Releaser concept, 33
Releaser mechanism, 9
Releasers, 4, 9, 49, 51
Releasing-stimulus–released-response interactions, 140, 141
Romanes, G. J., 95, 96

Index

Schiller, Claire H., 113
Schneirla, Theodore, 6, 105, 106, 109, 111
Search-image formation, 153
Self-termination of song-learning, 255–258, 260, 261
Seligman, Martin, 130, 131, 133
Sex ratios, 57
Sex-related differences in behavior, 216
Sexual dimorphism, behavior development, 216
Sexual imprinting, 5, 7, 24, 137, 161–163, 245–247, 253, 263, 265
Sexual preference, 250, 251
Sexual selection, 51
Signature cell and recognition of individuals, 162, 163
Skinner, B. F., 125, 131, 133
Social facilitation and food recognition, 142, 143, 145
Social interactions
  and development of imprinting, 250
  and development of song-learning, 259
Social learning, 137
Social relationships, mediating role in immune function, 201–203
Social stressors and immune function, 184, 185
Social variables in development of sexually dimorphic behavior, 224, 225, 227, 228, 230, 231, 235
Sociobiology, 50–52
  and ethology, 1–37
    contrasting ethology and sociobiology, 13–20
    definitions and introduction, 1–5

Sociobiology (*cont.*)
  ethology in retrospect, 5–13
  evolving continuity between ethology and sociobiology, 21–31
  whither ethology, 31–35
Song acquisition, 6, 8
Song copying, 254
Song-learning, 50, 243, 264, 265
  interaction with tutor, 252, 253–256, 259, 260
  interactive model, 257–261
  listening behavior, 262, 263
  timing process, 253–261
Spatial learning, 127
Species recognition, development, 247
Stereotyped chunks of behavior, 132; *see also* Modal action pattern
Stimulus filtering, 4
Stone, Calvin P., 104, 107
Stress
  control as determinant of effects, 187–189
  and immune function, 186, 187, 198
  and immunoenhancing effects, 189, 191
  and susceptibility to disease, 175, 180
Stress response, 196

Temporal decoupling of sexual behavior and sex hormones, 23, 26, 221
Testosterone and behavioral organization, 218, 221, 222
Thorndike, Edward, 95–97, 99, 125
Thorpe, William, 6, 7, 111, 112
Tinbergen, Lucas, 6

Tinbergen, Nickolas, 1–6, 8–11, 14, 16, 111–113
Traditional clock model, song-learning termination, 255, 260, 261

University of Chicago, 98

von Frisch, Karl, 2, 14

Watson, John Broadus, 93–99, 107
Weinland, David Fredrich, 88
Wheeler, William Morton, 90, 91, 98–100
Whitman, Charles Otis, 90, 98–100
Wilson, E. O., 2, 3

Yerkes, Robert, 97–99, 104, 111
Young, William C., 102